Genetic Improvement of Field Crops

Other Related Books

- Crop Improvement and Mutation Breeding — *A.K. Sharma et al.*
- Plant Breeding - Theory & Practice (2nd Enlarged Ed.) — *N.C. Stoskopf*
- Tree Mortality: Assessment and Mitigation — *J.D.S. Negi & P.S. Chouhan*

Genetic Improvement of Field Crops

Edited by

C.B. Singh
Dean
College of Agriculture
J.N. Agricultural University
Jabalpur, India

and

Dhirendra Khare
Scientist
Plant Breeding and Genetics
College of Agricultur
J.N. Agricultural University
Jabalpur, India

SCIENTIFIC PUBLISHERS

Published by:

Scientific Publishers (India)
5 A, New Pali Road, P.O. Box 91
Jodhpur 342 001 (India)

E-mail: info@scientificpub.com
Website: www.scientificpubonline.com

Print : 2026

ISBN: 978-81-7233-301-0 (HB)
eISBN: 978-93-8610-231-7 (EB)

Laser Typeset : Rajesh Ojha
Printed in India

FOREWORD

The quality of seed is the major component for augmenting crop production. The extent to which a plant can harness the natural resources viz., solar radiation, carbondioxide, water and nutrients in terms of biological yield depends on its genetic constitution. Also, the degree to which the potential is executed depends on the plant population and the environment. The use of genetic and physical pure seed of selected variety therefore is obligatory to efficiently utilize the soil fertility, water resource and climatic conditions. Seed production and its testing is one of the most significant advances in agriculture of this century. A book on seed technology is much needed to acquaint producers with the latest technique of production and testing.

This book assembles the recent techniques of seed production and its testing with comprehensive set of instructions for important crops. Authors have well organized the available information and presented it in a systematic and how-to-do-it format. The figures and drawing used in the book make the text more apprehensible. This book will be of immense value to seed production specialists, quarantine officers, seed analysts, field inspectors, teachers and extension workers apart from the students. I am sure the book comprising of both theoretical and practical ramifications will render valuable service to the promotion of a better understanding of the subject.

Dr. Panjab Singh
Vice-Chancellor
Jawaharlal Nehru Krishi Vishwa Vidyalaya
Jabalpur - 482 004 (M.P.)

PREFACE

Opportunities exist for increasing food production in a sustainable manner through the genetic improvement of field crops. One of the potential tools for improvement of the crop performance is the genetic alteration and selection. The ability to modify plant characteristics or select plants with desirable attributes, or both, has enormous implication for the potential productivity of a crop, sustainability of the crop and the cropping system along with the sustainability of agro-ecological environment as a whole. It requires detailed knowledge of individual crop.

Books on practical in plant breeding are meager. Students are normally unable to find the behaviour, problem, husbandry and solution of individual crop at one place. Therefore, the attempt is made to collect up-to-date information and write in how-to-do-it fashion by the scientists of repute working on the improvement of particular crop with vast experience.

The framework of the book is designed to serve as a ready reference in the classroom and laboratory and to bring the attention of the novice in plant breeding. The book comprises of 22 chapters and covers 16 crops. It brings all the information of field crops of Indian interest at one place. The problems of individual crop and methods to resolve are described in detail with clear objectives for the students of plant breeding. The text is designed in such a way that the instructor can insert his own experience and illustration throughout the course.

In persuasion of this book we acknowledge, the writers of different chapters, which provided information and base to this attempt in valuable manner. The manuscript has profited greatly from access to the very rich resources of Jawaharlal Nehru Krishi Vishwa Vidyalaya archive.

We thank to the authorities of JNKVV, Jabalpur for providing facilities to complete this endeavor and Messers Scientific Publishers (India), Jodhpur for their efficient help. Special appreciation is due to our family members for their support and patience.

We are sure the information provided shall be of practical use to many including researchers, teachers, students and people engaged in plant breeding. Readers must judge our success and we shall be honoured to have comments and criticisms.

Jabalpur
1st March, 2002

C.B. Singh
Dhirendra Khare

CONTRIBUTORS

Bhadauria, S.S.

Senior Scientist
(Rape seed and Mustard)
Plant Breeding and Genetics,
JNKVV, Jabalpur,
Zonal Agriculture Research Station,
Moorena

Billore, M. (Mrs.)

Associate Professor
College of Agriculture
Indore

Dabholkar, A.R.

Ex-Incharge
Sorghum Improvement Project,
JNKVV, Jabalpur
Flat 104, Dhanwantrinagar,
Indore, 452 012

Duhoon, S.S.

Project Coordinator
Sesame and Niger, ICAR,
JNKVV campus,
Jabalpur, 482 004

Gaur, P.M.

Senior Scientist, (Chickpea)
Plant Breeding and Genetics,
Visiting Scientist
ICRISAT, Hyderabad

Hegde, D.M.

Project Director,
Directorate of Oil Seed Research,
Rajendra Nagar,
Hyderabad 500030

Khare, D.

Scientist, (Seed Technology),
Plant Breeding and Genetics,
College of Agriculture, Jabalpur,
JNKVV, Jabalpur 482 004

Khire, A.R.

Scientist, (Groundnut)
Plant Breeding and Genetics,
JNKVV, Jabalpur
Zonal Agriculture Research Station,
Khargone

Koutu, G.K.

Scientist, (Cotton)
Plant Breeding and Genetics,
JNKVV, Jabalpur
B.M. College of Agriculture,
Khandwa

Kumar, A.

Scientist, (Rice)
Plant Breeding and Genetics,
IGKVV, Raipur
Regional Agriculture Research
Station, Bilaspur

Mandloi, K.C.

Senior Scientist and Zonal
Coordinator (Cotton)
Plant Breeding and Genetics,
JNKVV, Jabalpur,
B.M. College of Agriculture
Khandwa

Mishra, P.C.

Senior Scientist (Wheat)
Plant Breeding and Genetics,
JNKVV, Jabalpur
Zonal Agriculture Research Station,
Powarkheda

Mishra, R.

Scientist, (Rice)
Plant Breeding and Genetics,
College of Agriculture, Jabalpur,
JNKVV, Jabalpur 482 004

Moghe, M.

Pulse Breeder
College of Agriculture,
Indore

Nimbkar, N.

President,
Nimbkar Agriculture Research
Institute, Phaltan 415 523

Rao, S.K.

Professor and Head,
Plant Breeding and Genetics,
College of Agriculture, Jabalpur,
JNKVV, Jabalpur 482 004

Rastogi, V.K.

Senior Scientist (Maize)
Plant Breeding and Genetics,
JNKVV, Jabalpur
Zonal Agriculture Research Station,
Chhindwara

Raut, N.D.

Senior Scientist and In charge
(Seed Technology Research Unit)
Plant Breeding and Genetics,
College of Agriculture, Jabalpur,
JNKVV, Jabalpur 482 004

Satpute, R.G.

Senior Scientist, (Pigeonpea)
Plant Breeding and Genetics,
College of Agriculture, Jabalpur,
JNKVV, Jabalpur 482 004

Sharma, R. P.

Scientist (Wheat)
Plant Breeding and Genetics,
JNKVV, Jabalpur
Zonal Agriculture Research Station,
Powarkheda

Sharma S.M.

Ex-Project Coordinator
Sesame and Niger, ICAR,
Alok Nagar, Adhartal,
Jabalpur, 482 004

Shrivas, S.R.

Senior Scientist (Oilseeds)
Plant Breeding and Genetics,
College of Agriculture, Jabalpur,
JNKVV, Jabalpur 482 004

Shrivastva, A.N.

Senior Scientist, (Soybean)
Plant Breeding and Genetics,
College of Agriculture, Jabalpur,
JNKVV, Jabalpur 482 004

Shrivastva, M.K.

Technical Assistant, (Soybean)
Plant Breeding and Genetics,
College of Agriculture, Jabalpur,
JNKVV, Jabalpur 482 004

Shrivastva M.N.

Professor and Head,
Plant Breeding and Genetics,
IGKVV, Raipur

Singh C.B.

Dean,
College of Agriculture, Jabalpur,
JNKVV, Jabalpur 482 004

Singh, P.

Head,
Division of Crop Improvement,
Central Institute for Cotton
Research,
Nagpur

Singh, S.B. (Mrs.)

Scientist (Cotton)
Division of Crop Improvement,
Central Institute for Cotton Research
Nagpur

Singh, S.S.

Principal Scientist and Head,
Wheat Programme,
IARI, New Delhi 110012

Singh, V.

Scientist (Safflower),
Nimbkar Agriculture Research
Institute,
Phaltan 415 523

CONTENTS

CHAPTER - 1

WHEAT

P.C. Mishra, R.P. Sharma and S.S. Singh

1. TAXONOMICAL CLASSIFICATION, ORIGIN AND CHROMOSOME NUMBER

Wheat is the most important crop of the world. It belongs to the genus *Triticum* of family Gramineae. Out of sixteen species, *Triticum aestivum* L.Em. Thell and *Triticum durum* occupy large area while *Triticum dicoccum* is scantly grown.

The genetic origin of wheat is of polyploid nature (Table 1) in which related species have combined to form three natural groups namely diploid ($2n = 2x = 14$), tetraploid ($2n = 4x = 28$) and hexaploid ($2n = 6x = 42$). The tetraploid (AABB) wheat have originated with the combination of two diploids while the hexaploid (AABDD) species have originated with the addition of third genome (DD) in tetraploid species. Bread wheat *Triticum aestivum* L. Em. Thell and *Triticum durum* were widely grown in fertile crescent area of the middle east around 7000 years ago. They gradually became the principal types of wheat in western Asia and Europe and spread eastward to India, China and Japan. It was carried to the Western Hemisphere and to Australia and New Zealand by explorers and colonizers and thus encircled the whole world.

Wheat evolved into two large gene pool-the winter habit (life cycle in 9 to 11 months) and the springs habit (life cycle 3 to 6 months) and a third smaller pool, the facultative wheat. In India spring wheats are grown in winter season.

2. CLIMATIC REQUIREMENT, AREA AND DISTRIBUTION

Wheat is grown in 26.5 million hectares in all the states of India except Kerala. Its cultivation ranges from below 9°N (Palni hills in Tamilnadu) to above 35°N (Srinagar valley in J & K). Thus the crop is exposed to variable climates. The duration in which the wheat crop can grow is highly dependent on the temperature which comes to average 22°C. Thus in extreme south the duration of wheat crop is reduced to 100 days and goes to more than 200 days in Srinagar Valley. Wheat growing regions of the country are classified into 6 agroclimatic zones viz., North western plain zone, North eastern plain zone, Central zone, Peninsular zone, Northern hills zone and Southern hills zone.

Table 1. Genome formula of species of *Triticum**

Species	Chromosome number 2n	Genome formula
T. monococcum	14	AA
unknown (possibly an extinct species)	14 Diploid species (2n =2x)	BB
T. dichasians (formerly *Aegilops caudata)*	14	CC
T. tauschii (formerly *Aegilops squarrosa)*	14	DD
Secale cereale	14	RR
	Tetraploid species (2n = 4x)	
T. turgidum (formerly *T. dicoccum, T. durum, T. polonicum, T. durgidum*)	28	AA BB
T. timopheevi	28	AAGG
	Hexaploid species (2n = 6x)	
T. aestivum (formerly *T. vulgare, T. Compactum*, *T. spelta*)	42	AA BB DD

*Adopted from John Milton Poehlman (1986) New names for same species and combinations of several former species into single species.

3. SALIENT FEATURES OF CROP CULTIVATION

In India wheat is grown under three agro-climatic conditions viz., rainfed, (without providing irrigation but with conservation of moisture), irrigated timely sown condition and late sown irrigated condition.

When the mean daily temperature drops down to 22-23°C then the sowing is done in well-pulverized soil with adequate moisture in reading zone. Usually the desired level of temperature is observed in the first half of November but the sowing should not be delayed beyond 25th December. Seed should be treated with Vitavax or Bavistin or Raxil at the rate of 2.5g/kg of seed and placed at 5-6 cm dept with the detail given in table 2. The fertilizer should be applied as per the result of the soil test. Half dose of the nitrogen is applied as basal, whereas remaining half at first node stage. In rainfed conditions total quantity of the fertilizer is applied as basal dose. Zinc sulphate at the rate of 50 kg / hectare should be applied in zinc deficient soil.

Table 2. Prescribed sowing procedure

Production condition	Seed Rate [Kg/ha]	Date of sowing	Row to row distance (cm)	NPK [kg/ha]
Irrigated timely sown	100	Second fortnight of November	23	120:60:40
Irrigated late sown	125	11th to 15th December	18	80:50:30
Rainfed timely sown	100	October end to first week of November	30	40:20
Zero tillage	125			
furrow irrigated raised bed system	75			

Source : DAC – ICAR Interface.

On the basis of soil type and climatic conditions 4-6 irrigation are equipped. Irrigation should be applied at critical stages (Table 3) as per availability of water.

Table 3. Critical stages for irrigation as per availability of water

Irrigation available	Critical Stage					
	CRI	Tillering	Late joining	Flowering	Milk	Dough
Six	■	■	■	■	■	■
Five	■	■	■	■	■	
Four	■		■	■	■	
Three	■		■		■	

Weed control is important aspect in wheat. Pendimethalin may be applied as pre-emergence herbicide at the rate of 1000 g a.i. /ha up to three days after sowing for, whereas sulfasulfuron at the rate of 25.09 g a.i. /ha or Metribuzin @ 75 g a.i. /ha may be applied as post emergence herbicide after the first irrigation to control both grassy and broad leave weeds.

4. BOTANICAL DESCRIPTION

Root : During germination it produces both seminal and primary roots. The adventitious (coronal) root arises later from the basal node of the plant and remains functional throughout its life. Seminal roots are slender. The root of wheat spread 10 to 15 cm and penetrates in soil from one to two meteres.

Culms: The culms of wheat are erect, cylindrical and generally smooth with six solid nodes and hollow inter-nodes. Height of wheat plants is governed by the dose of one gene and classified as tall; one gene dwarf (Semi-tall); two gene dwarf (Semi–dwarf) and three gene dwarf.

Seed: Mature grain of wheat is caryopsis, a small dry, indehiscent, one seeded fruits with a thick and tight pericarp. It consists of an embryo and an endosperm. It is either ovate in shape or elliptical or oval. Prominent colour of wheat grain is amber or red to white. The texture of grain is either hard, semi-hard or soft. The scutellum is large lateral shield cotyledon of embryo. It is slightly longer than embryonic axis. The epidermis of scutellum is in close contact endosperm.

Coleoptile: It is a hollow, cylindrical structure, which completely encloses plumule (2 or 3 rudimentary leaves), with a small opening (*coleoptile pore*). The first green leaf of plumule eventually emerges through this opening.

Prophyll: The first leaf or prophyll of a lateral shoot is bi-nerved and has an open tip through which succeeding foliage leaves emerge.

Foliage leaves: Foliage leaves are arranged on culms alternatively in two ranks with each leaves at a 180^0 angle from one above and one below. Each leaf consists of a blade, sheath and legule with a pair of auricles at base of blade. The blade is linear and parallel veined. The pale green or pinkish claw like auricles is farmed in juncture of blade and sheath, where they loosely clasp sheath and stem from opposite side. The legule a thin, colourless, membranous structure with irregular edges and fringed with small hairs is located at juncture of blade and sheath.

Inflorescence : Inflorescence of wheat plant is known as spike or head. The main axis or rachis is a sinuous, notched structure composed of a number of short nodes and internodes. Each inter-node of rachis is narrow at the base and broad at the apex. One side of each inter-node is more or less convex and other is flattened or slightly concave. Margins may be fringed with hairs of variable length. The unit of spike is spikelet a reduced modified shoot. Each spikelet consists of two sterile bracts or empty glumes and form two to five florets. The florets are enclosed by glumes with alternate arrangement in two ranks (*distichous*) upon a central axis i.e., the rachilla. Each floret consits of a flowering glume called lemma and a thin glume called palea. Between lemma and palea sexual organs i.e., three stamens and a pistil as well as two small ovate lodicules are present. At anthesis the lodicules swell at the base and force lemma outward and open the flower. Lodicules collapse after anthesis.

Each anther is attached with a cylindrical filament. Two terminal styles are covered with hairs at the apex, each of which has numerous delicate stigmatic branches. The two glumes located at the base of spikelet, are shorter than the rest of glumes. Size, shape and texture of glumes; forms of keel are the characters used for verification of the varieties.

Shape of ear varies according to species and varieties, *fusiform* - taper towards apex or from middle towards both base and apex, *oblong* - uniform in width and thickness throughout length of spike but always several time longer then wide, *Clavate* - It is largely more dense at apex due to a shortening of rachis inter node in that part of spike and Elliptical - short and uniformly rounded at both base and apex and flattened on sides. The various varieties can be classified on the basis of awn as awnless, short tipped, long tipped, half-bearded and full bearded.

Glume shoulder shape is classified as wanting, oblique, rounded, square, elevated and apiculate.

5. SELFING AND CROSSING

Wheat is a self-pollinated crop. For genetic improvement in wheat hybridization is essential. The following steps are to be followed in hybridization. The spikelets for emasculation are selected when there is 50 % heading. The emasculation is done manually by removing all the three immature anthers from the spikelets with the help of forceps without injuring the ovary. The emasculated spike is covered with the parchment paper bag to avoid cross-pollination. it is labeled with a tag showing the name of the variety and date of emasculation.

After two to three days the spike of the desired pollinator with anther ready to dehisce are picked up and kept in sunlight for few minutes to extrudes anthers from the spikelets and pollen from the anthers. The spike is shacked gently over the emasculated spikes for pollination. Immediately name of the pollinator is recorded on the tag attached to crossing bag.

6. BREEDING OBJECTIVES

Presently, in India, emphasis is being given on the following objectives.

1. Evaluation and utilization of germplasm for diversification of genetic base.
2. Development of superior wheat suitable for various agro-ecological situation under irrigated timely and late sown conditions and timely sown moisture stress condition.
3. Incorporation of genes resistance to biotic stresses such as leaf and stem rusts and leaf blight.
4. Incorporation of genes for resistance/tolerance to abiotic stresses especially moisture stress.
5. Development of high quality *durum* wheat for export.

7. BREADING METHODS FOR WHEAT IMPROVEMENT

1. *Introduction:* The introduction of dwarf wheats during early 1960 from CIMMYT, Mexico resulted in heralding the green revolution in the country in 1966-67. The green revolution varieties Lerma Rojo 64 A, Sanora 63, Sanora 64 and Mayo 64 were developed by noble laureate Dr. Norman E. Borlaug and his associates in Mexico by successful introgression of *Norin* (dwarfing) genes in high yielding background of the tall spring wheats. In 1962, Indian Agricultural Research Institute, New Delhi introduced these dwarf wheats along with small samples of 630 advance lines developed at Mexico through the Rockefeller foundation. The year 1965 is considered as the landmark year for Indian agriculture as release of variety Lerma Rojo 64 A and Sanora 64 triggered off of the green revolution in country. These dwarf lines were highly responsive to input such as fertilizer and irrigation as they were lodging

resistant. The introduction of wheat germplasm from CIMMYT Mexico is helping the wheat breeders in selecting the good lines with enhanced yield potential and resistance to diseases.

2. *Hybridization:* Hybridization is the important process in order to get new recombinants.

3. *Selection* : The following selection methods are used to identify desirable genotypes from segregating progenies:

3.1 *Pedigree method*: For genetic improvement of wheat pedigree method of selection has been employed by most of the breeders. To achieve their objectives the plant breeder undergo hybridization, between two (some times more) selected parental lines carrying genes for desirable traits. It creates segregating population from which pure lines could be selected and few of them advanced as varieties.

Single plants with desirable gene combinations are selected from F2 generation onward up to F6 generation progeny rows are planted with seed harvested from previous generation. The segregating populations of F2 to F6 generation are surrounded by spreader rows constituted of rust susceptible wheat varieties Lal Bahadur, Kathia local and pissi local for creation of artificial epiphytotic condition.

3.2 Modified Pedigree bulk selection method : In the pedigree method of selection a breeder has limitation of handling few process in the load of work in keeping records of selected plants is too much. However, in the present situation a breeder is always looking for vast genetic variability for effective selection, therefore, he attempts large number of crosses. It becomes imperative to go for modified pedigree – Bulk selection method that gives opportunity for handling segregating progenies of large number of crosses. The method is described below :

F_1	Based on simple cross or back cross
F_2	2000 plants per cross are space planted. The individual plant is selected on the basis of plant type with agronomic base, disease resistance and grains.
F_3	The progenies selected from F_2 are planted individually as at a normal seeding rate of 100 kg./ha. in three rows of two meter per line, in order to observed and evaluate competitive ability with in the line. The selection of lines is thus based on visual assessment of agronomic performance and disease resistance in a plot rather than on an individual plant. Ten to fifteen ears are selected per selected F_3 line, threshed in bulk, and promoted to the F_4 generation.
F_4-F_5	Same as F_3
F_6	Planted as in the F_3 but five to ten ears are selected and threshed individually from selected F_6 plots.
F_7	Individual ears are planted, and plots are selected for subsequent complete bulk harvest.

Yield trial — The bulked lines are included in preliminary yield trials and subsequently replicated yield trials. Finally the superior lines may be included in state or national varietal evaluation trials.

3.3 Back cross : For introducing certain specific genes into the otherwise desirable commercial varieties, the back crossing technique has been used in wheat by earlier wheat breeder and secondly reducing the height of an inbred or a cultivar by back crossing can be accomplished much faster and with less likelihood of character loss or shift than by recurrent selection. Three to five years are required to complete a back crossing operation.

Five back crosses plus one generation to segregate out recessive or homozygous dominants should combine a donated gene with 98.4 percent of gene from recurrent parent. Linkage can not reduce the percentage much. All but the last generation could be carried this percentage much. A minimum of 10 hybrid seeds of each back cross will be adequate for wheat. Mixed pollen from a number of plants of a recurrent variety back crossed to more F_1s will be required in later generation to retain the variation in highly variable variety.

3.4 Hybrid varieties: In several crops hybrid varieties have shown its superiority over conventional varieties. In wheat attempt have already been made to exploit heterosis but success was not recorded on commercial basis, so for. However a well organized research on hybrid wheat was initiated about five years ago by the directorate of wheat research. Various gametocides have been tested and isolated in inducing reliable male sterility. It was observed that 20% yield advantage could be gained if there is no attack of disease like rust.

3.5 Shuttle breeding Shuttle breading methodology was proposed and implemented by Borlaug in 1968. This methodology has been responsible for the production of photoperiod insensitive and otherwise widely adopted germplasm. This approach involving contrasting location has proven most efficient way to introduce and select genes for photo period insensitivity. The photo period insensitivity genes, Ppd-1 and Ppd-2 along with the dwarfing genes Rht-1 and Rht-2 resulted into a new plant type which not only tolerant to lodging but also very high yielding with high biomass due to pleiotropic effects/close linkage (Hoogendoorn *et. al.*, 1988). When superimpose, with rust resistance the new gene combination provides adoption to most irrigated wheat growing areas of the sub-tropics (Borlaug, 1968). The success of Indian wheat breeding programme is mainly due to the shuttling of breeding materials between the plains and hill regions of Himanchal Pradesh (Lahaul-Spiti) and Nilgiri hills of Tamilnadu (Wellingaton).

8. ORGANIZATIONS INVOLVED IN CROP IMPROVEMENT

Indian wheat programme organized by directorate of wheat research is having close linkage with various International organizations in order to update the wheat research in the country. These organizations are CIMMYT [Exchange of breeding materials, nurseries and genetic stocks], ICARDA

[Exchange of wheat research materials] USDA [Exchange of research materials, advance generations material of winter spring wheat crosses], IRRI [Technological exchange for Rice/wheat crop rotation], ACIAR [Exchange of wheat research materials] and Asian Development bank [Rice/wheat rotation project]. The Indian wheat programme is being benefited by national MOU with countries like Mongolia, Ethiopia, Bangladesh, Nepal, Russia, Pakistan etc. Indian Agricultural Research Institute New Delhi is the lead centre for wheat research in the country. Its regional research stations in various states of the country are involved in wheat improvement.

9. MAJOR ACHIEVEMENT IN CROP THROUGH BREEDING

It is important to quote the statement from report of the project Director wheat in order to understand the progress in wheat programme "during 1999-2000 India created a land mark, as never in the history of this country 74.25 million tons of wheat was harvested. Available statistics indicate that during this decade wheat production increased from 55.1 mt. in 1991 to the current level of 74.25 mt. in 2000 AD, an increase of 19.11 mt over ten years. There has also been a gradual increase in the area under wheat during the last decade. Area gain in M.P. and Rajasthan was higher order and even in states like Punjab and Haryana, there has been a marginal gain. In Chhattisgarh area of MP there has been an increase under wheat during the last few years and variety GW 273 and Vidisha have been found to perform better. The gain of half a million ha in northern MP and use of high yielding (Table 4) varieties are the main reasons for producing more than 9.0 mt for the first time.

Varietal improvement has brought significant improvement in productivity. A new series of wheat varieties like UP 2338, PBW 343, WH 542 and HD 2687 have been developed and released in recent years, which resulted in increased yield potential. All these varieties can yield over 6.0 tonnes per hectare under high fertility condition.

Durum wheats occupy nearly 1.2 million-hectare area in central and peninsular zone. Increasing global demand, value addition potential, better price market and resistance to Karnal bunt are the factors that make durum an export commodity. Semi-dwarf, high yielding durum varieties like PBW 34, PDW 215, PDW 233, WH 896, HI 8498, HI 838, MACS 2846 etc. have been developed.

India is the only country in the world growing *T. dicoccum* over an approximate area of 50,000 ha. The *dicoccums* have high protein and thousand-grain weight. For the first time semi-dwarf *dicoccum* varieties i.e. DDK 1001 and DDK 1009 have been released under All India Co-ordinated Wheat Improvement Project.

10. HYBRID WHEAT

Since it is believed that the wheat productivity is reaching a plateau it becomes necessary to find new avenues for breaking yield barrier. The wheat yield scenario in India between 1975 to 1999 showed that only one percent (1%) yield gain per year is possible through conventional breeding method.

Table 4 Important Varieties of Wheat and their Special Features

Variety	Production condition	Maturity class	Grain size
NHZ			
	Irrigated timely sown /Rainfed timely sown	Medium	Bold
VL 738	Irrigated timely sown/Rainfed timely sown	Late	Medium
HS 240	Irrigated timely sown /Rainfed timely sown	Early	Medium bold
HS 295	Rainfed timely sown	Early	Medium
HS 365	Rainfed timely sown	Early	Medium
NWPZ			
PBW343	Irrigated timely sown	Medium	Medium
WH 542	Irrigated timely sown	Medium	Medium
PBW 373	Irrigated late sown	Early	Medium, Bold
HD 2687	Irrigated timely sown	Medium–early	Medium
UP 2338	Rainfed timely sown	Late	Medium, Bold
UP 2338	Irrigated timely sown/Irrigated late sown	Medium–Early	Bold
UP2425	Irrigated late sown	Early	
NEPZ			+
HP 1761	Irrigated timely sown	Medium	Medium
HDR 77	Rainfed late sown	Medium–Early	Medium
HD 2643	Irrigated late sown	Early	Bold
HP 1633	Irrigated late sown	Early	Medium
HP1744	Irrigated late sown	Early	Medium
NW 1012	Irrigated timely sown	Medium	Medium
K9107	Irrigated timely sown	Medium	Medium Bold
K9465	Rainfed late sown	Early	Medium
HD2933	Irrigated timely sown	Medium	Medium
CZ			+
GW173	Irrigated late sown	Medium	Bold
GW 190	Irrigated timely sown	Early	Medium
GW 273	Irrigated timely sown	Medium	Medium
HI8498(d)	Irrigated timely sown	Medium	Bold
DL 803-3	Irrigated timely sown	Medium	Medium
HW2004	Rainfed timely sown	Medium	Medium
JWS 17	Rainfed timely sown	Medium Late	Medium
RAJ 1555(d)	Irrigated timely sown	Medium Early	Bold
DL 788-2	Irrigated late sown	Medium	Medium
PZ			
MACS 2496	Irrigated timely sown	Medium	Medium
DDK 1009 (dic)	Irrigated timely sown	Medium	Bold
DWR 162	Irrigated timely sown	Medium	Medium
DWR 195	Irrigated late sown	Early	Medium
MACS 2846 (d)	Irrigated timely sown	Medium	Bold
K9644	Rainfed timely sown	Medium Late	Medium

d- *durum*, dic - *dicoccum*
Source – DAC –ICAR Interface

Directorate of wheat research has projected that in order to meet out target of wheat production of 95 to 109 million tones by 2020 AD from our present level of 65 million tones, the hybrid wheat approach can provide a valuable answer. The winter and spring wheat hybridization is also under investigation in the country. History indicates that the work on hybrid wheat was initiated in early fifties with screening of parents exhibiting heterosis followed by the use of 2 or 3 lines system using different sources of male sterility for seed production. However, in seventies efforts were shifted towards the use of gametocide for seed production. The first wheat hybrid was released in Australia for feed. The significant efforts have been made for commercial exploitation of hybrid wheat through the use of gametocide during last one decade. France and Italy are having commercial wheat hybrids using chemical hybridizing agents. France's National Agricultural Institute INRA registered first wheat hybrid. Literature indicates that harnessable heterosis is available in wheat and it can be better exploited through the use of double haploid.

REFERENCES

Borlaug, N.E. 1968. Public lecture; Wheat breeding and its impact on world food supply. PP 1-36 in proceeding of 3rd International wheat genetic symposium. K.W. Finley and K.W. Shephard (Eds.). Australian Academy of Sciences, Canberra, Australia.

Poehelman J.M. 1986. Breeding Field crops.

CHAPTER - 2

RICE

M.N. Shrivastava and A. Kumar

1. TAXONOMICAL CLASSIFICATION, ORIGIN AND CHROMOSOME NUMBER

Family Graminae; Sub-family-Oryzoidae; Tribe-Oryzeae; Genus- *Oryza*. The genus *Oryza* consists of 26 species, (Table-1) of which 24 are wild species whereas *Oryza sativa* L. and *Oryza glaberrima* Steud are cultivated species. *Oryza sativa* is grown worldwide and *Oryza glaberrima* is grown in parts of West Africa. Basic chromosome number of rice (X) is 12 and both the cultivated species are diploid.

Survey and re-examination of the geographic distribution of the wild and weed races of the cultivated rices alongwith the distribution of other taxa in the genus clearly indicate that the genus *Oryza* originated from the Gondwana-land continent. Judging from findings on the evolutionary process of the grasses and of crop plants, the general process of evolution can be visualized as; wild perennial- Wild annual- Cultivated annual. The annual weed race is a conglomerate of the natural hybrids of all the inter-grades among the above three. The parallel and independent trends in the evolution of two cultivated species i.e. *Oryza sativa* and *Oryza glaberrima* have been illustrated (Fig. 1).

The choice of *O. rufipogon* and *O. longistaminata* as the two wild perennials and *O. nivara* and *O. barthii* to represent the wild annual progenitors follows taxonomic conventions for description and nomenclature. *O. rufipogon* are characterized as perennial or largely asexually propagated, weakly *rhizomatous*, extravaginal in branching, having long anthers and adapted to continually flooded habitats, while, *O. nivara* is characterized as annual or essentially seed propagated, non-rhizomatous and adapted to shallow water. Accordingly, the weed races of Asia and Africa are designated

Table 1. Different Taxa in the genus *Oryza*

Complex Taxon	Genome group	Chromosome No. (2n)	Distribution
***Oryza sativa* complex**			
O. sativa (Cultigen)	AA	24	Worldwide
O. glaberrima (Cultigen)	AA	24	West Africa
O. barthii	AA	24	Africa
O. longistaminata	AA	24	Africa
O. nivara	AA	24	Tropical Asia
O. rufipogon	AA	24	Tropical Asia
O. meridionalis	AA	24	Tropical Australia
O. glumaepatula	AA	24	South America
***Oryza officinalis* complex**			
O. officinalis	CC	24	Tropical Asia to New Guinea
O. eichingeri	CC	24	East and West Africa
O. rhizomatis	CC	24	Sri Lanka
O. minuta	BBCC	48	Papua New Guinea
O. punctata	BB	24	Africa
O. latifolia	CCDD	48	Latin America
O. alta	CCDD	48	Latin America
O. grandiglumis	CCDD	48	South America
O. australiensis	EE	24	Australia
***Oryza meyeriana* complex**			
O. granulata	Diploid	24	South and Southeast Asia
O. meyeriana	Diploid	24	Southeast Asia
***Oryza ridleyi* complex**			
O. longiglumis	Tetraploid	48	New Guinea
O. ridleyi	Tetraploid	48	Southeast Asia
***Oryza angustifolia* complex**			
O. angustifolia	Diploid	24	Tropical Africa
O. perrieri	Diploid	24	Tropical Africa
O. tisseranti	Diploid	24	Tropical Africa
O. brachyantha	FF	24	Africa
Oryza schlechterianae			
O. schlechteri	Diploid	24	New Guinea

as the *spontanea* and *stapfii* forms of *O. sativa* and *O. glaberrima* respectively. The prototype of *O. sativa* evolved primarily along the foothills of the Himalayas in the South Asia and its associated mountain ranges in the main lands of Southeast Asia and Southwest China. The widespread distribution of the wild races in insular Southeast Asia, Sri Lanka and Taiwan was most probably facilitated by the land bridges, which existed as the Pleistocene. The continuous and varied movements of people of Asia during the prehistoric

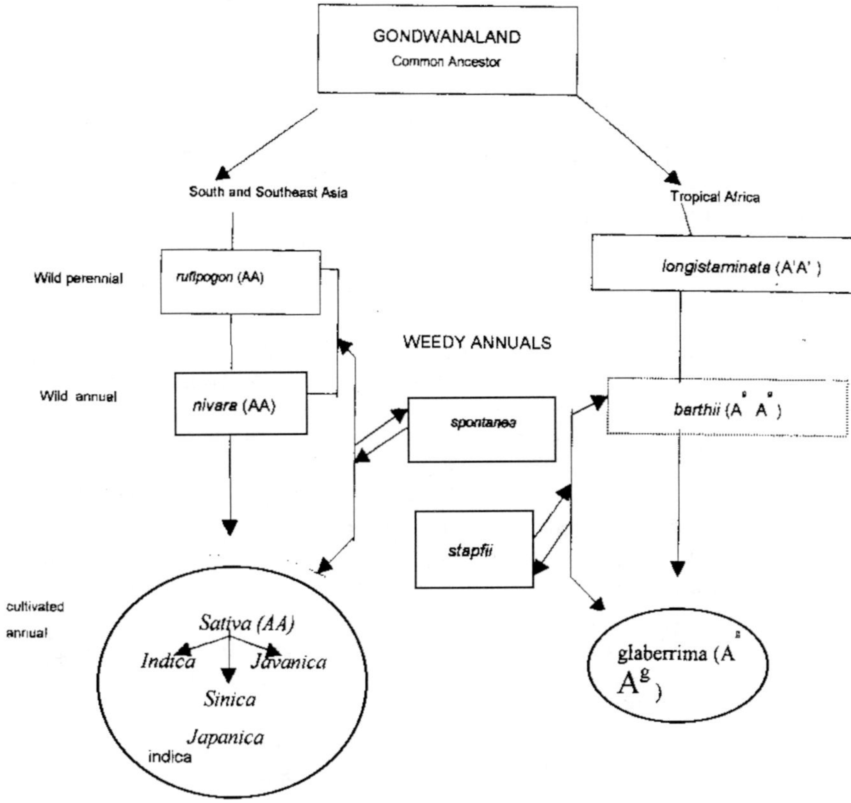

Fig. 1. Evolutionary pathway of the two cultivated species of rice

times was an important factor in the broad distribution of the early *O. sativa* forms which led to the formation of three ecographic races i.e. indica, japonica and javanica.

The most convincing archeological evidence for domestication of rice in Southeast Asia was discovered by Wilhelm G. Solheim II in 1966. Pottery sheds bearing the imprint of both grains and husks of *O. sativa* were discovered at NON Nok Tha in Korat area of Thailand. These remains date from at least 4000 B.C. *O. glaberrima* is postulated to have evolved in Niger River delta. The primary center of diversity for *O. glaberrima* is the swampy basin of upper Niger River and the two secondary centers to the southwest near the Guinean coast. The primary center was probably formed around 1500 BC while the secondary center were formed 500 years later.

2. CLIMATIC REQUIREMENT, AREA AND DISTRIBUTION

Rice is produced in a wide range of climatic conditions ranging from the area having annual rainfall of more than 5100 mm (Myanmar's Arkan Coast) to driest desert having annual rainfall of less than 100 mm (Al Hasa Oasis in Saudi Arabia). In upper Sind in Pakistan, where average rice season temperature is 33°C to Otaru, Japan having the mean rice season temperature of 17°C. The crop is grown at sea level on coastal plains and in delta regions throughout Asia to a height of 2600 meter on slopes of Nepal's Himalaya. Similarly, the variation in solar radiations ranges from 25% of potential during the main rice season in portions of Myanmar, Thailand and Assam State of India to approximately 95% of potential in southern Egypt and the Sudan.

Rice being a tropical and sub tropical plant, requires a high temperature, ranging from 20°C to 40°C. The optimum temperature appears to be in the neighborhood of 30°C for the day- time maxima and 20°C for the night time minima. Low temperature depresses the rate of germination and prolongs it beyond the desirable span of 6 days. During the seedling stage, low temperature may promote faster root development and better penetration but less prolification. It also depresses the internodal elongation and may result in partial emergence of panicles, all of which result in a low rate of photosynthesis and partial sterility. A high temperature of 35°C or more often shows many of the effects of low temperature, although the biochemical reactions are very different. Germination is halted because of high respiration rate. The pollen germination at the time of fertilization is slow and at ripening period, high temperature increases the thickness of bran.

The highest rice yields has been obtained from plantings in high- latitude areas that have long day length or in low latitude desert that have very high solar energy levels. Southwestern Australia, Hokkaido in Japan, Spain, Italy, Northern California and the Nile delta provides the best examples.

3. SALIENT FEATURES OF THE CROP CULTIVATION

Rice is produced in a wide range of locations and under a variety of climatic conditions with different methods of planting such as transplanting or direct seeded as broadcasted or in line. *indica* rice is grown as Upland (direct-seeded); *Aus* (summer rice, April to December); *Boro* (Winter December to May); *Aman* (transplanted autum, June to December) and *Aman* broadcasted (deep water, April to December). Similarly, Japonica is grown as Tropical lowland, upland (*bulu*) and temperate lowland and upland.

4. BOTANICAL DESCRIPTION

Cultivated rice plant may be defined as annual grass with round, hollow, jointed culm or stems, rather flat, sessile leaf blades and a terminal panicle. At maturity, the rice plant has a main stem and a number of tillers, each productive tiller bears a terminal flowering head or panicle.

The rice grain, may be defined as the ripened ovary, with the lemma, palea, rachilla, sterile lemmas, and the awn firmly adhered to it as hull. The rice grain is known as rough rice. Dehulled rice grain is called brown rice. The seed is enclosed by the pericarp. Next to pericarp, there are two layers of cells representing the remains of integuments, tegmen or seed coat. The bulk of the seed is the starchy endosperm. An aleurone layer encloses the endosperm, which lies below the tegnen. The embryo is located in on the ventral side of seed next to lemma. The embryo consists of plumule, mesocotyl, radicle, epiblast and the scutellum.

Seed absorbs adequate water at a temperature ranging from 10 to 40□C. Under aerated conditions the radicle, ensheated by the Coleorhiza emerges first followed by the emergence of plumule, which is enclosed in Coleoptile. Under anaerobic conditions, however, the plumule emerges first and the embryonic roots develop when the coleoptile has reached the aerated regions of the environment. If the seed develops in the dark (sown beneath the soil surface), a short stem (mesocotyl) develops, which lifts the crown of the plant to just below the soil surface.

The rice stem is known as culm. Culm is made up of series of nodes and internodes. Nodes are the solid portion of the culm, panicle axis, and panicle branches. Inter-node is the smooth, solid (when young), or hollow (when mature) part of the culm, between two successive nodes. The solid portion in the node separating the two adjoining internodes is called the nodal septum. The internodes generally increase in length from the lower to upper part of the stem, the uppermost (peduncle) bearing the panicle, being the longest. Each upper node bears a leaf and a bud, which can grow into a tiller. The number of nodes varies from 13 to 16 with only the upper 4 or 5 separated by long internodes. Under rapid increase in water level, some deepwater rice varieties can also increase the lengths of lower inter-node by more than 30 cm each. The leaf blade is attached at the node by the leaf sheath, which encircles the stem.

A branch of the rice plant, which is composed of roots, culm, leaves and with or without a panicle is called a tiller. The tillering stage starts as soon as the seedling is self–supporting and generally finishes at the panicle initiation. Tillering usually begins with the emergence of first tillers when the seedlings have five leaves. This first tiller develops between the main stem and the second leaf from the base (lower most node) of the plant. Subsequently, when the sixth leaf emerges, the second tiller develops between the main stem and the third leaf from the base. Tillers growing on main stem are called primary tillers. They may give rise to secondary tillers, which may in turn generate tertiary tillers. Although the tillers remain attached to the plant, at later stage they are independent because they produce their own roots.

Roots are the organs of absorption and anchorage, growing opposite the shoot. The rice root system consists of three major types: seminal roots, mesocotyl roots and nodal or adventitious roots. Seminal roots emerge from radicle and are temporary in nature. The radicle emerges after germination, elongates to about 15 cm and remains functional upto seventh leaf stage. Mesocotyl roots do not grow under normal condition. It develops under deep seeding condition from the axis between the coleoptile node and point of union of the culm and point of and secondary branches, pedicel, rudimentary glumes and the spikelets. The panicle axis extends from the panicle base to the apex; it has 8-10 nodes at 2 to 4 cm intervals from which primary branches develop. Secondary branches develop from the primary branches. Pedicels develop from the nodes of the primary and secondary branches and the spikelets are positioned on them. Since the rice has only one fully developed floret (flower) per spikelet, these are often used interchangeably.

Spikelet may be defined as the unit of rice inflorescence (panicle) consists of two sterile lemmas, rachilla and floret. Floret is the unit of spikelet that includes lemma, palea and the enclosed flower. Rachilla is a small axis that bears the floret. Flower consists of two lodicules, six stamens and a pistil. Pistil consists of a stigma, style and ovary with one ovule. Stamen consists of two celled versatile anthers borne on slender filaments. The flower is enclosed in the lemma and palea, which may be either awned or awnless. The sterile lemmas, lemma and palea are also called Bracts. Lemma is the five-nerved bract, which partly enclosed the palea. Palea is three-nerved bract, which closely fits to the lemma. The tip of the lemma and palea is known as apiculus. A thorn like extension at the tip of the rice hull is called awn. It is not present in all varieties. Two flowerless bracts at the base of the spikelet are sterile lemmas. Two scale like structures which are fused to the base of the palea are lodicules. Lodicules represent the perinath (calyx and corolla).

The leaves arise from nodes and borne at an angle in alternate phyllotaxy. A typical rice leaf includes leaf sheath, auricle and leaf lamina (blade). Leaf sheath is the lower portion of the leaf blade, originating from a node and clasping the culm. The swelling at the base of the leaf sheath i.e., just above its point of insertion on the culm is called sheath pulvinus. The leaf sheath contributes little to photosynthesis, but provides mechanical support to the whole plant and serves as a storage site for starch and sugar before heading. Auricles are a pair of small, ear like appendages borne at the base of the blade and usually arising at the sides where the ligule and the base of the collar are joined. It may not persist on older leaves. Ligule is a thin, upright, membranous structure found on the inside of the collar at its base where the blade joins the leaf sheath. The expanded portion of the leaf is called the leaf blade. the prominent ridge on the lower surface of the leaf is called the mid-rib. The joint between the leaf sheath and leaf blade is called collar. The uppermost leaf of the rice plant just below the panicle is called flag leaf. The first rudimentary leaf at the base of the main culm is bladeless and two killed bract. It is called prophyllum. The margins of the prophyllum clasp the young tiller. Prophyllum is also present between each secondary and the tertiary tiller.

Stages of development

Growth stages can be broadly classified into three distinct phases: Vegetative (from germination to panicle initiation), reproductive (from panicle initiation to flowering) and the ripening (from flowering to full grain development). The vegetative phase is subdivided into germination, seedling, tillering. It terminates with the initiation of panicle primodia. A 120 days variety, when planted in a tropical environment, spends about 60 days in the vegetative phase, 30 days in reproductive phase and 30 days in the ripening phase.

The first leaf emerges within three days of sowing of pregerminated seeds. From emergence to just before the appearance of first tiller is the seedling stage. During this stage seminal root development takes place and absorption of endosperm food occurs. Two more leaves come by tenth day and then one every 3-4 days. Adventitious roots replace temporary seminal roots. During tillering stage, primary tillers appear first, followed by secondary tillers and then tertiary tillers appear. Tertiary tillers continue to emerge upto maximum tillering stage, after this some tillers die and there is decline in total number of tillers. Active tillering refers to the period when increase in tiller number per unit of time is high. Active tillering stage is followed by maximum tillering stage when the tiller number per plant or per square meter is highest. Stem elongation begins before panicle initiation in long duration varieties whereas in short duration varieties stem elongation and panicle initiation occurs together.

The reproductive phase begins with the initiation of panicle formation. Formation of panicle becomes visible on the tenth day. The process is initiated first on the main culm then on the other tillers. Panicle moves upward and when the panicle is of five cm i.e. on the seventh day after detectable panicle initiation, spikelets are determined. Booting stage with swelling of sheath starts on the 16th day after detectable panicle initiation. Meiosis occurs when the collar of the flag and that of the preceding leaf are at the same level. During heading, panicle emerges from flag leaf sheath. Heading occurs about 22 days after the visible panicle initiation. Anthesis or flowering occurs on the 25th day of visual panicle initiation regardless of the variety. During this process, opening of lemma and palea, elongation of filaments, exsertion of anthers and dehiscence of anthers occurs. At the beginning of anthesis, tip portions of the lemma and palea begin to open, filament elongate and the anthers begin to exert from lemma and palea. Anther dehiscence usually occurs just before or when the lemma and palea open. Consequently, many pollen grains fall on the stigma and pollination occurs. It takes about 7-10 days for all the spikelets within the same panicle to complete anthesis. Under normal conditions, anthesis starts at about 8.00 and ends at about 13.00 hours. Under low temperature conditions, anthesis may start late in the morning and continue late in the afternoon (4-5 P.M.) or in rare cases it may be completely inhibited and pollination may be cleistogamous. The range of atmospheric humidity over which blooming occurs is 40-95 percent, the optimum being 70-80 percent. Spikelets on the upper branches of the panicle have anthesis earlier than those on the lower branches and within the same branch, spikelets at the top flowers first. After emergence from anther, pollen grains are viable for five minutes only whereas stigma can remain receptive for 3-7 days.

The ripening phase refers to the actual process of grain development following fertilization. Though, it is a continuous process, distinct stages such as milky, dough and mature grain stages are recognized. During the milky stage, the contents of seed, which are first watery turns milky. During the dough stage, the milky portion of the grain turns from soft to hard dough. The mature grain stage is characterized by the gradual change of panicle from green to yellow, grain becomes hard. Normally, the ripening stage takes about 25 to 35 days. However, it is prolonged when exposed to low temperature.

5. SELFING AND CROSSING PROCEDURE

The rice flower consists of pistil, six stamens and two lodicules. Rice is a self-pollinated crop and crossing in rice involves Emasculation – process of removing the anther from flower before the anthers shed their pollen grains on floret's stigma. The emasculated plant serves as the female parent in crossing programme and Pollination – Process of placing the anthers from other plant to be used as male over emasculated spikelets.

Emasculation

Methods are used to emasculate the rice spikelets viz., hot air emasculation, hot water emasculation, vacuum emasculation and manual emasculation by tweezers. The hot air and hot water emasculation method is based on the different response of pollen and stigma to high temperature. At high temperature (40-43□C) the pollen loose viability whereas stigma remains viable. Panicles are treated with either hot air or hot water in a thermos. Individual spikelets remain open for several minutes after panicle is withdrawn from thermos and are therefore pollinated before they close. It is a slow, tedious method and seed set is relatively low.

The most efficient method of emasculation is to clip the spikelet and remove the anther by hand or vacuum emasculator. The steps involved may be summarized as —

1. Select the parent to be used as female (parents for emasculation). Select the plants where panicle have emerged 50-60%. Remove the plant from field along with the mud in earthen pots in the early morning before 8.00. Add soil and water to the plant in the pot and tag the plant with proper designation. Transfer the pot to crossing house.
2. The process of emasculation is started in afternoon at around 2.30-3.00. Remove the leaf sheath from panicle. Cut away very young florets from the bottom of the panicle where height of the anthesis is less than half of the floret.
3. With scissors give a slant cut to floret to remove one third to half of the floret without injuring the stigma.
4. When the anthers are to be removed by hand, with forceps, gently press the anthers against the side of the floret and lift them out care should be taken to remove all the six anthers.

5. When anthers are to be removed by vacuum emasculator, adjust the emasculator to about 500 mm Hg. Touch the tip of the extractor to the cut surface area of the floret so that anthers are extracted. Be sure to remove all the six anthers.
6. In similar fashion, remove anthers from all the florets. Once finished, all florets should be inspected for any left over anthers. Most breeder prefers to start emasculation and pollination from panicle tip and work downward, in this way there is less risk of accidental deposition of pollen on stigmas below.
7. Cover the emasculated panicle with butter paper bag and put a tag for the date of emasculation.

Pollination

The stigmas of the emasculated spikelets remain receptive for three to seven days. Best results are obtained if the emasculated spikelets are pollinated next day. In case, viability of the pollen is limited, pollination can be extended to two to three days. The steps involved are —

1. The next day of emasculation, check the emasculated panicle for any anthers that may be left.
2. On the day when pollination is due to be effected, panicles in which an appreciable proportion of the anthers has reached the spikelet apex are removed from plants to be used as male parent between 10.00 a.m. to 2.00 p.m., position of the anthers is clearly seen when the spikelets are held against the sun, anthers show a recognizable colour and texture and discharge their pollen when punctured with needle.
3. Place panicles of each male parent in separate earthen cups containing water and arrange panicles in cups in such a way that each panicle is removed without disturbing the other. Earthen cups are placed in a closed room at high temperature without any air flow (can be maintained by providing several electric bulbs).
4. Watch closely for anther extrusion as pollen grains are viable for only five minutes after emerging from the anther. When anther extrude panicle is ready for use in pollination. Remove the butter paper bag from emasculated panicle, very gently lift the blooming panicle from water and shake it over the emasculated panicle.
5. Replace the butter paper bag, fold bottom edge over, place the paper clip on fold against the peduncle to keep the bag secured. Attach crossing tag having the name of the cross and date of emasculation and pollination. Remove the pot containing the plant from room to designated place in the crossing block.
6. Whenever, there is heavy rains, butter paper should be inspected for damage and should be replaced with fresh one.
7. F_1 seeds are ready between 18-21 days and should be harvested as

soon as they mature, otherwise fungus attack are very frequent. Matured seed turns hard and yellow from initial green colour. Once harvested, each F_1 seeds should be carefully dehulled by removing the half hull formed.

Raising the F_1

1. F_1 seeds are very delicate and prone to fungus attack. before sowing F_1 of each cross should be treated with bavistin solution 0.3% (3g per 1000ml).
2. F_1 seeds are firstly sown in Petridish. Arrange the properly marked (Cross name and number) Petridishes, cut filter paper to the size of Petridish. Put the F_1 seeds in the lid of the Petridish and treat the seeds with bavistin by pouring bavistin solution in the lid containing seeds, each seed should be properly dipped in solution with the help of brush and then washed thoroughly thrice with fresh water.
3. Arrange the seeds on the other part of the Petridish containing the filter paper, provide water to Petridish with water bottle so as there is sufficient moisture in the Petridish. Close the lid of the Petridish and place it in germinator at 28-30°C.
4. Inspect the seeds for germination and water every day. By fourth-fifth day, all the viable seeds will be germinated.
5. The germinated seedlings will be transferred to wooden tray having well prepared soil. Seeds of each cross are transferred in properly labelled individual row. Wooden trays are placed on raised surface and with protection against rats and ants. Initially the wooden trays are placed under shade for a week till the seedlings are well established and then transfer them to full sun light. When seedlings are well established, nitrogen in solution form is provided for the growth of the seedling.
6. Transplant when seedlings are 25 days old with one seedling per hill at spacing of 30X 30 cm.
7. Inspect the F_1 plants before and after flowering and remove the selfed plants.
8. When matured in the field, harvest each F_1 individually.

6. BREEDING OBJECTIVES

Rice is grown under the diverse agroclimatic situations ranging from upland to lowlands and on a wide range of soils i.e. acidic, alkaline, peat acid sulphate etc. Rice is host of more than 100 insects and many diseases. Thus rice ecosystem is divergent in nature. Rice growing environment is broadly classified as irrigated and rainfed. Rainfed ecosystem is very complex and is further subdivided based on land situations and water depth as upland, lowland and deep water. Problem soils and tidal wetlands are the other important ecosystem in which the rice is grown. Each ecosystem needs a particular plant type for yield maximization thus the breeding objective for each ecosystem will be different.

Irrigated ecosystem

The release of modern varieties led to tremendous increase in rice production. Rice production increased from 30.6 million tones in 1965-66 to 84.74 million tones in 1999-2000. T(N)1 for irrigated condition was released in Taiwan in 1956 and IR 8, was released by IRRI in 1966. IR 8 has yield potential of 10 t/ha in tropical latitude under dry season. In the last 28 years since IR 8 was released, rice yield potential has remained constant. This plateau is evident in the yield records of several long-term experiments. However, according to Bjorkman *et al.* (1971) potential production of C_3 plants is around 200 kg dry matter ha per day. Thus there is scope for yield improvement in irrigated condition. The economic yield in rice is the function of biological yield (biomass) and harvest index. The grain yield of rice can be increased beyond the current level by increasing the total dry matter and/or harvest index. The dry matter production can be increased through leaf area development, low maintenance respiration, adequate growth duration, efficient use of low CO_2 concentration and increased canopy photosynthesis. The harvest index can be increased through increased sink size (increased spikelets number per panicle and larger grains) and improved percentage of spikelet fertility. To improve the harvest index, genotypes or cultivars having rapid nutrient uptake and faster and more efficient translocation of photosynthetes from source to sink should be identified. The harvest index of the improved semi-dwarf cultivars is around 0.45 to 0.55 and it can be increased up to 0.60. The photosynthetic efficiency should be improved for raising the yield. The photosynthetic capacity per unit leaf area did not increase during the breeding programme. So breeding for high photosynthesis should be part and parcel of the breeding programme. Scientists at IRRI has developed a new plant type which should yield 12.5 t/ha on the irrigated land under direct seeding condition. These new plant-type have low tillering capacity with 3-4 panicle per plant; thicker, greener and more erect leaves; no unproductive tillers; thick and sturdy stem; 90-100 cm tall vigorous root system; 200-250 grains per panicle and 110-130 days growth duration and 0.6 harvest index.

Rainfed ecosystem

The modern semi dwarf rice varieties and technology have touched the lives of less than half of the world's rice farmers. Most of these farmers farm the irrigated lands and are reasonably certain of adequate and continued supply of water. Development of improved varieties for millions of farmers who depend solely on unpredictable monsoon to water their crops has not received much attention. Rainfed areas represent diverse ecological conditions ranging from rainfed upland to low-lying deep water. Rainfed rice is the crop grown with local rainfall. It includes the crop grown with supplemental irrigation, provided that the water is not imported from other catchments or with water from tanks and wells that depend entirely upon local rainfall. Rainfed environments are mostly unfavourable and characterized by variable water regime, occurrence of drought, submergence and water logging. Lack of water when needed is the widest constraint to higher yield. Yield reduction from drought at the vegetative and reproductive phase of the crop growth is significant. The drought at reproductive phase is more damaging and causes the high spikelet sterility. Rainfed ecosystem can be further subdivided int upland and lowland ecosystems.

Upland ecosystem

Upland rice may be defined as rice grown in rainfed, naturally well-drained soil without surface water accumulation and without phreatic water table. Based on pheratic water table, upland rice was further categorized as true upland rice and hydromorphic upland rice. In true upland rice water table never reaches the root zone whereas in hydromorphic upland rice water table may reach more or less frequently the root zone of the plant. The total upland in India is reported to be around 7 million ha with low productivity. This crop is grown on both the flat and sloppy land, which may be bunded or unbunded. Upland soils are poor in fertility, especially in nitrogen and phosphorus. They have poor water retentive capacity, which frequently leads to drought. The toxicity of manganese and aluminium is also encountered. Varieties should be tolerant to drought at vegetative as well as reproductive stage, highly responsive to low fertilization particularly nitrogen and phosphorus, tolerant to the soil problems like aluminium, manganese and iron toxicity depending upon the pH of soil, intermediate plant height with moderate tillering, good seedling vigour, droopy leaves to shade out weeds and conserve the soil moisture, erect upper leaves for efficient light penetration, deep, thick and dense root system and resistance against major diseases and insects.

Varieties with early maturity (less than 95 days) are desirable in areas with short and monomodal rainy season. However, long duration and /or semi-dwarf high yielding varieties may be useful in the areas where rainy season is long, favourable and distribution is uniform.

Rainfed lowland ecosystem

Rainfed lowland rice areas are characterized by bunds that trap rain-water. For working principle, three broad categories of rainfed lowland rice areas have been recognized as Shallow (water depth does not exceed 30 cm except during flooding, areas generally have one or more periods of drought), Intermediate (water depth between 31-60 cm and does not exceed 60 cm except during flooding, areas generally have consecutive deficits and excess of water) and medium (water depth 61-100 cm and does not exceed 100 cm except during flooding, areas generally have excess of water).

Nearly 38 million ha in the world is rainfed lowland of which 35 million ha is in South east Asia. There are four types of aberrant weather in rainfed lowland situation. Firstly when rainfall during vegetative and reproductive phase is adequate and assured. This is an ideal condition with no special problem. Secondally, rainfall during vegetative phase is adequate and at reproductive stage it is inadequate. Under this situation, there are two alternatives, either to provide supplemental irrigation by storing runoff water or to select shorter duration varieties that would ripen with limited moisture. Thirdly, rains inadequate during vegetative stage and adequate during reproductive stage. There are two alternatives under this situation, first is to dry drilling of rice followed by conversion to wet later in the season. The other is to use photoperiod sensitive varieties and to transplant them when rains are sufficient for puddling and transplanting. Fourth situation is when rainfall is inadequate during both vegetative and reproductive stage. Under this situation, rainfed lowland rice will be extremely risky. Crops other than rice should be preferred.

The improved lines for shallow and intermediate lowland situations should have drought resistance at different growth stages of the crop. For medium rainfed lowland condition, drought resistance at early vegetative stage is desirable. Flood is a common problem in north east India, affecting the crop at different growth stages. Flood tolerance would be a desirable characteristic. For shallow areas, submergence tolerance is needed. For intermediate and medium areas, submergence tolerance and internode elongation may be required. Rainfed lowland rices are grown from June to December. In countries like India, the low temperature during October-November coincides with the reproductive and ripening stages of the crop leading to high sterility of grains. For such areas cold tolerance will be a desirable characteristics. Early seedling vigour and competitive growth habit in the vegetative stage are considered essential for good establishment and compete effectively with weeds. For rainfed lowland situations, varieties with intermediate height (130-140 cm) and sturdy stem will be desirable as they will not lodge under moderate fertility and will be better suited to drought stress. Photoperiod sensitivity in rice is associated with timely floral initiation. If planted early, they complete the basic vegetative phase and pass through the photosensitive phase to wait for the appropriate day length in order to come to flowering. If planted late the early advent of appropriate day length causes initiation of flowering just after the minimum period required for basic vegetative phase. the growth duration of the varieties depend upon the rainfall distribution pattern. In areas where the rainfall distribution pattern is unimodal and of short duration, varieties of very short duration will be desirable for shallow water depth. For intermediate water depth, varieties with long growth duration will be desirable. In areas where rainfall distribution is bimodal, varieties with long growth duration would be preferred. rainfed lowland rice is grown mostly for local consumption and eating quality will play a very important role in farmers decision to adapt a new variety.

Deepwater ecosystem

In deep-water areas, the water depth exceeds 100cm. deep water area is around 12 million hectares in world, out of which 2 million hectares are in

India. The breeding objectives for deep water ecosystem will be- drought tolerance at the vegetative stage, submergence tolerance, elongation ability, photoperiod sensitivity, kneeing ability and resistance against major disease and pests.

Tidal wetlands

Tidal wetlands are found near seacoast and inland estuaries, which are influenced by tides directly or indirectly. Tide causes the submergence of the crop in all the wetlands. Therefeore, the varieties to be developed should have submergence tolerance, elongation ability, tall in stature, photoperiod sensitive, drought tolerance, salt tolerance as the acid soils of the tidal wetlands have the phosphorus deficiency, and iron toxicity, whereas, alkaline soil of tidal wetlands have boron toxicity, and sodium toxicity.

Drought

It is recognized as the primary constraint to rainfed rice production. Many adaptive trait has been proposed for use in breeding programme for drought prone areas such as leaf rolling, epicuticular wax, amino acid accumulation, root length, thickness and length density, spikelets fertility and harvest index. However, an important consideration to be followed will be selection of genotypes under stress condition. When genotypes are selected under favourable conditions, much useful genetic variation for stress tolerance may be lost. One of the outstanding successes of the breeding programme over decades has been development of modern varieties that resist damage from diseases and insects. It is an important component of national and regional breeding programme keeping in view of the prevailing pest problems.

7. BREEDING METHODS FOR CROP IMPROVEMENT

The common breeding methods used in rice breeding program are pure line selection, pedigree, bulk population, back crossing, mutation and heterosis breeding. Breeding method to be used depends upon the character to be improved and the difference lies in the handling of segregating generations and screening methods.

Plant height

It is a simply inherited character with dwarfness being monogenic recessive. Traditional tall varieties are non-responsive to nitrogen and tend to lodge with increased nitrogen application. There are excellent sources of dwarfing genes with simple inheritance such as DEE-geo-woo-gen (DGWG), 1-geo-tze (IGT) and Taichung Native-1, which should be used in breeding programme. Actually, dwarfing genes from these sources have been transferred to many improved breeding lines, which can be used as parents in crossing programme. However, for rainfed condition specially for rainfed lowland, plants with intermediate height (120-140 cm) are more appropriate. The recessive semi-dwarf gene sd-1 has been extensively used to develop plants with intermediate height. In crosses involving tall and dwarf, in F_2 dwarf plants with other desirable characteristics may easily be selected. It is necessary to fertilize the F_2 nursery with heavy nitrogen dose so that the differentiating characters can be expressed properly. Besides plant height, other factors that contribute to lodging resistance are – thickness of the culm, diameter of the culm, density of lignin. Scores for lodging resistance are affected by growth duration i.e. short duration entries tend to lodge more easily than long duration entries. A more objective measure of lodging resistance is the bending resistance. Many breeders evaluate bending resistance by bending the culm halfway over, releasing it and observing the speed with which it returns upright (Table 2).

Table 2. System of scoring for plant height, lodging resistance and culm strength (IRRI, 1996)

Character	Score	Description	Characteristics
Plant height	1	Semi-dwarf	lowland- less than 110 cm upland- less than 90 cm
	5	Intermediate	Lowland-110-130 cm upland- 90-125 cm
	9	Tall	Lowland-more than 130 cm Upland-more than 125 cm.
Lodging resistance	0		No plants are lodged
	1		Less than 20% of the plants are lodged
	3		20-40% plants are lodged
	5		41-60% plants are lodged
	7		61-80% plants are lodged
	9		More than 80% plants are lodged
Culm strength	1	Strong	No plants are bending
	3	Moderately strong	Most plants are bending
	5	Intermediate	Most plants are moderately bending
	7	Weak	Most plants are nearly flat
	9	Very week	All plants are flat

Vegetative Vigour

It is associated with several characters such as rapid seedling emergence and development, early and heavy tillering, moderately long droopy leaves. Under direct seeded condition, vegetative vigour ensures good stand and provides competition ability to plants with weeds. Vegetative vigour is polygenic trait with low heritability. Selection should start from F_2 generation and continue to later generations. In F_2 and F_3, selection pressure should be mild and from F_4 onwards, selection pressure should be strong.

In direct seeded rainfed lowland rice fields, weed competition is intense, non vigorous genotypes can be overcome completely by weeds. The direct seeded *aus* and *gora* rices of eastern India and Bangladesh are some of the most vigorous rice cultivars in the world. In transplanted condition, vegetative vigour helps rice plants to compete against weeds. Seedling height has proved to be related to submergence survival in many of the experiments. Both vegetative vigour and seedling height are closely and positively related to mature plant height, which implies that it may be difficult to combine shorter stature with seedling vigour. Cultivar such as Chandina of Bangladesh has vigour at seedling stage although it has short height at maturity. Seedling vigour is scored as (IRRI, 1996) extra vigorous (vary fast growing, majority of plants have two or more tillers at five leaf stage), vigorous (fast growing, seedlings with one to two tillers at four to five leaf stage), normal (seedlings are at four leaf stage), weak (seedlings are somewhat stunted, 3-4 leaves with no tiller formation) and very weak (seedlings are stunted and yellowing of leaves takes place).

Number of Panicles

Number of panicles per square meter is an important determinant of rice yields. The optimum number for transplanted, high yielding varieties is 300 panicles/m2 and for direct seeded it is $600/m^2$. Panicle number is largely determined tillering ability, thus breeding for cultivars that produce large number of tillers has become an important objective. Tiller number is inherited quantitatively and is highly influence by cultural practices. It is desirable to combine heavy tillering with compact and non-spreading culm arrangement, which allows increased penetration of solar radiation to tiller and cause less mutual shading. Tiller compactness is reported to be recessive as against spreading culm arrangement. Selection for tiller number and erectness is possible in F_2 but greater emphasis should be laid in the selection for this trait in F_3. Under direct seeded situation, a very low seed rate should be used so that plant is able to exhibit its tillering ability without any inter plant competition. However, the recent approach has been to breed for few tillers (4-6) with large number of filled grains (250-300) similar to the approach followed earlier in breeding for maize and sorghum. Scoring for tillering ability (IRRI, 1996) is very high (more than 25 tillers), good (20-25 tillers), medium (10-19 tillers), low (5-9 tillers) and very low (less than 5 tillers per plant).

Percentage of filled spikelets

The percentage of filled grains is determined largely by the number of spikelets that are fertilized and the ability of the plant to fill the fertile spikelets. Environmental stress such as drought, salinity, low and high temperatures and low solar radiation increase the number of unfilled grains. Lodging can also increase the unfilled grains especially if it occurs before flowering. Scoring for spikelet fertility (IRRI, 1996) is highly fertile (more than 90 percent), fertile (between 75-89 percent), partly sterile (between 50-74 percent), highly sterile (less than 50 percent to trace) and no fertility (0 percent fertility).

Leaf characters

Leaf characteristics which is of interest to plant breeders are leaf shape and size (narrow/ broad/ erect /droopy, leaf angle, leaf colour, toughness, senescence and pubescence). Erect leaf is considered a most important trait and associated with high grain yield as it allows greater and even penetration of of sun light in leaf canopy and promote higher photosynthetic activity. Erect leaf is highly heritable. Selection for erect leaf may be carried at flowering during F_2 generation. Flag leaf supplies photosynthate to panicles and play an important role in grain filling. Moderately long erect leaves protect panicles against possible bird damage. Selection for desired flag leaves should commence from F_2 and should continue for several generations. Small and erect flag leaves are preferred. Nitrogen deficiency and heavy plant density tend to shorten the flag leaves. Therefore, segregating generations should be raised in fertile soil with optimum plant density.

Other panicle and grain characteristics

Panicle exsertion is the emergence of the neck node from the flag leaf collar. When panicle does not emerge completely, some spikelets remain within the sheath, and these spikelets tend to have high sterility rates. Environmental factors such as drought, low temperature, and certain diseases can impair panicle exsertion. Breeders try to select for complete panicle exsertion in F_2 and pedigree nurseries. Threshability is an important consideration for farmers. A proper level of non-shedding is necessary. If it is more, than there will be problem during the threshing (japonica rice) and less threshability will result into grain shedding in the field.

Growth duration

Rice is a short day plant i.e. short day treatments can hasten the flowering and long day treatments can prevent or delays its flowering. Rice cultivars exhibit a wide range of variation in their degree of sensitivity to photoperiod and are classified into photoperiod sensitive and photoperiod insensitive types. The vegetative growth phase of the rice plant consists of Basic Vegetative Phase (BVP) and Photo-period Sensitive Phase (PSP). The BVP refers to juvenile growth stage of the rice plant not influenced by photoperiod variation. BVP ranges from 10 days to 63 days (shortest duration to flowering minus 35 days). The PSP is the growth stage of the rice plant sensitive to photoperiod. In photoperiod sensitive cultivars, the PSP determines the sensitivity of the rice plant. The PSP of the photoperiod insensitive cultivar ranges from 0 to 30 days while that of sensitive cultivar lasts from 31 days to much longer period (difference between shortest and longest duration). Optimum photoperiod is the day length that minimizes the growth duration. It varies between cultivars, but 10 hours often is used in experiments (photo-period for shortest duration). Critical photoperiod is the longest day length at which the strongly sensitive cultivar will flower or the day length beyond which it will not flower.

The presence or absence of photoperiod response usually is controlled by one or two dominant or recessive genes. A major gene controlling photoperiod sensitivity is designated *Se-1* and is located on chromosome 6 and seems to be present in many sensitive cultivars. A recessive gene *Se-2*, has been mapped to chromosome 7. Another major gene, *Se-3* has been identified in some very long duration sensitive cultivars. The two dominant genes seem to have a cumulative effect in controlling photoperiod sensitivity.

Very short duration photoperiod insensitive cultivars have been successful in areas where they can be direct seeded. For other conditions, they are not suitable because- they are sensitive to delayed transplanting, they mature during the period of heavy rainfall, which makes proper harvesting and drying difficult, they compete poorly with weeds as they tend to be very short, and they are not adapted to drought.

Short duration photoperiod insensitive cultivars have been found successful not only in irrigated but rainfed areas also because- It reduces the risk of drought injury in areas where the amount or duration of rainfall is limiting; in particular, plants are spared from terminal drought stress at reproductive stage; when rainfall is sufficient, it allows intensification of the cropping system. Photoperiod–sensitive cultivars have many advantages for typical rainfed lowland conditions. There appears no genetic barrier in developing sensitive rice cultivars with improved height and yield potential. The type of parents used for crossing must be chosen according to breeding programme objectives, particularly desired growth duration. At least one photoperiod sensitive parent with optimum flowering date should be used in crosses in order to select sensitive progeny with shorter duration than the sensitive parent because it is difficult to recover sensitive progeny that have longer duration than the sensitive parent (Table 3). If two or more photoperiod sensitive cultivars are crossed, most of the progeny are sensitive, making it easier to obtain desirable recombinants.

Table 3. Guidelines for photo period insensitive and sensitive parents in breeding photo period sensitive rices

Response to photoperiod	Consideration in selecting parents
Insensitive	Avoid short duration parents unless early flowering or weakly sensitive recombinants are desired. Productive, medium to long duration lines with intermediate height are desirable.
	Use long duration lines (more than 140 days) if very late flowering recombinants are desired.
Sensitive	Use traditional cultivars with wide adaptability, stress tolerance or other desirable traits.
	Choose a parent that has a flowering date the same or later than the desired flowering date.
	Use the traditional cultivar as female parent to help diversify the cytoplasm sources.

Once crosses have been attempted, the F_2 generation and the pedigree nursery can be handled normally and the insensitive progeny should be eliminated during the early stage (Table 4). sensitive progeny can be identified in several ways —

When seeded early (April to early June), insensitive plants or pedigree lines will flower earlier than the sensitive parents (before October) because the growth duration of insensitive plants do not exceed 150 or 160 days.

When two generations per year are grown, the sensitive pedigree lines have much longer growth duration in the wet season than in dry season, if wet season nursery is planted by July.

When a special nursery are planted in the dry season (January), sensitive lines are those that do not flower in 140-150 days.

Table 4. Breeding scheme for developing photoperiod sensitive rices

Year	Season	Generation	Activities
1	Wet	F2	Select for plant type
2	Wet	F3	Select for plant type
	Dry	F4	Advance generation, screen, select for plant type
3	Wet	F5	Select for plant type
	Dry	F6	Advance generation, screen
4	Wet	F7	Bulk best rows
	Dry	F8	Increase seed
5	Wet	Fixed lines	Conduct observation trials and replicated yield trials.

Drought resistance

Drought stress affects the rice crop directly by reducing the growth rate and tillering and when severe, killing the leaves and then the plants. It also affects the crop indirectly as it causes the plant to decrease their rate of transpiration, which results in reduced nutrient uptake. Drought stress is less damaging during the vegetative phase than during the reproductive phase. At vegetative phase it causes reduction in tiller number and leaf area. At reproductive phase, desiccates spikelets and anthers, reduces pollen shedding, inhibits panicle exsertion and increases sterility rates. Drought resistant plants are those varieties that are able to grow and yield satisfactorily in areas liable to periodic drought. Three types of drought resistance are known.

Drought escape

The ability of a plant to complete life cycle before being subjected to serious water stress. Crop duration is adjusted so that the plant completes yield determining processes while drought is not present. With the use of short duration varieties in areas having short rainy season will be able to complete the reproductive stage before drought stress becomes severe.

Drought avoidance

It represents plant's ability to maintain a high level of water status under conditions of increasing soil moisture deficit. Attributes of plant morphology and phenology may affect drought avoidance. Leaf area affects total evapo-transpiration. Reduction in leaf growth when water stress develops is a major mechanism by which plants reduce water requirement. Narrow leaves are considered a stress adaptive trait by some wheat breeders. Leaf orientation may affect leaf water status. Upright configuration of leaves in soybean causes lower leaf temperature. Narrow upright leaves should be selected as a drought adaptive trait instead of wide lax leaves. Root size, morphology, depth, length density are important in maintaining high leaf water potential against evapotranspiration. In areas where prolonged dry spells occur varieties with drought avoidance has significance.

Drought tolerance

It is the ability of plants to withstand internal water deficits with minimal injury. Plant maintains low internal water content during the period of drought but the ability to recover and grow when soil water is replenished. During the seedling stage, this type of drought tolerance has significance as the roots are smaller and has not penetrated the deeper layers.

Table 5. Adaptive traits for drought resistance in rice

Type of trait	Mechanism	Effect
Root related	Deeper or denser root system	Increase water uptake with large soil water
	Increased root penetration of	Increased water uptake under hardpan restricted soil water reservoir
	Root osmotic adjustment	Expanded crop available soil water under restricted soil water reservoir
Shoot related	Stomatal closure	Maintenance of leaf water potential
	Leaf rolling	Maintenance of leaf water potential
	Thicker epicuticular wax layer	Maintenance of leaf water potential
	Osmotic adjustment	Maintenance of turgor potential
	Accumulation of growth regulator	Physiological adjustment
Reproductive	Increased panicle diffusive	Flower at higher panicle
stage specific	Resistance	water potential
	Early morning anthesis	Flower at higher panicle water potential
	Tolerance for high temperature	Decreased sterility
	Emergence of new panicle from internodes	Recovery of lost yield component
	Remobilization and translocation of stored reserves	Maximization of potential harvest index
	Nonsynchronous tiller development	Lower probability of catastrophic yield loss

The best source of drought resistant donors is the local varieties that have proven their abilities in the field. these donors may have one or more of the adaptive traits for drought resistance. Heritabilities for yield associated traits tend to be lower when environmental conditions are stressful than when they are not. Some breeders propose therefore, selection of early generations under optimum conditions. However, using drought as a selection pressure during the early generations maximize the chance of identifying superior plants. In the later generation pedigree nurseries (F_4 to F_6), more favourable environmental conditions can be applied so lines that perform well when water is available can be identified (Table 5).

Screening for drought resistance: At IRRI, an upland nursery is used for screening at the vegetative stage. After the onset of dry season, the entries are dry seeded on a granular clay soil in well tilled upland field that drains freely and is not influenced by subsurface ground water. the rows are spaced at 30 cm apart and are 2.5 meter long. A total of 60 Kg/ha of nitrogen is applied in two equal doses by top dressing at 10 and 20 days after seedling emergence. Mercury tensiometer are installed at several locations in the field. the field is sprinkled irrigated at 4 days interval or when the soil metric potential at a 10 cm depth approaches –0.03 Mpa (0.3 bars). Irrigation ends 30 days after seeding. Each entry is monitored three times at SMP –0.2, -0.5 and –1.0 Mpa. After the final drought screening, the field is twice sprinkler irrigated to saturation and after 10 days, the entries are monitored for their recovery (Table 6).

Table 6. Scoring system for evaluating resistance to and recovery from drought (IRRI, 1996)

Response to drought	Score	Description
Resistance-Leaf drying	0	No symptoms of stress
	1	Slight drying of leaf tip
	3	Tip drying extended upto one fourth length in most of the leaves.
	5	One fourth to half of the all leaves fully dried
	7	More than two third of all leaves fully dried
	9	All plants apparently dead
Leaf rolling	0	Leaves healthy
	1	Leaves start to fold (shallow V shaped)
	3	Leaves folding (deep V shaped)
	5	Leaves fully cubbed (U shaped)
	7	Leaf margins touching (O shaped)
	9	Leaves tightly rolled
Recovery	1	90-100 percent plants have recovered
	3	70-89 percent plants have recovered
	5	40-69 percent plants have recovered
	7	20-39 percent plants have recovered
	9	0-19 percent plants have recovered

At reproductive stage, at IRRI, a three tired screening programme is followed i.e., Tier one–Vegetative stage screening- the first tier of screening takes place during the vegetative stage as described above. Tier two- mass screening- In it, genetic material is staggered planted during the dry season to synchronize flowering dates, irrigation is interrupted during the flowering period for generating water deficit and plants are scored for spikelet fertility. Tier three- advanced screening- It focuses on 20-30 cultivars that performed well during the mass screening. Five planting dates are used to make more evaluation possible. This method requires the use of two irrigation regimes, full irrigation for a control group and artificial drought stress for the trial groups. Each experiment is laid out in a block within replication design, replicated three or four times. records on flowering dates and estimates of spikelet fertility and grain yield are gathered for each experiment. Irrigation management is same as that in mass screening. Superior genotypes are those that have greatest deviation from the regression.

Disease and insect resistance

Among the hundred insect pest that affect the rice crop, stem borer, in particular the yellow stem borer (*Scirtpophaga incertulas*) are the most serious pest in India. gall midge, *Orseolia Oryzae* causing silver shoot damage comes next in importance. Among the plant hoppers, the brown plant hopper, *Nilaparvata lugens* and the White backed planthopper, *Sogatella furciferaare* the most important, sometimes causing, total devastation of the crop. The green leaf hopper, *Nephotettix viriscens* causes low damage to the crop as direct feeders but causes considerable damage indirectly as a vector of the tungro disease. Rice hispa, *Dicladispa armigera*, traditionally a sporadic pest,is appearing in many areas as a major pest. At times, the leaf folder, *Cnaphalocrocis medinalis*, thrips and gundhi bug also becomes serious.

Among the 85 diseases that affect rice crop, blast disease caused by *Pyricularia grisea* is considered a major problem. Bacterial blight caused by *Xanthomonas Oryzae* pv *Oryza*e is a major problem in irrigated and rainfed ecosystems.Sheath blight, caused by *Rhizoctonia solani*, assumed importance. Sheath rots has become problematic in uplands and rainfed lowlands of eastern India.

In 1968, Van der Plank recognized two basic types of resistance on the basis of genetic and epidemiological concepts namely —

(i) *Monogenic, complete, race specific resistance called as vertical resistance* – It is qualitative in nature. It can be either immunity or high level of resistance, which neither allows the pathogen to sporulate nor allows plant injury under any known conditions, but is short lived in nature and is often known to break down with serious economic consequences.

(ii) *Polygenic, incomplete, race non specific resistance called as horizontal resistance* – It is quantitative in nature and is also called as field resistance, generalized resistance, rate reducing resistance etc.

(iii) There is another term called Durable resistance, which has remained effective for long period in a cultivar that has been widely cultivated in an environment favouring the disease even though another resistant cultivar succumbs to an epidemic, the mode of inheritance is not fully understood.

Genetics of disease and insect resistance

Inheritance of resistance against blast and bacterial blight have been studied extensively in Japan and at IRRI. These studies has led to the identification of atleast thirty one genes for blast resistance and fourteen genes for bacterial blight resistance. Similarly, several studies have been done on genetic characterization of resistance against major insect pests such as BPH, WBPH, GLH, gall midge, stem borer etc (Table 7). Studies at IRRI have led to the identification of five dominant and five recessive genes against BPH; four dominant and four recessive genes against WBPH, six dominant and two recessive genes against GLH. Studies at Indira Gandhi Agricultural University has led to the identification of five dominant and one recessive gene for gall midge resistance.

Table 7. Genes conferring resistance against major pests of rice

Gene	Source	Gene	Source
Brown planthopper			
Bph1	Mudgo	*bph4*	Babawee
bhp2	ASD7	*bph5*	ARC10550
BPH3	Rathu Heenati	*BPH6*	Swarnalatha
bph7	T12	*bph8*	Chin saba
Bph9	Balamawee	*BPH 10(t)*	IR65482-4-136-2-2
White backed plant hopper			
Wbph1	N22	*wbph4*	Podiwi A8
Wbph2	ARC10239	*Wbph5*	N'Diang Marie
Wbph3	ADR52	*Wbph6(t)*	Giu-yi-Gu
Green leafhopper			
Glh1	Pankhari 203	*Glh5*	ASD8
Glh2	ASD7	*Glh6*	TAPL796
Glh3	IR8	*Glh7*	Maddai Karuppan
glh4	Ptb8	*glh8*	DV85
Gall midge			
Gm1	W 1263	*Gm5*	ARC5984
Gm2	Siam 29	*Gm6*	Duokang 1
gm3	RP2068-18-3-5	*Gm7*	RP2333-156-8
Gm4	Abhaya		
Blast			
Pi-a	Aichi Asahi	*Pi-I*	Ishikari Shiroke
Pi-k	Kanto 51	*Pi-kh*	HR22
Pi-kp	Pusur	*Pi-km*	Tsuyuake
Pi-ks	Norin6	*Pi-8*	Kasalath

Pi-6(t)	Apura	*Pi-9(t)*	WHD-IS-75-1-127
Pi-7(t)	Moroberekan	*Pi-10(t)*	
Pi-5(t)	Moroberekan	*Pi-11(t)*	
Pi-2	5173	*Pi-3*	Pai-Kan-tao
Pi-sh	Shin2	*Pi-1*	LAC23
Pi-zt	TKM1	*Pi-b*	Bengawan
Pi-ta	Tadukan	*Pi-ta2*	tadukan
Pi-z	Zenith	*Pi-t*	Tjahaja
Pi-12(t)	Moroberekan	*Pi-15(t)*	GA25
Pi-13(t)	Maowangu	*Pi-16(t)*	Aus373
Pi-14(t)	Maowangu	*Pi-17(t)*	DJ123
Bacterial blight			
Xa1	Kogyoku	*xa9*	Sateng
Xa2	Rantai Emas 2	*Xa10*	Cas209
Xa3	Wase Aikoku	*Xa11*	R944-102-2-3
Xa4	IR 20	*Xa12(t)*	Kogyoku
xa5	IR1545-339	*Xa12h(t)*	IR28
Xa6	Malagkit	*xa13*	BJ1
Xa7	DV85	*Xa14*	TN1
xa8	PI231129	*Xa21*	*Oryza longistaminata*

Donors for resistance

Donors for different diseases and insects have been identified (Table 8). Some donors have been source of broad based resistance in many breeding programmes, but in many cases, differences in pathogen races or insect biotypes require the use of donors with resistance genes that are specific to the location. Furthermore new donors are being identified continually. With the use of donors, several resistance varieties have been breed (Table 9).

Table 8. Resistant donors for different insects and diseases

Pest	Donors
Brown planthopper	ARC 5984, ARC 6650, Karivennel, Leb Mue, Nhang, Ptb 10, Ptb 18, Ptb 21, Ptb 33
White backed planthopper	Ptb 33
Gall midge	Eswarakora, CR 143, Leaung 152, Ptb 21, Siam 29, Abhaya, RP2068-18-3-5, ARC 5984, RP2333-156-8, Samridhi, Surekha, Phalguna
Stem borer	TKM 6
Blast	Etep, Tadukan, Zenith, Co4, Moroberekan, IRAT3, IAC25, Tride 1
Bacterial blight	BJ1, TKM6. Zenith, Nira, Java 14, Wase Aikoku
Sheath blight	T141, ARC 15368, OS4, BCP3, Saibham, Bhuhjan, Saduwee. Laka
Brown spot	T 141, BAM10,Ch13, Ch45
Rice tungro	Ptb2, Ptb18, ADT21, ARC 10599, ARC 14320, ARC 14766

Table 9 Rice varieties with resistance to diseases and insect pests

Pest	Variety
Blast	ADT 36, ADT 38, ASD 18, Athira, Amrut, Avinash, BK 79, Bhadur, Bhagoil, Bhavani, Birupa, Co 43, Co 45, CR 1002, GR 102, GR 101, GR 103, Gaur 1, Gauri, Govind, Ghanteswari, Indryani, IR 64, IET 7564, Janaki, Jaya, Jayasree, Jayanthi, Kairali, Kanchana, Karjat 1, Karjat 2, Khandagiri, Kitish, Kunti, Kushal, Laxmi, Manhar, Mahaveer, MTU 9993, Neeraja, Narendra 359, Netravati, Onam, Palghar 1, Pantdhan 4, Pantdhan 6, Pantdhan 10, Pantdhan 11, Pranava, Patel 85, Penna, Pusa 169, Pusa 205, Pavizham, Prabhat, Punshi, Phew Oibi, PR 106, Radha, Rangoli, Rajavadlu, RP 2421, Safri 17, Sagarsambha, Sakoli 6, Salivahna, Sambamahsuri, Sarjoo 52, Savithri, SKL 47-8, Shaktiman, Sneha, Sonamahsuri, Sujatha, Sarbani, Swathi, Swarnadhan, Swarnamukhi, Sye 1, TKM 9, TKM 10, TPS 10, TPS 3, Tulasi, VL Dhan 221, Vytilla 4, TRC Borodhan
BLB	Ajaya, PR 110, PR 111, Pantdhan 12
RTV	Nidhi, Viramarya, Bharani, Vasundhara, CORH 1
GLH	Vikramarya, Lalat
BPH	HKR 126, Nandi, Arvindar, PMK 2, ADT 42, TPS 3, PKU 2, Kanakam, Chaitanya, Krishnaveni, Vajram, Pratibha, Macom, Pavizham, Manasarovar, Chandana, Nagarjuna, Sonasali, Rasmi, Jyothi, Bhadra, Neela, Annanga, Daya, Aruna, Remya, Bharathidasan, Karthika, Nandi, Co 46
WBPH	HKR 120
Stem borer	Sasyasri, Vikas, Ratna
Gall midge Biotype 1	Samridhi, Phalguna, Surekha, Abhaya, Bhuban, Samlei, Shakti, Suraksha, Daya, Pratap, Mahamaya, Sarsa, Neela, Lalat, Dhanyalaxmi, Tara, Khira, Shaktiman
Biotype 2	Phalguna, Surekha, Abhaya, Suraksha
Biotype 3	Rajendradhan 202, Karna, Ruchi, Divya
Biotype 4	Abhaya, Suraksha
Biotype 5	Triguna

For blast, in areas where the incidence of blast is high, durable resistance seems to have an advantage over vertical resistance. Breeding for durable resistance requires the use of donor that has durable resistance. Early generations should be screened in an upland nursery with appropriate susceptible check (IR20) and resistant check (IR36). F_2 populations should be screened as single plants and F_3-F_6 populations as single rows. Select lines with susceptible type of lesions but with resistance similar to or greater than IR36. Evaluate elite lines for neck blast in three rows replicated trials and evaluate elite lines for neck blast in lowland field trials.

At least 15 genes have been identified that are associated with resistance to bacterial blight and breeding for resistance has been successful in many irrigated areas. Quantitative resistance to bacterial blight can also be exploited. The resulting resistance is partial but appears to be more durable than that conveyed by major genes. Early generations and breeding lines can be screened under artificial inoculation by clipping method using standard evaluation system of IRRI and resistant lines can be selected.

The inheritance of sheath blight resistance is not well understood. Heritability is low indicating that selection in early generations will not be effective in developing resistant lines. Advanced breeding lines can be artificially inoculated and screened and highly susceptible lines can be discarded. While screening, sclerotia from fungi grown in rice hulls can be applied to the leaf whorl to increase the level of infection.

Resistance against stem borer appears to be inherited polygenically and therefore, is difficult to transfer; however, breeders have been successful in transferring moderate resistance in improved varieties. Recurrent selection using diallel mating system and male sterile facilitated inter mating, has been used at IRRI to increase levels of resistance to yellow stem borer.

Resistance against gall midge is inherited simply. However, in choosing a donor for resistance, it is important to be sure that the resistance applies to the prevailing biotype. Presence of single silver shoot per plant can be the criterion of susceptibility while breeding for vertical resistance. Plants can be screened in the green house in the field for resistance to gall midge. If the development of new biotypes becomes a problem, more effort will be needed to look for the sources of horizontal resistance.

Resistance against brown plant hopper and green leafhopper is conferred by a series of major genes, most of which are dominant. Of the resistance genes against brown plant hopper, *Bhp1* and *bph2* are not effective in India but *Bph3* and later genes are effective. Resistance usually is determined through seed box screening using 7 days old seedlings infested with second or third instar nymph.

Grain Quality

Grain quality is determined by milling quality, cooking quality, grain size, shape and appearance. Quality factors that can be assessed visually are grain size, shape and chalkiness, chemical characteristics such as amylose content, gel consistency, gelatinization temperature are more difficult to measure but are equally important and a new variety must pass the ultimate test the taste test before it is ready to be released.

Grain size, shape and appearance

Preference for grain size and shape varies from one group of consumers to another. Some ethnic groups prefer short bold grains, some have a preference for medium long grains and long slender grains are preferred by others. In general, long grains are preferred in the Indian subcontinent, but in Southeast Asia, the demand is for the medium long grains. In temperate areas, short grains are preferred.

The length of the grain is determined after the grains are hulled. Grains are classified as (IRRI, 1996) extra long (more than 7.50 mm), long (6.61 to 7.50 mm), medium (5.51 to 6.60 mm) and short (less than 5.51 mm).

Grain shape is the ratio of length to width of brown rice grains. the classifications are slender, (more than 3.00), medium, (2.1 to 3.0), bold, (1.1 to 2.0) and round (less than 1.1)

Grain appearance is largely determined by the amount of chalkiness in the endosperm either on the dorsal side of the grain (white belly) or in the centre (white center) and the condition of the eye as in some varieties the grain tends to break more frequently at the eye when it is milled. In most regions, rice that has clear endosperm is preferred over rice that has opaque endosperm, even though all grains are equally translucent after cooking and clarity of endosperm does not affect the rice taste or texture. Chalkiness is caused by loose packing of starch and protein particles. Chalkiness is under genetic control and most breeders experience show that it is fixed in early generations. So strict selection against chalkiness should begin as early as possible i.e. F_3 grains from F_2 plants should be evaluated by recording average score of 5 or 10 grains per plant by classification as none (nil), small (less than 10 percent), medium (11 to 20 percent) and large (more than 20 percent).

Cooking and eating characteristics are determined by starch properties that make up ninty percent of the milled rice. It consists of amylose, which is the linear fraction and amylopectin, which is the branched fraction and classified as Waxy (0-2% amylose), low amylose (8-20 % amylose), intermediate amylose (21-25 percent) and high amylose (more than 25% amylose).

Low amylose rices are moist, sticky and glossy when cooked, they split and disintegrate when over cooked. Most Japonica rices have low amylose content. High amylose rices, which are widely grown in Asia, cook dry and fluffy but become hard when cool. Rices with intermediate amylose content approach the ideal for many consumers, they are fluffy when cooked and remain soft when cool. Breeding for specific amylose content is not difficult because the trait is simply inherited. However directly determining the amylose content is difficult because the procedure is expensive, slow and delicate. However, based on alkali test for gelatinization temperature, rices with different amylose content can be identified as varieties that have high gelatinization temperature always have low amylose content and rices that have intermediate amylose content, rarely have low amylose content and can be retained.

Cooking quality can be tested by determining a rice gelatinization temperature, the temperature at which water is absorbed and starch granules swell irreversibly. It ranges from 55 to 79°C and rices can be classified as low (below 70□C), intermediate (70-74□C) and high (Above 74□C).

Gelatinization temperature is estimated by the extent of alkali spreading and clearing milled rice soaked in 1.7% protein hydroxide for 23 hours at room temperature. Rices with low gelatinization temperature disintegrate completely, where as those with intermediate gelatization temperature shows partial disintegration and those with high gelatinization temperature remain largely unaffected in the alkali solution. Rices with intermediate gelatinization temperature is preferred as rices with high gelatinization temperature becomes excessively soft and disintegrate when over cooked. They also require more water and time for cooking.

Crosses using a high gelatinization parent always produce many high gelatinizing F_2 plants. Crosses between japonicas (which has low gelatinization temperature and indicas also produce high gelatinizing segregants. Because high gelatinization temperature is associated with low amylose content, such high lines can be discarded.

Gel consistency is also an important characteristics associated with quality. Rices with soft gel consistency are preferred. Rices that have less than 24% amylose, usually have soft gel consistency.

Milling quality is the ability of the rice grain to resist breaking when mechanically hulled. Milling recovery of the rough rice is the estimate of the quantity of head rice and total milled rice that can be produced from a unit of rough rice. It is generally expressed as percentage. To determine the milling yield, a given sample of clean rough rice is passed through a shelling device to remove the hulls from kernels. The dehulled or brown rice is then milled to remove the bran and the embryos. Generally, the hulls form 20 to 22% of the rough rice although variations from 18 to 26% have been recorded and bran and embryos constitute another 8 to 10%. Thus from a given sample of rough rice., about 70% milled rice is obtained. The milled rice is then separated into broken and whole grains with a sieving device. The proportion of whole grains is known as head rice recovery. The head rice recovery may vary from as low as 25% to as high as 65%. head rice recovery is dependent upon grain size, shape and appearance. In general, varieties with long and long bold grains and those having chalky grains give lower head rice recovery. Varieties with medium long, slender and translucent grains give the best head rice recovery. Direct measurement for milling quality should begin when elite lines are bulked for preliminary yield trials. A 1 kg sample of grain dried to 14% moisture content is dehulled and milled with laboratory equipment. The results produced by a 1 kg sample generally correlates with those produced by large commercial mills. Smaller sample produces less reliable results.

Grain size, appearance, amylose content, gelatinization temperature, gel consistency are useful for classifying lines into different groups. But cooking quality still varies between lies within a group. It is therefore essential for the breeder to cook and taste samples of hot and cold rice of promising lines. Rough rice should be stored for atleast four months after harvest before taste test are conducted. Storage helps rice to absorb more water and expand more during cooking, resulting in flaky cooked rice. One important aspect of cooking quality is volume expansion. The volume of rice grain expands 300-400% during cooking. Most varieties produce grains that expand evenly, but some lengthen by as much as 100% without any expansion in girth. Prominent examples of such variety are Basmati rices of Punjab.

The scent or aroma of rice is a minor characteristic. Still aromatic rices are preferred in some parts of Asia such as Basmati rices of India and Pakistan. Basmati rices are characterized by intermediate amylose and gelatinization temperature, soft gel consistency, strong aroma and high degree of grain elongation. In some areas, however, aromatic rice is not preferred. Therefore breeders must be cautious of preferences of the area before taking the breeding programme.

8. INTERNATIONAL AND NATIONAL ORGANIZATIONS INVOLVED IN CROP IMPROVEMENT

(A) International Organizations

1. International Rice Research Institute (IRRI), Philippines – IRRI was established in 1960 by the Ford Foundation and Rockefeller Foundation with the support of the Government of Philippines and consists of laboratory and training facilities on a 252 ha. experimental farm in Los Banos, Philippines.
2. Centro Internacional de Agricultura Tropical (CIAT) - (International Center for Tropical Agriculture), Cali, Colombia – CIAT was established in 1967 by the Rockefeller Foundation and Ford Foundation with support from Government of Colombia. The Center has 5000 ha. area at its headquarter Palmira in the Cauca Valley of Colombia. CIAT works for improving the nutritional and economic well being of the rice growers and low-income consumers in Latin America and the Caribbean.

(B) National Organizations

Central Rice Research Institute, Cuttack, Directorate of Rice Research, Hyderabad and Indian Agricultural Research Institute, New Delhi. Different State Agricultural Universities situated in different states of India have established research stations for research in rice improvement and it is known that more than 160 research stations are engaged in rice research work in India.

9. LIST OF VARIETIES

Table 10. Popular rice varieties for different ecosystems in India

Ecosystem	Variety
Rainfed upland	Tulasi, Aditya, Prasanna, Rasi, Satya, Rudrama, PNR 381, Vandana, Birsadhan 101, Birsadhan 104, Birsadhan 105, Birsadhan 201, Birsadhan 202, Heera, Kanchan, Kalinga III, Annada, GR 2, GR 5, GR 6, Himdhan, Nagardhan, VLK Dhan, VL Dhan 221, RP2421, Himalaya 741, Himalaya 2216, Amrut, Mukti, IET 7564, Suvama, Modan, Onam, JR 75, R281-31-1, Ratnagiri 73-1-41, Kranti, Ngoba, Parijat, Pathara, Kalyani II, Sattari, Neela, Rudra, Vanprabha, Khandagiri, Nilagiri, Ghateshwari, Sneha, Patnai 23, Aswani, Renu, Saket 4, Narendradhan 18, Narendradhan 80, Narendradhan 97, Narendradhan 118, Khitish, Kiran, Bhupen, Poornima.
Rainfed shallow	Manasarovar, Swarnadhan, Phalguna, Mandya, Vijaya Nandi, Swarna, Sambamahsuri, Pinakini, Krishnaveni, Thikkana, Chaitanya, Pothana, Orugallu, Sri Ranga, Sagarasamba, Simhapuri, Lakshmi, Salivahana,

	Bahadur, Kushal, Ranjit, Manoharsali, Mahsuri, Moniram, Rangili, Bhogali, pankaj, Savitri, Sita, Jayashree, Janaki, Radha, CR 1002, Rajashree, Kanak, Vaidehi, Hemavathi, Nethravathi, IET 7191, Abhilash, Intan, KPH 2, Neerja, Rashmi, Kanyamkulam, Shyamla, Safri 17, Kranti, Surekha, Mahamaya, Ratnagiri, SYE 75, SYE-ER-1, Ratnagiri 3, Jagannath, NEH Mega Rice 1, NEH Mega Rice 2, Rajeswari, Seema, Parijat, CR 1014, Jajati, Urbashi, Samlei, Pratap, Saradhi, Gauri, Daya, Mahalakshmi, Lakshmi, Savitri, Madhukar, Jayalakshmi, Suresh, Dinesh, Biraj, Bipsa, IR 42
Semi deep	Amulya, Nalini, Jogen, Sabita, Biraj, Jaladhi 1, Suha, Jaladhi 2, Vaidehi, Jalmagna, Rambha, Utkal Prabha, Manika, Mahalaxmi, Kanchan, Panidhan, Jallahri, Jalnidhi, Jalapriya, Jitendra, Madhukar, Chakia 59, Natina, Mandira, Matangini, Purendra
Irrigated	Rasi, Vikas, Prabhat, Abhaya, Rajendra, Divya, Sasyashree, Bhaddrakali, Suraksha, IR 64, Ajaya, Vikaramarya, Vibhaya, Satya, Saleem, Surekha, Kavya, Erramallelu, Puas 2-21, Pusa 44-23, Gautam, CR 1002, Archana, Rajendradhan 201, Rajendradhan 202, Jaya, Vikram, Karjat 2, Karjat 3, Ratnagiri 3, Sugandha, Ambika, Jaya, Narmada, Ratna, IR20, IR 36, GR 4, GR 6, Gaur 10, GR 102, GR 103, HKR 102, Haryana Basmati, Taroari Basmati, Basmati 370, Pusa 44, Pusa Basmati 1, Kasturi, PR 103, HKR 126, RP732, Himalaya 2, Himalaya 732, Ranbir Basmati, Jhelum, K78-13, SKAU 23, SKAU 27, Akash, Avinash, Mahaveer, Prakash, Vikram, Vibhava, Sonasali, Annapurna, Mandya Vani, Triveni, Jyothi, Kartika, Makom, Sabari, Jayathi, Kanchana, Swarnaprabha, Aishwarya, Aruna, Remya, Kanakam, Mata Triveni, Kairali, Ruchi, Mahamaya, Patel 85, Madhuri, Pawana, Sugandha, Punshi, Maniphouibi 1, Maniphouibi 2, PR 106, PR 109, PR 108, Basmati 385, PR 111, Bharatidasan, Jawahar, Aravinder, Punithavathi, Savitri, Puduvai Ponni, BK 79, BK 190, Mahisugandha, IR 50, TKM 9, PMK 1, ADT 17, IR 20, Co 41, Co 43, Co 44, Co 42, Co 45, TRC Boro Dhan, Saket 4, Sarjoo 52, Narendra Dhan 359, Govind, TPS 1, MDU 2, MDU 3, ADT 39, ADT 42, JJ 92, ASD 17, White Ponni, Ponni, Paiyur, ADT 40, Madhukar, Jayalakshmi 10, Pantdhan 12, Manhar, Narendradhan 2, Kunti, Lakshmi, Munal
Saline area	CST 7-1, Lunishree, CSR 10, Panvel 2, Vytilla 2, Vytilla 3, Vytilla 4, CSR 5
Alkaline area	Vikas, CSR-5

Table 11. Description of important varieties

Situation	Soil type	Varieties	Duration (Days)	Yield (q/ha)	Resistance
Upland	All	Kalinga	85	30	LS grain
		Aditya	105	35	Blast
		Vanprabha	90	30	Semi tall
		Tulsi	105	35	Blast
		Poomima	105	40	LS grain, Drought
Lowland Rainfed					
	Matasi	Annada	105	35	Drought
		Tulsi	105	35	Blast
		Poomima	105	40	LS grain, Drought
	Dorsa	IR 36	115	45	Blast, BLB, GM
		IR 64	112	45	BLB, BPH
		Kranti	128	55	Drought
		Mahamaya	126	60	BLB, GM, Sh.R. BS
	Kanhar	Kranti	128	55	Drought
		Mahamaya	126	60	BLB, GM, Sh.R. BS

Behra	Safri 17	140	45	Photosensitive, Dr
	Swarna	150	60	LS grain
	Shyamla	135	45	Purple leaves
Irrigated situation	IR 36, IR 64, Abhaya, Kranti, Mahamaya, Madhuri, Pusa Basmati-1, Tarori Basmati			

10. REQUIREMENTS FOR SEED PRODUCTION

As in other crops, there are four classes of seed that are used in seed production programmes- Breeder seed, Foundation seed, Registered seed and Certified seed. Seed has to meet certain requirements before it will be certified for distribution as improved seed. These requirements include genetic purity, physical purity, freedom from weeds, freedom from disease and pests, high germination capacity etc.

Genetic purity means absence of seeds of other varieties of the same crop species as well as that of other crop species. It ensures that the seed is of the variety under certification. Physical purity implies freedom from inert matter (sand, pebbles, soil particles, straw) and defective seed (broken, shriveled, diseased, insect infested). Rice crop should have minimum of 98% purity. A high percentage of germination is necessary to obtain a good crop stand. Rice crop should have minimum of 80 percent germination. Freedom from weed seed is necessary to prevent weeds from spreading through seed and to reduce losses caused by weeds. The maximum amount of weed seed permitted in rice is 0.1%. The certified seed must be free from seed borne diseases and should have moisture content of 13%.

11. MAJOR ACHIEVEMENTS IN CROP THROUGH BREEDING IN M.P. INDIA AND ABROAD

Varietal improvement work

Rice breeding in India, started in 1911 in undivided Bengal with the appointment of Dr. G.P. Hector as economic botanist with his headquarter at Decca. In 1912, Madras province, had the first crop specialist fully devoted to rice. Indian Council of Agricultural research was established in 1929, it initiated the rice research projects in many states and by 1950, there were 82 research stations devoted to rice in 14 states. These research stations released 445 improved varieties mainly by pure line method of selection. Some of these varieties like Manoharsali, SR 20B, FR 13A and Nonasali Basmati 370 and NP 370 are still being grown. Other popular varieties were MTU1, MTU 15, HR 19, Chinsurah 7 7, Kolamba strains, GEB 24, Co2, Co 25, Co 26, ASD 1, T 141, SR 26B, T 136, N 22 etc. The broad objectives of the early breeding work were earliness, Deep water and flood resistance, lodging resistance, drought resistance, Non-shedding grains, Dormancy of seed, control of wild rice, disease resistance and higher response to heavy manuring. Central Rice Research Institute was established in 1946. The F.A.O. sponsored *japonica* X *indica* project was started in South East Asian Countries in 1956 with aim of transferring the high yielding ability and response to heavy fertilizer inputs from *japonicas* to *indicas* which had tolerance to prevalent diseases and pests of the region. The project produced a total of 710 *japonica* x *indica* hybrids, however very limited success was achieved as only four varieties were released from the project. Mahsuri and Malinja in Malaysia, ADT 27 in India and Circna in Australia. The All India Co-ordinated Rice Improvement programme was started in India in 1965. The development of Taichung (Native) 1 from semidwarf mutant Dee-geo-woo-gen was the major event in rice in Asia. IR 8 was released by IRRI in 1966 and Jaya was released from hybridization programme of AICRIP in 1968. The post Jaya period of varietal improvement in India may broadly be grouped into four phases —

The first decade of varietal improvement for yield, adaptability and quality

The second decade of breeding for resistance to pests and diseases

The third decade of breeding for adaptability to diversified rainfed ecologies

The development and use of hybrid technology

The AICRIP identified 17 centrally released and 73 state released high yielding varieties of short and medium duration for irrigated ecosystem till 1979,. Some of the popular varieties among them were Jayanthi, IR 20, Padma, Ratna, Pusa 2-21, Saket 4, Annapurna, Jagannath and Pankaj.

Pest disease syndrome causes 25-30% yield loss in rice. The problem became serious with the introduction of high yielding semi-dwarf varieties. Appearance of new pests like bacterial blight, rice tungro virus, grassy stunt virus, sheath blight, brown planthopper, green leaf hopper, leaf folder and hispa in addition to the traditional ones like blast, stem borer, gall midge increased their virulence. The intensive breeding efforts led to the development of a large number of resistant varieties against these diseases and pests (Table 9). Similarly, a large number of varieties suited to different rice growing situations have been identified (Table 10).

To break the yield barrier of popularly grown dwarf plant types, research on hybrid rice was initiated in 1989 and has led to the release of more than 11 hybrid varieties. APHR 1, APHR 2, DRRH 1 in A.P.; MGR 1, CORH 1, ADTRH 1 in Tamilnadu; KRH 1, KRH 2 in Karnataka; CNRH 3 in W.B.; UPHR 17 in U.P. and KJTRH 1 in Maharashtra were released by respective state variety release committee and PHB 71 by central variety release committee. These hybrids yield 15% higher than the prevailing popular varieties.

REFERENCES

Annonymous (1994). Rice research in Chhattisgarh pp. 10-13. Indira Gandhi Krishi Vishwavidyalaya, Raipur, Chhattisgarh, India.

Bjorkman O., Pearcy, R.W. and Nobs, M.A. (1971). Annual Reports, Deptt. Plant Biol, Carnegie Institution, Stanford, California. P. 53.

Chandraratna, M.F. (1964). Genetics and Breeding of rice. Longmans, Green and Co. Ltd. London.

Chang, T.T. (1976). Evolution of rices. Euphytica 25 :426-440.

Chaudhary, R.C. (1999). Breeding rice for resistance to diseases and insect pests. In: Breeding and Genetics- Research priorities and Challenges edited by J.S. Nanda. Oxford and IBH Publishing co. Pvt. Ltd., New Delhi. pp. 143-167.

Gangadharan, G. (1985). Breeding pp. 77-109 in Rice research in India, Indian Council of Agricultural Research, New Delhi.

INGER (1996). Standard evaluation system for rice. International Rice Research Institute, Los Banos, Philippines.

IRRI (1997). Rice Almanac. International Rice Research Institute, Los Banos, Philippines.

Krishnaiah, K. (1998). Rice (*Oryza sativa*) research in India. Indian Journal of Agricultural Science 68 (8, special issue) : 385-95.

Macalinga M. Victor and Nordland, H. Flyod (???). Morphology of the rice plant. Module GM-1. Rice production training and research. International Rice Research Institute, Los Banos, Philippines.

Mackill, D.J.; Coffman, W.R. and Garrity, D.P. (1996). Rainfed lowland rice improvement. International Rice Research Institute, Los Banos, Philippines.

Nanda, J.S. (1997). Manual on Rice Breeding. Kalyani Publishers, New Delhi.

Paroda, R.S. and Tomar, J.B. (1997). Ideotype concept for different ecosystems in rice. *Oryza* 34: 94-101.

Reddy, A.P.K and Bentur, J.S. (1999) Insect and disease resistance in rice. In: Breeding and Genetics- Research priorities and Challenges edited by J.S. Nanda. Oxford and IBH Publishing co. Pvt. Ltd., New Delhi. pp. 143-167.

CHAPTER - 3

HYBRID RICE BREEDING FOR INCREASING YIELD

Rajesh Mishra

1. INTRODUCTION

Among the various options available to break the yield ceilings, exploitation of hybrid vigour through the development of F_2 hybrids is widely acknowledged as the potential tool.

In rice, heterosis was first reported by Jones (1926), which was confirmed by several scientists (Ramiah 1935, Idsumi 1936, Kadam *et al.,* 1937, Capinpin and Singh 1938, Ramiah and Ramaswamy, 1941, Borwn 1953, Oka 1957, Sen and Mitra 1958). None of these researchers, however, showed interest in commercial exploitation of this phenomenon. During the 1960's suggestion for commercial exploitation of heterosis in rice were made in India (Richharia 1962), USA (Stansel and Craigmiles, 1966), Japan (Shinjyo and Omura 1966) and China (Yuan 1966). Chinese scientists pursued the research intensively enough to attain practical application in rice breeding even with the difficulty in producing bulk quantities of hybrid seed. Chinese scientists developed and used commercial rice hybrids which increases yields by 20% over semi-dwarf varieties (Yuan *et al.,* 1994) without any significant different in input use (He *et al.,* 1984, 1987). In 1991 hybrid rice were cultivated about 17 million ha in China which constituted 55% of its total rice area and contributes 66% of total rice production of the country and 20% of the total rice production of the world. Virmani (1994) reviewed the literature and noted several reports showing significant heterosis, standard heterosis for yield, grains per panicle, grain weight and panicles per plant in different crosses.

2. HETEROSIS

2.1 Combining ability in relation to heterosis in rice

A number of reports on combining ability in rice indicating that both additive and non-additive gene action were important in the inheritance of yield components and several other agronomical traits. Yield components usually showed preponderance of GCA variance compared to SCA variance. Heterosis crosses showing high SCA for yield are usually derived from parents possessing high x high on high x poor GCA. Sometimes, average x poor general combiners show heterosis. Rahman *et al.,* (1981) reported high SCA value of certain crosses derived from poor combining parents. As a rule parents possessing high GCA also possess good *per se* performance, although exceptions to this rule are not uncommon.

2.2 Genetic basis of heterosis

Genetic basis of heterosis is agreed that various genetic mechanisms viz., accumulation of dominant genes, over dominance effects of genes and complementation between cytoplasmic and nuclear genetic factors, play an active role in the manifestation of this phenomenon. Biochemically, heterotic effect is the result of efficient integration and the timely operation of the essential individual enzymes and the optimum balance of the enzymatically catalyzed metabolic systems.

2.3 Prediction of heterosis

Several methods viz., *per se* performance, genetic diversity as determined through geographic origin, multivariate analysis using morphological and agronomic traits, isozyme and RFLP polymorphism, combining ability, mitochondria complementation etc., tried with variable success. Molecular markers (isozyme and RFLPs), make it possible to identify and tag heterotic gene block in different inbreds to transfer in the desired through appropriate breeding procedures.

3. MALE STERILITY SYSTEMS FOR HYBRID RICE BREEDING

Rice, being a self-pollinated crop requires use of an effective male-sterility system to develop and produce F_1 hybrid varieties. Major male-sterility systems used for hybrid rice breeding are cytoplasmic genetic and chemically induced male sterility system. Recently, photo period sensitive and thermo-sensitive genetic male sterility systems are also being tested for their usability.

Three line hybrid involves cytoplasmic-genetic male sterility system with CMS (A) line, maintainer (B) line and restorer (R) line This is caused by an interaction between genetic factor(s) present in cytoplasm and the nucleus. Absence of a sterility-inducing factor either in the cytoplasm or in nucleus makes a line male fertile. Presence of certain dominant restorer gene(s) in the nucleus makes a line capable of restoring fertility in the hybird derived from it and a CMS line.

A CMS line is multiplied by crossing it with its maintainer line, either by hand-crossing (to produce small quantities of seed) or by out-crossing in an isolated plot (to produce bulk quantities of seed). The CMS line is maintained by crossing it with maintainer line, which is morphologically, similar to 'A' line except male sterility. Restorer or 'R' line possesses dominant fertility-restoring genes and restorers fertility when crossed with the CMS line. Since this system involves the use of three lines (A,B and R lines), the hybrid developed by using this male sterility system are known as three line hybrid.

Procedures for identifying a CMS

It can be identified in inter-varietal reciprocal crosses and inter-specific crosses.

Identification of CMS sources by inter-varietal crosses

Differences in reciprocal crosses between varieties with respect to male sterility indicates cytoplasmic-genetic interaction (Fig. 1). The occurrence of high frequency of completely male sterile plants in BC_2 generation indicates that variety B is a donor of cytoplasmic factor inducing male sterility e.g., chinsuarah Boro II, ARC 13 and 24 etc.

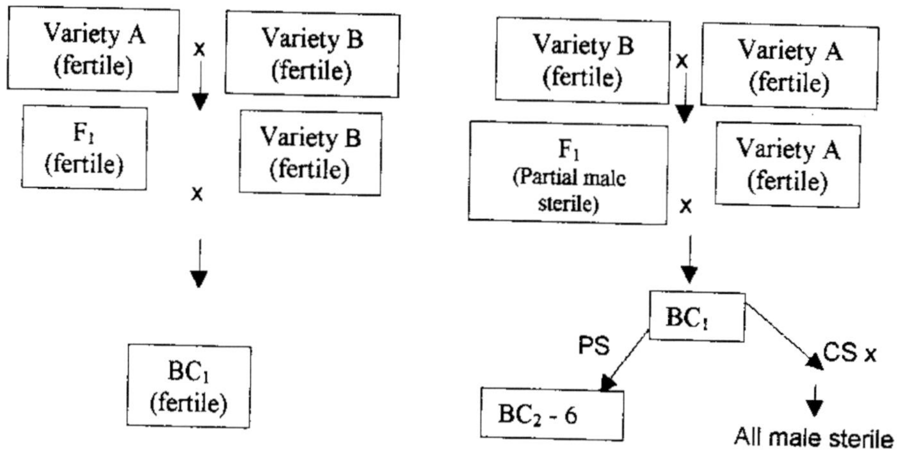

Fig. 1. Identification of new CMS sources in inter varietal crosses

Identification of CMS sources in inter-specific crosses

Crossing between wild species and cultivated varieties identify new CMS sources (Fig. 2).

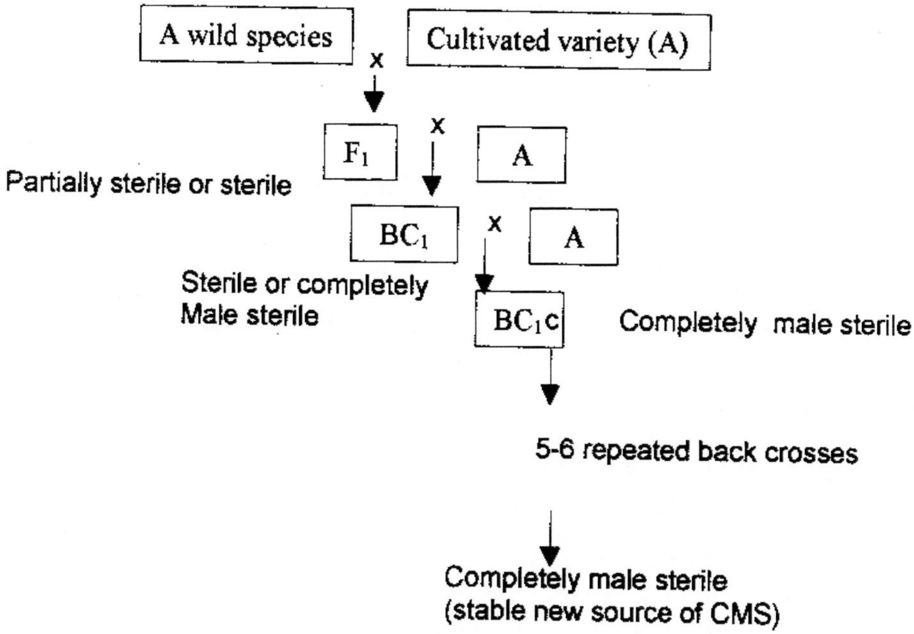

Fig. 2. Identification of new CMS sources in inter-specific crosses

Need for diversified CMS sources

Most of the rice hybrids cultivated are based on the 'WA' system of cyto-sterility (Table 1). Such over dependence on a single cyto-sterility source may be disastrous. Therefore, diversification of CMS sources should be an important component of a strong hybrid rice-breeding program.

Table 1. Some sources of male sterility inducing cytoplasm in rice

Designation	Cytoplamic Source	First Nuclear Donor
CMS – WA	Wild rice, abortive pollen	Zhen Shan 97, V20, V41
CMS – DA	Dwarf wild rice, abortive pollen	Xue Qin Zhao
CMS – IP	Indonesial Paddy	II-32
CMS – DI	Dissi-type	297
CMS – HL	Hong Lian	Lian – Tana Zhao
CMS –KR	*Oryza rufipogon*	Taichung 65
CMS – BT	Chinsurah boro II	Taichung 65
CMS – TN	TN 1	Pankhari 203
CMS – GAM	Gambiaca	Chao Yang 1
CMS – ARC	Assam Rice Collection	IR 10179-3-2-1
CMS–*O. parennis*	O. perennis	IR 64 R

Characteristics of a commercial usable CMS line

Although several cyto-sterility sources have been identified but the most commonly used cytosterility sources are WA, BT, DI, DA and IP because they have —

Stable male sterility over environments.

Adaptability to target environment.

Easy restorability, so that many elite lines can be used as male parents.

Good out-crossing ability to result in higher seed yield.

Good combining ability.

Good grain quality, so that rice hybrid can be developed with acceptable grain quality.

Restoring ability

Availability of a wide range of restorers is an essential prerequisite for exploitation of heterosis. The frequency of restorers varies between ecotype and geographic regions. *Indica* rices have higher frequency of restorers than *japonica* types with higher frequency of restores in *aman* and *boro* than *aus* types. *Bulu* rices are weaker restorers than *tejereh* cultivars of *Jawa*. Frequency of restorer line is generally higher in rice varieties originating from lower altitudes and south and South-east Asia and southern China than those from higher altitudes and in Northern China and Far Eastern Asia.

Environmental - sensitive-genic-male sterility system (EGMS)

Although the CMS system has been found most effective for developing commercial rice hybrid, but it is cumbersome because —

Seed is produced in two steps i.e. A x B multiplication and Ax R for F_1 production.

The Choice of male parents is limited to only those genotypes, which are identified as restorers.

Sometimes the sterility inducing cytoplasm exerts adverse negative effect on the expression of agronomic traits.

A CMS system may cause a genetic vulnerability of the derived hybrids if this system get associated with susceptibility to abiotic stress.

Used restricted to those germplasm in which maintainers and restorers are abundant.

A novel photoperiod sensitive genic male sterility (PGMS) and temperature-sensitive-genic male sterility (TGMS) in rice which reverts back to fertility under certain photoperiods or temperature regimes were reported by Shi, (1981) and Zhou *et al.,* (1988) respectively. Based on these systems, Yuan (1987) put forward a new strategy of hybrid, which did not involve a maintainer line, hence called the two line breeding method. Any fertile line could be used as pollen parent to develop rice hybrids through this system.

PGMS – Photoperiod sensitive genetic male sterility includes genic male sterile line, which responds to the photoperiod or duration of day length for expression of pollen sterility and fertility behaviours. Most of the PGMS lines remain male sterile under a long-day (>13.75^h) condition and revert back to fertility under short-day (<13.75^h) conditions.

TGMS – Thermo-sensitive genic male sterile lines are genic male sterile lines with alteration in expression of male sterility by change in temperature regimes. Most of the TGMS lines remain male sterile at a high temperature (maximum 30□C) and revert back to partial fertility at a lower temperature (maximum 30□C). Theoretically, thermo-sensitive sterility/fertility points vary from genotypes to genotypes. The critical thermo-sensitive stage for fertility alternation in the TGMS line varies from 15 to 25 days before heading or 5-15 days after panicle initiation.

In India, the TGMS (Fig. 3) system is more useful than PGMS, because of non-availability of long photoperiod, which is prerequisite for PGMS system. Number of TGMS lines viz., MDTG-4, DRITG-1, DRITG-2, DRITG-3, TS-16, TS-29 identified in India.

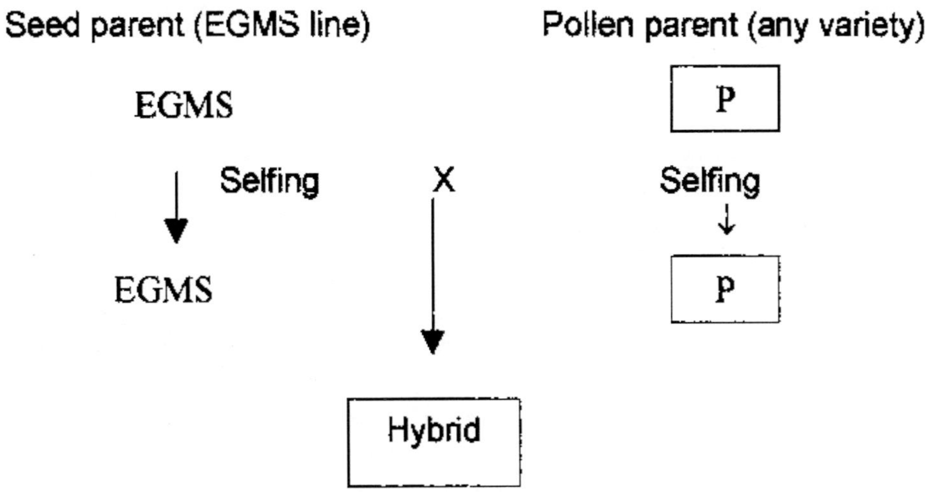

Fig. 3. Use of EGMS line for developing two-line hybrids

4. CHEMICALLY INDUCED MALE STERILITY

Male sterility is induced by spraying a chemical gametocide that can kill pollen grains without affecting the pistil. In hybrid seed production, two parents are planted in alternate strips. One is sprayed with chemicals at appropriate growth stage, and the other is used as pollen source to produce the hybrid seed. The ideal gametocides (Table 2) should have —

Selectively induce male sterility without adversely affecting the female fertility.

Systematic effects so as to sterilize both early and late panicles.

Broad range of effectively to withstand adverse environmental conditions.

Minimum side effects on plant growth and panicle development.

For developing hybrids by using gametocides, the female line must have a synchronous flowering habit even in the tillers, it should respond to chemical treatment and the parents should possess good out-crossing traits. The efficiency of chemical gametocides depends upon the correct dosages, appropriate stage of treatment and uniform coverage of chemical. In rice, chemically induced male sterility is of limited use due to availability of effective CMS and EGMS systems.

Table 2. Gametocides found useful in rice

Chemical	Concentration	Growth stage for application
Ethrel	8000-10,000 ppm	Pre-boot and boot stage
Monosodium Methyl Arsenate (MGI)	0.02% or 2000 ppm	Uni-nucleate, pollen stage
Sodium methyl	0.02% or 2000 ppm	5 days before heading

5. APOMIXIS

The limiting factor for wide-scale adoption of hybrid-rice in tropics and subtropics is the complexity of the seed production system based on cytoplasmic male sterility and influence of environment for EGMS system. Apomixis will increase the efficiency of hybrid rice breeders in producing many new true breeding hybrids compared with the hybrid produced by three or two-line breeding system apart form the use of seed every year. The availability of large number of hybrids will help increase genetic diversity and reduce genetic vulnerability. Apomixis is indicated by —

Occurrence of identical maternal progeny from plants of cross-pollinated species, or progeny of F_1 crosses

Limited or no genetic variation in F_2 population of a cross between two distinct parents

Recessive genotypes from a cross of parents with recessive genes pollinated with a parent possessing a dominant marker gene.

In aneuploids, triploids and wide crosses sterile aneuploid chromosome number or structural heterozygocity remains constant from parent to progeny.

Unusual high seed formation.

Multiple stigma, ovules per floret, fused ovaries and seedlings per seed.

Chen *et. al.,* (1988) have claimed apomixis in rice in China from an interspecific cross but was not confirmed. IRRI has initiated a search for apomixis in rice (Khush *et al.,* 1994) by analysis of the teleraploid wild germplasm of oryza, induction of mutations for apomixis and use of molecular approaches to engineer apomixis.

For arresting the development of the sexual embryo Kathiresan *et. al.* (2000) reported generation of an inhibitory protein under the control of an egg-or zygote–specific promoter. They adopt two genes approach (Cre-lox recombination system) because such promoter is not available in plants. The Cre recombinase is delivered through meiotic lineage to the zygote where it excises the lox box intervening the coding region of the gene that can cause ablation (Fig. 4). It necessitates isolation of a gene that is specifically expressed in meiotic cells and a favourable gene that expresses in the embryo but not in the embryosac or endosperm. REE 5–specific PCR and southern analyse indicated that the rice genome contains two REE 5 genes. By screening the IR 36 genomic library, they isolated clones containing the two REE 5 genes. To isolate the REE 5 gene that is expressed during early

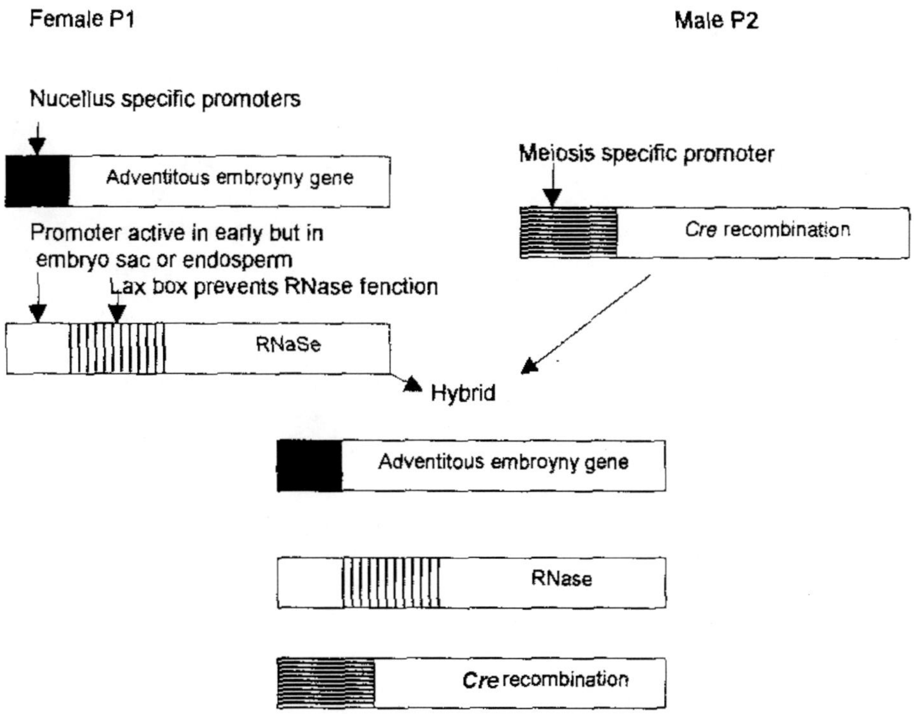

Fig 4. A Schematic diagram showing the constructs that can be used to derive an apomictic hybrid

embryogenesis, they performed 3' rapid amplification of C DNA ends (3' RACE) using spikelets at 2 days after fertilization. Screening of the genomic clones containing the two REE 5 genes using 3' untranslated regions derived from the 3' RACF clone revealed that REE 5 encodes a novel phospho-protein. The expression pattern of REE5 genes using RT-PCR, Northern and in situ hybridization techniques is under study.

6. ORGANIZATION OF HYBRID RICE BREEDING PROGRAMME USING CMS SYSTEM

Procedures for developing rice hybrids are quite distinct from those employed for breeding conventional varieties. In hybrid breeding productivity genes are assembled and exploited under a heterozygous condition for only one generation. On the contrary, conventional breeding involves the accumulations of productivity genes that perform well under a homozygous condition year after year. Hybrid rice breeding broadly covers (i) development of parental lines (ii) seed production of parental lines and experimental rice hybrids and (iii) evaluation of hybrids. For the efficient development of parental lines, breeding materials should be grouped into separate nurseries i.e. source, test cross, retest cross and back cross-nurseries. The genetic base of the material used should be as wide as possible. The evaluation of heterosis is the most crucial part of hybrid rice breeding and experimental hybrids should pass through each stage and finally tested in the farmer's field before their release.

7. DEVELOPMENT OF PARENTAL LINES

In order to develop parental lines for breeding rice hybrids it is extremely useful to establish a close linkage with inbred rice breeding programmes to have continuous access to genetically diverse elite lines possessing good per se performance, multiple disease and insect resistance and acceptable grain quality. In a three-line hybrid rice breeding system this step involves development of male sterile (A), maintainer (B) and restorer [R] lines.

(A) Source Nursery (SN)

This nursery includes elite lines, which have the potential to become parents of commercial hybrids. The best available CMS and TGMS lines are also included.

(B) CMS line maintenance and evaluation nursery (CMSN)

It is the breeding nursery in which the CMS lines, both developed locally and those introduced from outside, are maintained and evaluated.

(C) Test Cross Nursery (TN)

It is the breeding nursery, where F_1s of cytoplasmic male sterile lines and test varieties from the source nursery are screened for pollen sterility/fertility, spikelet fertility, and other agronomic traits to identify the potential maintenance and restorers and heterotic hybrids.

(D) Restorer Purification Nursery (RPN)

This breeding nursery comprises the progenies of the cross line and individual plants of restorer lines, which are selected for purification and seed multiplication process.

Restorers identified among released varieties are elite breeding lines have not been bred for fertility restoration. Therefore, they may show different restoration between and within test cross progenies with differential restoration, the progenies segregate for fertility restoration which would adversely affect the yield heterosis of the hybrid. Therefore, purification of promising restorers becomes extremely important before their use in producing rice hybrids.

Restorer purification involves re-test-crossing of a large number of single plants of a restorer line with the CMS line and evaluation of re-test cross F_1 progenies for uniformity or segregation for splikelet fertility. The progenies of the single plants of the restorer used in the re-testcrosses are grown to observe uniform looking restorer progenies, i.e. those that give normally fertile and uniform re-test cross F_1 progenies, are bulked to obtain a nucleus seed of restorer line.

(E) Back-Cross Nursery (BN)

A back-cross nursery is the back bone of an effective and efficient hybrid rice breeding programme. Breeder needs a wide range of CMS line to develop desired hybrids. CMS lines introduced from elsewhere may not be adaptable and/or suitable for a given target area. For example, Chinese CMS lines introduced to the tropics were susceptible to major diseases and insect pests besides poor in grain quality. Therefore, unsuitable for development of commercial rice hybrids for tropics. Thus, it is necessary to transfer the available CMS systems into elite lines well adapted to the tropics. This is feasible only when the elite lines are effective maintainers. After identification of maintainer, back crossing is initiated in the test-cross nursery. About 5-6 back crosses are required to transfer a CMS system in to genetic background of elite maintainer line. Back-cross progenies from BC_1 to BC_6 are handled in the back-cross nursery.

(F) Combining ability Nursery (CAN)

Assessment of the combining ability of parental lines is extremely useful in hybrid breeding programme, especially when a large number of prospective parental lines are available. Line x tester method (kempthorne, 1957) is commonly used for the purpose.

Nucleus and Breeder seed production of 'A', 'B', 'R' and TGMS lines

The availability of a genetic pure seed is a prerequisite for exploiting the full potential of hybrids. Lack of purity in parental lines and improper isolation in seed production are the major causes of poor quality of hybrid seed. With every 1% decrease in purity of the hybrid seed, the eventual yield loss in the F_1 hybrids is about 100 kg/ha.

Procedure for A, B and R line seed production

Nucleus and breeder seed of A, B and R lines can be produced simultaneously by the method described below —

Select about 50-100 typical and completely male sterile single plants of the selected CMS line.

Make 50-100 crosses of the selected CMS plants with corresponding single plants of the maintainer and restorer lines of the promising hybrids.

Plant few seeds of A x B crosses and all the seeds of A x R crosses in an identification nursery to select the best pairs that produce typical, uniform, and stable progeny.

Plant, balance seeds of A x B crosses and their corresponding B line progeny in an isolated multiplication block 21 days after the planting of A X B crosses in an identification nursery.

Based on the observations made in A x B crosses grown in the first plantings mark those which lack uniformity in growth and flowering and show lack of stable male sterility. Such A x B crosses and their corresponding B lines are removed from the A x B multiplication block before flowering. The remaining 'Ax B' pairs are allowed to get cross-pollinated to produce the nucleus seed of 'A' line.

Among the pairs of A x R crosses, identify those, which exhibit poor restoration and lack of uniformity.

Plant all the maintainer and restorer progenies of respective A x B crosses and A x R crosses in isolated plots for multiplication. Such lines whose F_1 progenies failed to meet the set standards based in the observations made in identification nursery are discarded and the remaining lines are bulked to form the nucleus seed of 'B' and 'R' lines.

The nucleus seed of 'A' and 'B' lines are used for producing breeder seed of the 'A' line. Plant the 'A' and 'B' lines in strictly isolated plots (preferably 100 m away from other rice varieties) to produce the breeder seed of 'A' line.

Plant the nucleus seed of 'B' and 'R' lines in isolated plots as per certification standard for producing the breeder seed of B and R lines.

Procedure for TGMS line seed production

TGMS lines multiplied continuously for several generations without any selection may segregate for critical sterility point. It causes major problems in maintaining purity. Therefore, nucleus and breeder seed production has to be taken up on a continual basis.

Nucleus seed production of a TGMS line is initiated in the fertility-inducing environment. Seeding of TGMS lines is arranged in such a way that the sensitive stage occurs when the temperature is favourable for higher seed set. At the time of flowering, about 100 typical plants are selected from the population of a TGMS line and their panicles are bagged. Selection process is completed within a week. After the harvest, the selected plants are scored for spikelet fertility (based on the main panicle) and 50 plants with higher spikelet fertility are selected.

Progenies of the selected plants are grown in the sterility inducing-environment. About 30 seeds are taken from each of the selected plants to grow single row progenies and the remaining seed is stored carefully. Progenies showing uniform and complete male sterile are marked. The balance seeds of such selected progenies are bulked to form the nucleus seed.

Breeder seed is produced in the fertility-inducing environment by adapting the method recommended for production of nucleus seed. Fresh breeder seed should be used by seed production agencies to produce foundation seed of TGMS lines. The latter should be used to produce hybrid seed.

Seed production of hybrid rice

The success and sustenance of hybrid rice technology depends economic and efficient production of adequate quantity of pure hybrid seed and economically viable and high level of heterosis. Significant cross-pollination occurs on male sterile plants, depending on their flowering behaviour, floral traits, synchronization of flowering, amount of pollen available from male parent and prevailing weather conditions. Both genetic improvement of parental lines for flowering behaviour and floral traits and modifications in seed production practices are very essential to boost the seed yield in seed production plots and reducing the seed cost.

Extensive seed production research on isolation distances, determination of seedling interval, synchronization and prediction/adjustment of flowering of parental lines, planting geometry, row ratio, application of GA_3, flag leaf clipping and supplementary pollination were carried out in India. An optimum seed production package with ensures seed yield of 1.5 to 2.0 ha^{-1} has been developed (Table 3) and popularized by Directorate of Rice Research (DRR), Hyderabad.

Table 3. Optimum package for seed production of hybrid rice and multiplication of CMS line

Activity	Particulars
Seed rate	Seed parent : 15 kg $ha.^{-1}$, pollen parent : 5 kg $ha.^{-1}$
Nursery	Sparse seeding (20 g/m^2) to ensure multi-tillered (4-5) seedlings in 25 days.
Row Ratio	2 B : 8 A for CMS multiplication; 2R: 10A for hybrid seed production.
Spacing	Male : Male = 30 cm; Male : Female = 20 cm; Female : Female = 15 cm, Plant : Plant = 15 cm.
GA_3 Application	45-60 g ha^{-1} in 500 lit. of water at 5% heading in two split dose on consecutive days.
Supplementary pollination	Two-three times a day at peak anthesis during flowering phase.
Roguing	Twice during vegetative phase: Based on morphological characters.

Agronomic management of hybrids

In order to realize the full yield potential of hybrids, there is need for an appropriate agronomic package. High grain yield of hybrid rice is associated to high vegetative biomes production, high leaf area and increased number of spikelets and filled grains. DRR carried out a extensive studies on agronomic management (Table 4) at different net work centers of hybrid rice.

Table 4. Recommended Agronomic package for cultivation of hybrids

Component	Recommended
Seed rate	15-20 kg ha^{-1}
Seed density (in nursery)	20 g/m^2
Spacing	15 x 15 on 20 x 15 cm.
Planting	01 seedlings $hill^{-1}$
Nitrogen dose	120-150 kg ha^{-1}(depending upon soil test)
Nitrogen Management	3 splits (1/2 basal + ¼ panicle initiation + ¼ booting)
Water Management	Cyclic of continuous submergence.

8. FUTURE OUTLOOK

Hybrid rice, a great success in China and during the next few years India, Vietnam and Philippines will be benefited from it. The technology is useful due to its higher yield potential and its capacity to increase rural employment opportunities through a hybrid seed industry. The potential of this technology is higher in countries like India and China, where a high proportion of irrigated rice area and a high land : labour ratio exist.

The current level of heterosis (15-20% or 0.75 to 1 t/ha^{-1}) in *indica* rice hybrids is economically viable, but a higher level, observed in *indica / japonica* hybrids, will be more attractive. This will tackle certain associated problems, viz., inter-varietal hybrid sterility, taller stature, late maturity and grain quality. Some *indica/japonica* rice hybrids are being developed in China whereas in tropical countries such hybrids can be developed during the next 5 years from *indica*/tropical *japonica* crosses.

Significant heterosis, observed for vegetative vigour and root characteristics, suggests that hybrid rice should be explored for abiotic stress environments, where transplanting is practiced or some seeding equipment is available for direct seeding with a reduced seed rate. The CMS system is most effective and practical for developing rice hybrids, but only few sources have been deployed to develop commercial rice hybrids. It is important to critically analyze the available sources for their utility in developing commercial rice hybrids and more genetically diverse CMS sources should be deployed for breeding commercial rice hybrids.

Rice hybrids with resistance to biotic stresses and acceptable grain quality can be developed by an appropriate choice of parental lines. Therefore, while selecting parental lines, their level of resistance to biotic stresses and grain quality should be monitored critically. Apomixis is the ultimate genetic tool to develop true breeding hybrids and facilitate commercial exploitation of heterosis by even resource poor farmers. So far, there is no confirmed report of occurrence of apomixis in rice, but research is in progress to discover, induce, and/or genetically engineer such rices.

Thus, hybrid rice holds great promise to increase rice production and contribute significantly in meeting the projected global rice demand in the 21st century.

REFERENCES

Ramaiah, K. and Ramaswamy, K. (1941). Hybrid vigor in rice, Indian J. Genet., 1: 4-12.

Brown, F.B. (1953). Hybrid vigor in rice. *Malay Agric. J.* 36 : 226 – 256.

Burton, G.W. (1983). Utilization of hybrid vigor In : Wood D.R. (ed), Crop Breeding. Am. Soc. Agron. Crop. Sci., Am., Madison WI, pp 89-107.

Capinpin, J.N. and Singh, P. (1938). A study of varietal crosses and hybrid vigour in rice Philpp. Agric. 27 : 255-277.

Chen, J.J., Lin, Y. and Jun, C.W. (1988). A study of apomixis in rice. Genet. Manipulation Crops Newslett., 4(2): 3-10.

Comstock, R.E. and Robinson, H.F. (1952). Estimation of average dominance of genes. In Heterosis '(J.W.Gowen ed.), pp. 494-516. Iowa State college press, Ames, IA.

Darwin, C. (1877). The effects of cross and self-fertilization in the vegetable kingdom. Dappleton, New York.

East, E.M. (1908). Inbreeding in corn. Rept. Connecticut Agric. Expt. Station for 1907, pp. 414-428.

East, E.M. (1936). Heterosis. Genetics, 21 : 375-397.

He, G.T., Te A., Zhu, X., Travens, S.L., Lai, X. and Herdt, R.W. (1984) The economics of hybrid rice production in China. IARI Res. Paper Sen., 101:14

He, G.T., Zhu, X., and Flinn, J.C. (1987) A comparative study of economic efficiency of hybrid and conventional rice production in Jiangsu Province China, Oryza 24: 285-296.

Idsumi, Y. (1936). Investigations in heterosis of rice plant. Proc. Crop Sci. Soc., Japan 8 (4): 504-515.

Jinks, J.L. (1983). Biometrical genetics of heterosis. In : Frankel, R. (ed). Heterosis – reappraisal of theory and practice. Springier Berlin Heidelbery, New York.

Jones, J.W. (1926). Hybrid vigor in rice. J. Am. Soc. Agron., 18: 424-428.

Kadam, B.S.,Patil, G.G. and Patankar, V.K. (1937). Heterosis in rice. Indian J. Agric. Sci., 7 : 118-126.

Kathiresan, A., Khush, G.S. and Bennett, J. (2000). Molecular tools for zygotic embryo ablation in rice, IRRN 25 (3) : 16.

Kempthorne, O. (1957). An introduction of genetical statistics. I edn, New York, pp 458-571.

Khush, G.S., Brar, D.S.; Bennett, J. and Virmani, S.S. (1994). Apomixis for rice improvement. In "Apomixis : Exploiting Hybrid Vigor in Rice", (Khush, G.S., ed.) pp. 1-21. Int. Rice Res. Inst. Manila, Philippines.

Mendel, G. (1865). Versuche uber pttlanzen hybriden. Naturuf. Ver. in Brunn. Ver. 4: 3-47.

Oka, H.I. (1957). Phylogenetic differentiation of the cultivated rice. XVIII. Sterility and vigor of lines derived from hybrid between distant varieties of rice in crosses with parental varieties. Jpn. J. Breed., 1: 7-11.

Rahman, M., Patwary, A.K. and Miah, A.J. (1981) Combining ability in rice. Indian I. Agric. Sci., 51:543-546.

Ramiah, K. (1935) Rice genetics. Proc. Ass. Eco. Biol., Coimbatore 3: 51-61.

Richharia, R.H. (1962). Clonal propagation as a practical means of exploiting hybrid vigor in rice. Nature, 194: 598.

Sen, P.K. and Mitra, G.N. (1958). Inheritance of photoperiodic reaction in rice. Nature, 182: 119.

Sharma, H.L. and Virmani, S.S. (1990). Some varietal sources of twin seedlings trait in rice. Rice Genet. Newsl., 7:134-135.

Shi, M.S. (1981). Preliminary reports of later japonica nature 2- lines and application. Hubei Agric. Sci., 7.

Shinjyo, C. and Omura, T. (1966). Cytoplasmic male sterility in cultivated rice, Oryza sativa L.I. Fertilities of F_1, F_2 and off-springs obtained from their mutual reciprocal back crosses and segregation of completely male sterile plants. Jpn. J. Breed., 16 (suppl. 1) : 179-180.

Sprague, G.S. and Tatum, L.A. (1942). General and specific combining ability in single crosses of corn. Am. Soc. Agron., 34: 923-932.

Stansel J.W. and Craigmiles, J.P. (1966). Hybrid rice: Problems and potentials, Rice J., 69(5): 14-15.

Suh, H.S. (1985). Studies on multiple pistillate male sterile rice. Korean J. Breed., 17: 439-443.

Suh, H.S. (1988). Multiple pistillate male sterile rice for hybrid seed production. In "Hybrid Rice", pp. 181-187. Int. Rice Res. Inst., Manila, Philippine.

Virmani, S.S. (1994). "Heterosis and Hybrid Rice Breeding". Springier verla, Berlin.

Virmani, S.S.(1996). Hybrid Rice. Adv. Agro., 57 : 378-462.

Yuan, L.P. (1966) A preliminary report on male sterility in rice. Sci. Bull., 4:32-34

Yuan, L.P. (1987). Strategy conception of hybrid rice breeding. Hybrid Rice 1,: 1-4.

Yuan, L.P., Li, Y.C. and Deng, H.D. (1990). Progress of studies in rice twin seedlings. Paper presented at the Fourth Annual Meeting of the Rockefeller Foundation's International programme on Rice Biotechnology, May 9-12,1990. Int. Rice Res. Inst. Manila, Philippines.

Yuan, L.P., Yang, Z.Y. and Yang, J.B. (1994). Hybrid rice in China. In "Hybrid Rice Technology : New Developments and Future prospects (S.S.Virmani, Ed.), pp. 143-147. Int. Rice Inst., Manila, Philippines.

Zhou, T.B., Xiao, H.C., Lei, D.Y. and Duan Q.X. (1988). The breeding of indica photosensitive male sterile line. J. Hunan Agric. Sci., 6 : 16-18.

CHAPTER - 4

MAIZE

V.K. Rastogi

1. TAXONOMICAL CLASSIFICATION, ORIGIN AND CHROMOSOME NUMBER

Corn is the only species [*mays]* of genus *Zea* of family Gramineae and tribe Maydeae, which, possesses naked kernels. It has following seven sub species/groups

Zea mays indurata Stuart. : Flint Corn

(Well-developed starchy tissue enclosed by the horny endosperm)

Zea mays indentata Stuart : Dent Corn

(Outer surface depression is caused by the shrinkage of the starch matter in drying)

Zea mays everta Staurt : Pop Corn

(Size of kernels and ears are small and having horny endosperm. It excludes starchy tissues, which gives the property of popping.

Zea mays amylacea Staurt : Soft Corn of Flour Corn

(No corneous endosperm)

Zea mays tunicata Staurt : Pod Corn

(The kernels of ear are enclosed in husk constituting together a pod for each single kernel)

Zea mays saccharata Staurt : Sweet Corn

(Wrinkled and more or less translucent seed, Starch is absent and is replaced by sugar dextrines)

Zea mays ceratina Kulesh : Waxy Corn

Zea mays has 2n = 20 chromosomes. The individual ten chromosomes comprising the gametic number are morphologically distinguished. Maize and teosinte (*Euchlaena mexicana*) each with 2 n = 20 cross readily and produce fertile progenies. *Tripsacum* species have 2n = 36 and 72 chromosomes.

According to Mangelsdorf and Reeves (1959), first cultivated form of maize should have derived by mutation from a wild form of pod maize. Teosinte (*Euchlaena*) is a natural hybrid of *Zea* and gamma-grass (*Tripsacum*) which occurred after cultivated maize was brought to Central America. From America, it had been brought to South East Asia about the sixteenth century by Portuguese traders. South American centre has been suggested as the possible origin of maize.

2. CLIMATIC REQUIREMENT, AREA AND DISTRIBUTION

The maize is a warm weather plant and is grown generally at all the latitudes except where it is too cold or the growing season is too short. It can be successfully grown under varying environmental conditions especially *kharif*, *rabi* and spring at 10 to 2680 msl with pH from 5.4 to 8.5. The soil type, where maize is generally cultivated, varies from sandy, clay, medium black to silty clay loam. In addition, about 80 per cent area to maize belt being rain dependent remains exposed to vagaries, either excess water or drought.

3. SALIENT FEATURES OF CROP CULTIVATION

The maize is emerging as third most important cereal after only wheat and rice, in the world agriculture with respect to both area and productivity. Maize being a C-4 plant, it has best physiological efficiency and has a productivity as high 3300 kg/ha against 2200 and 2500 kg/ha of wheat and rice, respectively. Maize is traditionally grown during monsoon. In winter 6-8 t/ha average grain yield is harvested to the absence of problems viz., water logging, drought, disease etc. Sowing of maize crop in rows with a pre-emergence application of Atrazin @ 0.5 kg a.i. per ha followed by 1 to 2 inter-culture operations control weeds. Fifty to seventy cm of well-distributed rain is sufficient for proper growth of maize. The deep medium textured soil preferably high in organic manure, well drained with good water retentive and pH 6.5 to 7.5 is considered to be the most desirable.

Seed should be planted 3-5 cm deep and across the slope. Planting on flat bed followed by earthing up is found best in the plains while the planting on the side of ridge is recommended for high rainfall area. Dry sowing with *palewa* irrigation is found better over late sowing. The placement of compost/FYM in furrow along with basal dose of 50 : 60 : 40 NPK kg/ha is necessary. Row to row and plant to plant distances of 60 x 25 cm in composite and 75 x 20 cm in hybrids may maintain 75000 and 65000 plants per ha respectively. Thinning after 10 days; first top dressing of urea (25 kg N/ha) at knee high stage followed by *dora* operation and weeding; second top dressing of urea at pre tasselling stage followed by *dora* with sarpati for earthing up are essential.

Irrigation is necessary at long moisture stress especially at flowering. Early composite varieties for low rainfall area and double cropping rainfed, while hybrid seed with irrigation under good management are recommended. Urid and cowpea can be successfully inter-cropped in between the two rows of maize (75 cm. apart). One spray of 0.1% Endosulphan 35EC or application of 4% Endosulphan granules @ 8-10 kg/ha (2-3 granules per plant) on 10-15 days old, minimise the loss due to borers.

4. BOTANICAL DESCRIPTION OF THE PLANT

Plants are herbaceous tall (up to 4 meter) and often single stalked. Roots are adventitious. Seminal root (main) develops from the radicle with its lateral roots. Stilt and fibrous, developing from the lower nodes of the stem thick and often pigmented. The main roots may penetrate to depths of 2 metre. Two to four cm thick, cylindrical, solid, slightly furrowed stem possess an average 14 internodes (8-20), those are thick at the base and tapering to-wards the tassel. Leaves distichous, each with leaf sheath, ligule, auricle and lamina. Sheath entire below and split above; ligule is membranous, fringed with hairs and closely invests the stem; auricle broad; lamina long and ribbon like, with a strong mid rib and waxy.

Inflorescence: The maize is a monoecious plant. The male inflorescence called *tassel* is terminal on the main axis. The spikelets are arranged in pairs, one sessile and another pedicelled both being identical in size, shape and structure. The female inflorescence known as the *'cob'* or the *'ear'* may be considered as a modified lateral branch derived from an axillary bud of the main stem. The internodes become short from a central stout axis and the overlapping sheaths of the modified leaves cover the inflorescence, forming the "*husk*" of the ear. The soft and succulent spikelets in 8 to 12 longitudinal rows are closely attached to the thickened central axis. The lemma and paleas are thin, membranous and broad. The fertile pistillate flower is found with the second lemma. The ovary has a long slender style (36 to 40 cm), emerging from the top of the husk as *silk*.

Floral biology

Anthesis: the maize plant is protandrous. In staminate inflorescence, spikelets closer to the tip of the tassel open first then anthesis proceeds downwards and complete in 14 days. The receptive style of each ovary grows in length and exerts out of the husk and after fertilization, the style begins to wither.

Pollination: About 95% of the ovules on a shoot are formed by cross-pollination. Pollen may be carried up to one kilometre by wind. A single tassel may produce as many as 2,50,00,000 pollen grains. Each shoot is composed of a shank from which the husks arise and terminates in the ear on which the pistillate flowers are borne. Fresh silks function as both style and stigma and are receptive to the fresh pollens through out their entire length. Fertilization of the ovules usually occurs with in 12 to 28 hours after pollination.

Stages of plant development

During germination, the primary root or the radicle bursts through the *coleorhiza* and the *coleoptile* emerges later. In about 45 days root, stem and leaf are fully developed. With the initiation of tassel at the top of the plants and simultaneously the silk appears on the mid node. After fertilization, grain development stage starts, which continued upto dough stage then maturity and drying of grain starts. The seed to seed cycle completes in 90 to 120.

5. SELFING AND CROSSING PROCEDURE

System of controlled pollination: The ear (spadix) is covered with a transparent bag before emergence of silk (Silking). While tassel with anthesis in $1/4^{th}$ part is also covered with a transparent bag (Tassling). The date of pollination will be one day later than the date of tasselling. After tasselling and silking, pollination can be done according to the programme of selfing or crossing.

Self-Pollination: The pollen is collected after 4-6 days from the first day of anthesis in tassel bag. Next day, silk of the same plant is pollinated by the pollen collected from the tassel of the same plant. Then the spadix is covered by the same tassel bag. The tassel bags are stapled in the back of the stem of the plants.

Sib-Pollination: Sib mating (sister-brother mating, half sib (one parent common) and full sib (mating of plant within same inbred in all possible combination) population is developed by making crosses between desirable parents through silking and tasseling.

Cross-pollination: Depending upon the number and arrangement of the parent inbred lines single [(AXB)], double [(AXB) (CXD)], three way [(AXB) XC], top cross [(A X open pollinated variety)], back cross [(AXB) AX (CXD) XC] hybrids are formed.

6. BREEDING OBJECTIVES

1. *Yield per unit area:* Yield per unit area can be increased by introducing early maturity and number of ear per plant.
2. *Seed production of hybrid* : Freedom from leaf firing and early silking in relation to time of pollen shedding and its duration will be most beneficial attributes for consideration in the breeding of hybrids and composites.
3. *Industrial use:* Varieties with waxy endosperm (Amylopectin), high amylose and oil (7.5%) are developed for maize-based industry. Maize starch [72%] is used in the production of medicines, paper absorbents, adhesives and textile.
4. *Hard endosperm:* Improvement of *Opaque-2* with soft kernel, to meet the increasing demand for sweet corn, baby corn and pop corn cultivars and suitable forage varieties.

5. *Resistance to water logging:* Maize plants can not withstand water logging condition. Development of varieties, possessing reasonable degree of resistance would minimize the losses. Variety VL-42, developed at Almora has considerable resistance to water logging

6. *Early maturity:* Varieties with early maturity fit well in traditional system under rainfed conditions and can escape the dry spell at any stage of plant life.

7. *Resistance to drought:* Drought resistance is a complex concept and can be divided into mechanism of *drought escape, drought tolerance* and *drought avoidance.* Maize, a rainfed crop faces drought at many stages of crop establishment. There is a need to identify progenies with minimum time interval between pollen shed and silking or larger pollen shed period, which may prove helpful for stability in yield under drought conditions. Smaller tassel size with short, narrow and upright leaves are helpful in increasing grain yield especially at the time of lack of moisture and low soil fertility. These characters can be exploited together to maximize yield under rainfed situation.

8. *Resistance to biotic stresses* :

9. Development of heterotic pools for hybrid development programme.

7. BREEDING METHODS

1. *Breeding for prolificacy and yield into early duration maize:* Phenotypic recurrent selection has been proposed by Paterniani (1974) to transfer prolificacy into early duration non-prolific composite. The F1 is back crossed with the non-prolific parent followed by sibbing. Main emphasis during back crossing and subsequent sibbing was given on early maturity, total plant yield and ear number. The population thus developed has shown marked increase in ear number per plant.

2. *Breeding for drought tolerance:* Full sib family selection procedure (Fisher *et al.*, 1981) is used for improving yield under drought in a typical maize population "Tuxpeno". Families were selected for stable high yield and drought index at CIMMYT. Families selected for drought index have been used for developing experimental varieties in the recent years.

3. *Modified Gametic Selection:* For improvement of maize varieties gametic selection is modified as —

Step 1. Pollinate the tolerant inbred line from bulk pollen of susceptible population.

Step 2. Raise progeny of population x inbred cross, select the desirable plants under disease free condition and sib-mat selected plants. Select and harvest only pollinated plants.

Step 3. Grow the progeny from the harvest, select and sib-mat moderate and tolerant plants. Select and harvest only pollinated plants.

Step 4. Raise the progeny from selected ears. Self the selected plants select and harvest only selfed plants.

Step 5. Raise the progeny and continue to self, to develop inbred lines until acceptable level of tolerance is achieved.

4. *Population Improvement:* Population improvement in maize involves following steps —

Step 1 Incorporate monogenic (Necrotic area surrounded by yellow hollow, increased and collapsed) and polygenic type (Retained their tolerance until maturity) of tolerance through bulk pollen from selected tolerant lines during "*Mass-Sibbing*".

Step 2 Only tolerant plants again sib-mated.

Step 3 Selection and "*Mass-sibbing*" is continued until acceptable level of tolerance into the population.

5. *Breeding for resistance to biotic stresses:* Two available sources of resistance viz., Antigua Gr. 1 and H-207 are used to develop population incorporating resistance in the *ear-to-row* progenies. An integrated half-sib (HS) and full-sib (FS) system of breeding is proposed to evolve a resistant high yielding population. Promising FS families may also be developed to evolve experimental varieties for release. HS progenies can be recombined with new yielding and resistant germplasm. This may be a better substitute population for exploitation by FS families selection procedure. The balanced composite of the selected resistant families may serve as a source for resistance.

Examples

(a) Resistant to leaf blight and stalk rot (Raj. R. Benner, 1987) *Helminthosporium turcicum*: CM103, CM105, CM019, CM114, CM115, and B-49.

(b) Late wilt (*Cephalosporium maydis*) CM103, CM104, CM105m, CM106, CM111, CM113, CM202

(c) Charcoal rot (*Macrophomina phaseolina*) Orange: CM 103, CM104, CM107, CM111, CM112, and CM202 White: CM303, and CM400

(d) Turicum leaf blight PhDMR 61; Hi: 36: TZI 30, CM118, CM401, KUI 1411: KUI 1414 A.

(e) Late wilt (%) Inbred CM 111 = (9%); SS 111 = (7%) Adec = 913%); R N6 = (16%)

(f) Resistant to SDM (Sorghum Downy Mildew)
Accession 7936 and 84 33; Pozarica 8336 and 8149.

(g) Resistant to TLB:
PTR; A dec; RWF

Hybrid DHM [(CM104 x CM105) x (CM202 x CM201)] resistant to leaf blight; DHM-103 [(CM120 c CM118) x (CM119 x CM208] and DHM-101 [(CM202 x CM206) x (CM115 x CM114)] and DHM-105 [(SS 111 x RN6) x (Adec x CM111)] resistant to leaf blight and stalk rot were developed by utilizing the above mentioned sources with good yielding ability.

6. *Breeding maize for starch:* Starch is composed of amylose and amylopectin. Two mutant genes i.e., Waxy (WX) and amylose extender (ae) are transferred by back cross method results in a starch with 100% amylopectin [Waxy gene], where as 'ae' increases the relative proportion of amylose. Hi-starch prepared with the use of this source was released in 1963 for industrial purpose.

7. *Modified ear-to-row selection scheme (Lonnquist, 1964):* It involves following steps —
    1. Select 190-250 ears from O.P.V.
    2. Plant ear to row in 7 x 7" tripple lattice design, one replication each at three locations with one or two check varieties.
    3. 4 female: 2 male rows are planted in an isolated area.
    4. The male rows are the composite made up of an equal number of seeds from each 190 ears. The female rows are detesseled including check entries.

It provides a high degree of recombination among the progenies grown each year. Phenotypically best five plants in each 190 female rows are marked, harvested and placed in separate bags.

8. ORGANIZATIONS INVOLVED IN THE CROP IMPROVEMENT

International Maize and Wheat Improvement Center (CIMMYT), Mexico and Directorate of Maize Research, IARI, New Delhi, Indian Maize Development Association, 75N, Connaught Circus, New Delhi and many private sectors are involved in maize improvement.

9. MAJOR ACHIEVEMENTS

Development of the varieties and hybrid: Many varieties hybrids and composites have been released for cultivation in different parts of India mainly based on crop duration (Table 1). The characters of the hybrids and composites are presented in the Table 2.

Development of broad based pools/Population: The material is grouped based on their heterotic pattern viz. maturity and there after diallel may be made with the material falling in one group (Majumdar and Singh, 1987). Example Development of Jawahar maize-12 (E) and Jawahar maize-8 (EE) during 1995-1999.

Table 1. Varieties under cultivation in India

Full season		Medium duration		Early duration	
Hybrid	Composite	Hybrid	Composite	Hybrid	Composite
>30	3	5	5	10	5

Table 2. Description of the hybrids and composite under cultivation

Hybrid/ composite	Year of release	Season	Grain colour and type
Full season hybrids			
Kph- 528	1994	*Kharif*	Yellw, semi dent
Deccan-105	1990	*Rabi*	Yellow-orange, semi-dent
Trishulata	1991	*Kharif*	Yellow-orange, semi-dent
Ganga-11	1988	*Kharif, Rabi*	Orange-yellow, flint
Deccan-105	1997	*Rabi*	Orange-yellow, semi dent
Pro-311(4640)	1997	*Kharif*	Yellow flint with caps
Pac-705	1997	*Rabi*	Yellow semi flint
X 1382 (w)	1997	*Rabi*	Yellow semi flint
Bio-9681	1997	*Kharif*	Yellow semi flint
Full season composite			
Prabhat	1988	*Kharif*	Yellow flint
Dhawal (w)	1988	*Rabi*	White flint
Medium duration hybrids			
Dhm-107	1993	*Kharif*	Orange-yellow, semi flint
Ssm-510	1994	*Kharif*	Orange-yellow, semi flint
Medium duration composite			
Navjot and Arun	1988	*Kharif*	Orange-yellow, semi flint
ShaKharifti-1	1997	*Rabi*	Orange, flint with chalky base
Early duration hybrids			
Dhm-109	1994	*Kharif*	Yellow, semi flint
Him-129	1997	*Kharif*	Yellow, flint
Pusa early hybrid maKharifKharifa-1	1997	*Kharif*	Yellow, flint
Pusa early hybrid maKharifKharifa-2	1997	*Kharif*	yellow- orange flint
X- 1123 g (3342)	1997	*Kharif*	Yellow, semi flint
Early composites			
Mahi-Kanchan	1995	*Kharif*	Yellow, flint
Jawahar maize - 8	1995	*Kharif*	White, flint
Jawahar maize-12	1998	*Kharif*	White, flint
Kiran	1988	*Kharif*	Orange, semi flint

Development of maize for high oil content : A total of 25 inbred lines derived from 'CIMMYT' with oil content ranging from 7.7 to 10.6 percent are being used in breeding high oil content maize cultivar. The oil content of the original *Burr white* variety of maize was 4.7 per cent. This was increased to 7.37 per cent after 10 generations of selection and 15.3 per cent after fifty generations.

Development of Maize cultivars using 'b m3' gene for fodder: The discovery of 'bm3', a recessive gene reduces lignin content. Gene 'bm3' is transferred in African Tall and Kisan composites through VC-80. Genotypes "bm3 bulk (y) bm3 bulk2 (Y and WY), bm3 bulk-3 (W), bm3-4 and bulk-4 (W) are selected and used for production of low lignin content and high digestibility fodder cultivars.

Development of Popcorn and sweet corn varieties: High yielding varieties of popcorn with high popping volume (321 Cu inch) have been developed. Example 'HAWA ii AN SUGAR' (37 q/ha), H 254 (28 q/ha), Amber pop (18 q/ha) and Mushur.

Development of Hard endosperm composite with superior protein: Significant improvement was made in changing soft endosperm into hard one in a few *opaqe-2* materials. Rattan and Protina were released in 1971 with soft endosperm. For the first time in the country, high lysine nutritionally superior hard endosperm composite variety "SHAKTI' was released for commercial cultivation. This may be used directly as food for infant like *'Balahar'*.

Hybrid with early maturing: Single cross [DES-33 (57 q/ha)], double cross [E H F-1121 (43 q/ha)] and top-cross [EH-50536 and EH-50526 (47 q/ha)] hybrids with early maturity were developed.

Population Improvement in the development of maize hybrid: Development of heterotic pools for hybrid development programme is an achievement. Inter-varietal hybrids are developed by reconstituted improved varieties and synthetics. it is possible to extract lines and families, which may be used as a parent for non-conventional hybrids (Mukherjii and Ahuja, 1996). Several inter family hybrids, full sib x S1 line hybrids; inter-population and superior lines from identified S1 lines are developed.

Developing better performing hybrid oriented heterotic groups : White and Yellow full season are synthesized after pooling all the elite available germplams. Ludhiana centre have developed heterotic pools using *stiff stalk* and *Lancaster sure cropper lines* as testers to identify lines for two hererotic pools developed by them. Introgression to temperate germplasm into tropical materials may be extremely useful to improve the efficiency traits as the temperate materials are known to be efficient in nitrogen utilization and have better harvest index, but are susceptible to leaf diseases (Singh, 1996).

10. HYBRID, COMPOSITE AND SYNTHETIC

Hybrid : It is the first generation progeny from a cross-involving inbred lines. The breeding of hybrid maize involves —

1. Development of Inbred lines by controlled self-pollination.
2. Evaluation of inbred lines for GCA and SCA and which of them may be combined into productive crosses.
3. Combining the inbred lines into heterotic hybrids.

It is of following types —

Inbred lines: There are the "Pure lines" developed by five to seven generations of selfing/inbreeding. The pollens collected in the tassel bags are transferred to the silk and the tassel bag is then placed over the ear shoot to avoid contamination.

Single cross hybrid: It is a progeny from a cross between two hybrids. e.g. Pusa Early Hybrid 1(IPA-3 x IPA-4) tolerant to excess water and Pusa Early Hybrid 2 [IPA-9 x IPA-21] Tolerant to diseases.

Double-cross hybrid: It is a hybrid between two heterozygous single cross parent lines. The pedigree of a double cross hybrid shows the four inbred going into the double cross. e.g. Ganga hybrid Makka-3: (CM 109 x CM 110) x (CM 202 x CM 11).

Three-way cross hybrid: It is the progeny between a single cross and an inbred. Three inbred lines are involved in a single cross which is pollinated by an O.P.V. e.g. Ganga safed-2: (CM 400 x CM300) x CM 600; GANGA -5: (CM 202 x CM 111) x CM 501.

Inbred-variety cross hybrid: It is a cross between an inbred and an OPV and is often known as a *top cross*.

Composite varieties : It is the varietal crosses made in advance the generations. As there are number of germplasm involved in the heterozygous population hence it is sometimes called as *germplasm complexes*. Composite varieties are usually developed from the varieties or other heterozygous populations or germplasm which have originally not been subjected to inbreeding (Singh, 1965). According to Dhawan, (1965), a wide spectrum of genetic variability exists and is available to the maize breeders all over the world. Heterozygous effect of a very high order can be obtained by crossing certain races of genetic diversity. Advance generations of such crosses or of population developed from multiple crosses of elite races or germplasm complexes, have been shown to give stabilized yields at a high level of productivity.

Synthetic varieties: Synthetic varieties may be produced by making all possible cross combinations. Equal amount of seed of each cross of known inbred is composited to produce the synthetic variety. It refers to the open-pollinated increase from a multiple hybrid. The greatest progress has been made in breeding high yield synthetics where lines with good combining ability were chosen to enter, into the synthetic. Synthetic VARUN (1991) is the result of chain crossing of 5 donors x 5 potential inbreds.

11. SEED PRODUCTION

Maize seed production can be undertaken in areas of assured irrigation.

Quality seed is produced in A.P. and Karnataka in *rabi* season whereas, in *Kharif* it is difficult due to incidence of downy mildew. Zinc spray increases yield of maize inbred CM 202, used in seed production of many hybrids by 4-5 quintals. An isolation of 300-500 metre. Alternatively, a time isolation of 8-10 days restricted contamination to one percent in seed production of hybrid maize (Sharma and Dadlani 1995). Planting ratio of 2 male : 6 female gives maximum yield of hybrid seed (Anonymous, 1991).

The success of seed programme depends upon the nicking of male and female parents. It has been observed that flowering of both the parent is influenced by the time of planting location of seed production programme. The best way of nicking problems is the planting of another row of male with slight higher density after 5-7 days of first planting. In the event of late silking both male and female rows should be irrigated after application of urea. Flowering in female lines can be delayed by irrigation of female line after top dressing of urea. During winter, high seed rate of male parent followed by frequent irrigation with urea is advocated because of reduction in pollen production and receptivity of stigma. During summer top dressing of urea improves the receptivity of stigma.

Roguing and detasseling from vegetative stage to pre-flowering stage are very important measures to avoid genetic contamination. All the tassels from rows of female parent are pulled out when they are well out of the *'boot'* i.e., prior to shedding or silk emergence but never pre-maturely.

Since the performance of the hybrids (F_1) is the direct function of the purity of the parents, the maintenance of genetic purity (gene and genotypic frequency) of the parents is very important and essential. Inbreds are maintained by hand pollination preferably by sib-mating or mass-sibbing. Open pollinated and synthetic varieties are maintained by open-pollination in strict isolation with large population size. An OPV will seldom be as uniform as a single-cross hybrid for most morphological traits. Therefore, certification standards for OPV's should be flexible, realistic and appropriate for the conditions prevailing in a given region.

Composite varieties, which are mass reservoirs of desirable alleles, can be effectively managed without any loss of desirable alleles, if appropriate population size is kept. Following points are kept in mind for maintenance of the composite varieties —

1. To check mechanical mixing or cross pollination with other varieties. Plant the seed in isolation of 200 m in all direction.
2. Plant one acre and then selecting about 3-5 thousand ears from central portion of the field leaving 30 feet boarder all around.
3. After 3 years (70% germination) sib-mating procedure be followed. Selection of cobs led to new genotype hence reject, which do not fit or conform to the original.

Seed Production procedure for starch types : White colour is controlled by a recessive gene and can be handled like *WX* and *ae*. The gene for white colour does not adversely affect the starch yield. The special types of maize (Waxy, amylo and white) differ from normal in being homozygous recessive at one locus. Thus, contamination arising any foreign pollen would mask the expression of the desired trait. Seed production procedures are same as those used for normal maize. The produce of maize cultivated during winter season has bolder kernel than that of monsoon season.

Drying of seed upto 7 per cent moisture content and packing in 700 gauge polythene bags maintain germination above 80 per cent for upto 3 years when stored in ambient conditions (Anonymous, 1995).

REFERENCES

Anonymous, 1991. Report of QRT Seed Technology Research (1985-90), ICAR, New Delhi.

Anonymous, 1995. Research report (1992-94) NSP on Seed Tech. Res. ICAR New Delhi.

Lonnquist, J.H. 1967. A modification of the ear-to-row procedure for the improvement of maize population. *Crop Sci.* ***4***: 227-28.

Mangelsdorf, P. C. and R. G. Reeves 1959. The origin of corn. *Botanical museum leaflets. Harvard Univ.* **18:** 329-440.

Paterniani, E. 1974. Phenotypic recurrent selection for prolificacy in maize. *Maydica* ***23***: 29-34.

Sharma, S. P. and M, Dadlani 1995. Seed technology research in cereals. Hybrid Research Development, 189-98.

Singh, N. N. 1996-97. Seed Research and production of maize in India. (Compilation of Research papers of maize) Maize Winter Nursery, DMR, Hyderbad. 1-20

CHAPTER - 5

SORGHUM

A.R. Dabholkar

1. INTRODUCTION

Sorghum, *Sorghum biocolor* (L.) Moench, commonly known as jowar (Hindi), jonna (Telgu), cholam (Tamil), jola (Kannada), jawari or jondhala (Marathi), is one among the five major cereals of the world (wheat, rice maize, barely and sorghum), It is grown extensively in tropical and subtropical environments and is most suitable for rainfed farming. Sorghum is a C_4 type of plant and hence is capable of producing more dry matter per unit of time/per unit of land as compared with other cereal crops except maize.

2. TAXONOMICAL CLASSIFICATION, ORIGIN AND CHROMOSOME NUMBER

Taxonomical classification: Sorghum belongs to family Gramineae and genus Sorghum. Sorghum has been variously classified. Pliny was the first to clearly give the identifiable written description of sorghum (House, 1985). In 1794 Moench distinguished genus *Sorghum* from genus *Holcus* (Celarier, 1959; Clayton, 1961). Clayton (1961) considered the name *Sorghum biocolor* (L.) Moench. Our present generic and species concepts of *Sorghum* agree with the definition established by Moench, and all the specific names now described above are now referred to as synonyms of *Sorghum biocolor* (L.) Moench.

Snowden (1936) divided the genus sorghum into two sections, viz., *Para sorghums* and *Eusorghums*. *Para sorghums* have bearded sheath nodes, simple primary branches of the panicle and simple racemes. The chromosome number is 2n = 10. The *Eusorghums* has glabrous sheath nodes, finely pubescent and not bearded. Primary branches of the panicle are divided and racemes lateral and terminal. Section *Eusorghum* has two sub sections, viz., *Arundinacea* and *Halepensia*. *Halepensia* consists of wild species.

Arundinacea is further subdivided into two series *Spotanea* and *Sativa*. *Spontanea* comprises of 10 wild species while series *sativa* comprises of 31 cultivated species.

Murty *et al.* (1967) classified about 4000 entries of sorghum world collection into 8 sub series and 70 preliminary working groups. Harlan and de Wet (1972) developed a simple and informal classification useful to plant breeders, for cultivated sorghums and their closest wild relatives. The cultivated texa, covering 28 (out of 31) species of Snowden's series of *Sativa* were partitioned into following races under S. *bicolor* subspecies *bicolor.* The fifteen races of cultivated sorghum can be classified by looking at the mature spikelets alone based upon five fundamental spikelet types described by Harlan and de Wet (1972).

Basic races	Hybrid races
1. Race – bicolor (B)	6. Race – guinea – bicolor (GB)
2. Race – guinea (G)	7. Race – caudatum – bicolor (CB)
3. Race – caudatum (C)	8. Race – kafir-bicolor (KB)
4. Race – Kafir (K)	9. Race – durra. bicolor (DB)
5. Race – durra (D)	10. Race – guinea-caudatum (GC)
	11. Race – guinea-kafir (GK)
	12. Race – guinea-durra (GD)
	13. Race – kafir-caudtum (KC)
	14. Race- durra-caudatum (DC)
	15. Race-kafir-durra (KD)

The origin and domestication : Greatest variability in both wild and cultivated sorghums is found in north-east quadrant of Africa. It appears that the people, who occupied central and northern parts of Ethiopian plateau played major role in the development of African plants, including sorghum (Murdock, 1959). From Ethiopia cultivated sorghum moved to West Africa at an early date. From this region it was carried to Sudan and then to the regions of upper Niger river. Here the Mande people developed a diversified agriculture and numerous varieties of sorghum were produced. Movement of sorghum out of Africa to the east was most probably brought about through trading route between east Africa and India via Arabia. According to Snowden (1936) *S. arundinaceum, S. verticilliflorum and S. sudanense* were the progenitors of sorghum.

Chromosome number: The basic number of chromosomes for sorghum is five, and this number is found in *parasorghums* and *stiposorghums.* All the remaining genera or subgenera have chromosome number which are multiples of five. The chromosomes of *parasorghums* and *stiposorghums* are quite large, measuring some 6 at metaphase I, whereas those of other genera are small. The cultivated *sorghum (Sorghum bicolor* (L.) Moench), have chromosome number, 2n=20. Polyploid origin has been confirmed by cytological investigations of *S. bicolor* as well as *S. halepense* (Endrizzi, 1957).

3. CLIMATIC REQUIREMENTS AND AREA AND PRODUCTION

Sorghum is grown in a wide range of environments. India is the largest sorghum producing country in the world, contributing 34 percent of the semi-arid tropics total. Sorghum adapts to many environments with 90 to 140 days of maturity. Highest yields are usually obtained from varieties maturing in 100 to 120 days. Floral development and seed set are normal at temperature between 40 to 43□C with 15 to 30 percent relative humidity. Sorghum grows slowly at 20□C.

Almost all the core rainy season sorghum growing areas are located between 800 and 1000 mm per year rainfall isohytes. Sorghum requires less moisture for growth than other cereals. 332 kilograms of water per kilogram of accumulated dry matter which is less than maize, wheat and barley require 368, 434 and 514 kilograms water, respectively.

In India sorghum has been a predominant crop cultivated during *Kharif* season. Large acreages were found in peninsular India, the Malwa plateau and the Punjab region. Due to low demand for food, lower price relative to oilseeds (such as soybean), there has been a drastic decline in sorghum acreage in the country. Despite considerable decline in the acreage under sorghum, the production has registered increase due to adoption of high yielding varieties accompanied with their package of practices. Maharashtra shares largest sorghum acreage of the country followed by Karnataka and Madhya Pradesh. Productivity of sorghum is highest in Maharashtra during kharif while during rabi it is highest in Tamil Nadu.

4. SALIENT FEATURES OF CROP CULTIVATION

In most of the sorghum growing states of India, it is grown during rainy season (Kharif). However, in the states like Maharashtra, Karnataka, Andhra Pradesh, Gujarat and Tamil Nadu it is also cultivated in post-rainy season (rabi).

Sowing: Sowing of 8 kg seed /ha should be done about a week before the on set of monsoon otherwise, low temperature and high relative humidity makes congenial situation for shootfly (*Atherigona soccata),* which causes dead heart formation resulting in poorer grain yield. In post-rainy season, optimum time of sowing is middle of September. The seed must be treated with fungicide, like Thiram 2.5 g per kilogram seed, before sowing. For kharif season, the optimum plant population is 1,80,000 per hectare. The seed is drilled 45 cm apart in rows. After 8 to 10 days of germination thinning is done to maintain 10-12 cm plant to plant distance. Seedlings damaged by shoot fly, should be removed. In rabi season, optimum plant density is 1,35,000 per hectare.

Fertilizer management: The indigenous varieties are not management responsive. However, recently developed hybrids and high yielding varieties can respond high dose of nitrogen (150 kg per hectare). 40 kg Nitrogen and 40 kg P_2O_5 per hectare as basal dose, whereas 40 kg N/ha at 30-35 days old (crop differentiates from vegetative to reproductive stage) crop is applied as side dressing. In rabi season 40 kg Nitrogen and 20 kg P_2O_5/ha are applied as basal dose.

5. BOTANICAL DESCRIPTION

Plant description: It is a single culmed i.e., a single solid erect stem supported by strong adventitious root. Buds, from lowest node forms tillers depending on the cultivar and environmental conditions. The centre of the stem may be sweet or insipid, juicy or dry.

Leaves are arranged alternately usually in two ranks. Each leaf originates from a node and has a sheath, blade or lamina. The leaf sheath has overlapping margins and encircles the inter-node. The number of leaf per plant can vary from 7 to 30. The leaf blade may be 1.5 to 13 cm in width and 30 to 135 cm in length. The uppermost inter node of the plant bears the inflorescence called peduncle. Leaves are lanceolate or linear lanceolate. Leaf margin may be flat or wavy. The mid-rib may be white or yellow in dry, pithy varieties while in juicy varieties the midrib is opaque-green, often with a fine white strip down it. There is often a deposit of wax near the base of mid-rib at the junction of leaf.

Duration for maturity in sorghum is governed by four major genes (Ma_1, Ma_2, Ma_3, Ma_4,) called maturity genes. The peduncle may be well exerted above the flag leaf. Poor exsertion encourages insect attack and grain mould development in the hidden portion. Under extreme drought conditions such panicles may fail to emerge from the boot. The peduncle may be erect or recurved (goose neck). Panicles may be short or long, compact to very lax open, ovate, conical, long and slender or pyramidal. The length of panicle range from 4 to 25 cm with 2 to 20 cm width. Inflorescence has a central axis or rachis from which primary branches arise. They bear secondary and sometimes tertiary branches. The secondary branches carry the racemes of spikelets.

Only temporary root in sorghum is a single radicle of germinating seedling, remainder of the fibrous root system develops from lowest nodes of stem. The sorghum seed is a 'caryopsis' which varies in colour, shape and size. Large variation for pericarp colour (white, cream, lemon yellow, red or brown), shape of seed (spherical, oval, ellipsoidal, pear shaped or turtle shaped with a flat surface on one side) and seed size (0.75 to 7.5g/100 seed wt) has been observed.

Stages of development: Vanderlip (1972) distinguished nine growth stages of early maturing hybrids (Table 1).

Of the various schemes proposed for recognition of growth stages, that of Eastin (1972) has been widely accepted. This scheme distinguishes the period from germination to panicle initiation as growth stage 1 (GS1); from panicle initiation to blooming as GS2, and from blooming to physiological maturity, identified with the formation of black layer at the base of grain (Eastin *et. al.* 1973) as GS3. Secondary roots begin to develop when plant is 3 to 7 days old.

Plant remains in vegetative phase for about 30 to 40 days, during which all the leaves are formed. Flowering initiates from 30 to 40 days after germination. It is an important growth stage because maximum potential seed number set during this stage.

Table 1. Growth stages of sorghum

Stages	Approximate days after emergence	Identifying characters
0	0	Emergence, coleoptile visible
1	10	Collar of 3rd leaf visible
2	20	Collar of 5th leaf visible
3	30	Growing point differentiation (approximately 8 leaf stage)
4	40	Final leaf visible
5	50	Boot, head extended into flag leaf sheath
6	60	Half-bloom, half of plants in some stage of bloom
7	70	Soft dough
8	80	Hard dough
9	95	Physiological maturity

Floral biology: The inflorescence of sorghum is a panicle. The raceme bears spikelets in pairs. Each pair consists of a sessile and hermaphrodite spikelet and a pedicelled spikelet which may be male or sterile. The terminal sessile spikelet is accompanied by two pedicelled spikelets. Sorghum spiklet has five different types of expressions (Fig. 1).

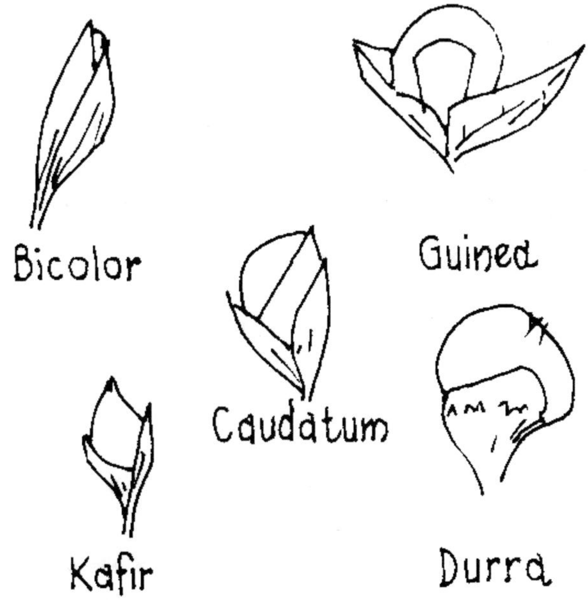

Fig 1. Classification of sorghum spiklets

Sessile spikelet is fertile 3 to 10 mm long and oval, elliptic or obovate in shape. It has two leathery or papery glumes (Fig. 2). The colour of mature glume varies from cream to almost black. Lower glume has a coarse keel like nerve on each side.

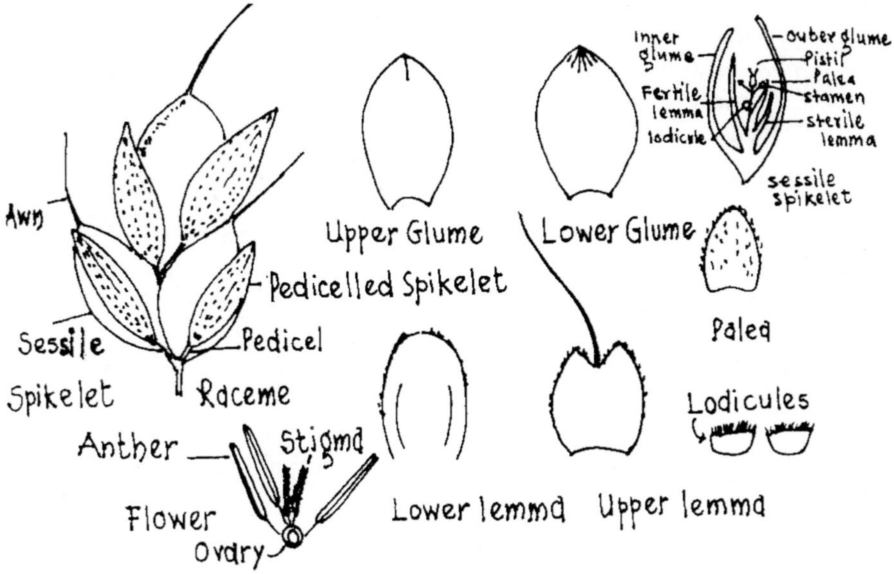

Fig. 2. Parts of sorghum raceme and spiklet

Two glumes enclose two florets i.e., lower one sterile whereas the upper one fertile and perfect. The fertile lemma has a cleft at the apex and may bear an awn. Adjacent to fertile lemma, short, broad, fleshy and truncate lodicules with ciliate margin are present.

It has three anthers each with four lobes. Ovary is single celled with two pistils and long style, ending in feathery stigma of white, light yellow, or pink colour. In general, a typical panicle may possess as many as 1500 to 4000 sessile spikelets.

Pedicelled spikelets are much narrower in shape than the sessile spikelets, usually lanceolate and persistent or deciduous. These are usually sterile and consist of only two glumes. Frequently the upper lemma is fertile and bears stamens, which produce functional pollen. Very rarely the pedicelled spikelets may contain a rudimentary ovary that could produce relatively smaller seeds than those of sessile spikelets.

Anthesis: Flowering starts with complete elongation of peduncle. Usually a few spikelets at the tip of the panicle open on the first day and blooming continues successively downwards for several days. Blooming of florets occurs in a fairly regular manner with florets in a horizontal plane across the panicle opening at same time. Flowering of a panicle may be completed in 4 to 9 days. In cooler climates the panicle blooming period may be extended. Anthesis begins from 2 a.m. and opening continues till 8 am and may prolong to 10 am. Protogynous flowers are pollinated by the higher flowers of same panicle; though anemophilous flowers are visited by insects, chiefly the bees.

Flowers open due to the pressure exerted by swelling of the two lodicules. As the glumes spread stamens emerge, rotate and spread out with rapid filament elongation. The anthers, as a consequence fall and become pendant and dehisce. The time taken from the commencement of glume opening to completion of closing is about 1-2 hours. In tropics, on clear sunny day, dehiscence of anthers takes place around sunrise which could be delayed on damp and cool days upto 10 am.

Sorghum is classified as a predominately self-pollinated crop. Amount of cross-pollinationis usually 2 to 40 per cent, and is normally 2-4 folds higher in the top quarter of the panicle. Pollens are driven by wind and remain viable for 3-6 hrs. For cross- pollination, pollens are collected in paper for a few hours after dehiscence. Pollen is usually available for 4 to 10 days. Each anther may contain approximately 5000 pollen- grains.

The stigmas are receptive for one to two days before blooming and remain receptive for 8 to16 days after blooming.

Fertilization of egg cell occurs two hours after pollination. The embryo is formed 12 days after fertilization (Doggett, 1970). Development of embryo and endosperm continues alongwith starch accumulation for about 30 days.

6. SELFING AND CROSSING

Selfing: Sorghum is predominantly a self-pollinated crop. Due to out-crossing, crop must be protected from contamination by maintaining isolation distance of 400 m from Johnson grass and grassy sorghums and 300 m from field of other sorghum varieties. In breeding nurseries, where thousands of diverse cultures of sorghum are to be maintained, protection from contamination is accomplished by enclosing sorghum panicles inside craft paper bags of 42 x 12 x 6 cm size.

For selfing all the plants within a row are usually put inside sorghum pollinating bags as soon as panicles emerge out. The flag leaf is removed before selfing a panicle. If the panicle has started blooming, the flowering portion is removed and then covered with a pollinating bag. The sorghum pollinating bag is fastened at the bottom, around the peduncle by means of staple or paper clip. Date of selfing is written on pollinating bag. Ear-heads left for long period inside the pollinating bags are damaged. Hence, pollinating bags should be taken out after complete blooming (about 8 days) and panicle should be treated with insecticides like Malathion 50 EC.

Crossing: The emasculation work is carried out, in the part, which will be ready to flower on the next day. Remove one fourth top and base spikelets from the panicle chosen for emasculation. About 4 to 5 primary branches in the middle portion of panicle are selected for emasculation, rest are trimmed off. It is convenient to emasculate florets facing away from the panicle axis. From these primary branches, remove all the pedicelled spikelets. Grasp the sessile spikelet (which is bisexual) between the thumb and forefinger and then insert a forceps (Fig. 3) just below the middle of floret and move it towards back side of glume rotate slightly to lift all the three anthers. Emasculate 10-15 spikelets per branch. Occasionally pistils come up with anthers during emasculation, which can be pushed down into the floret immediately after the anthers are removed.

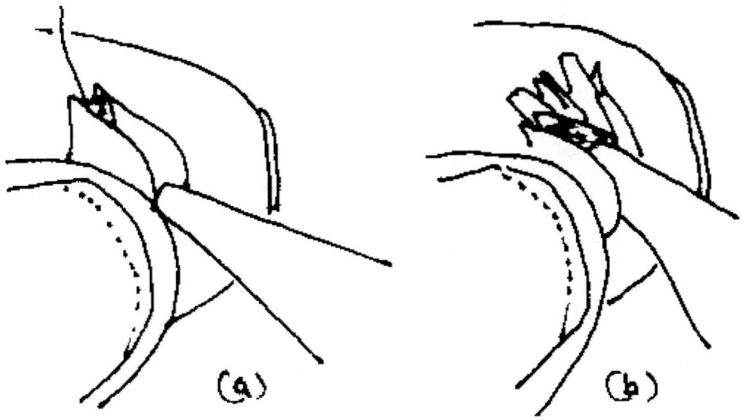

Fig. 3. Procedure of emasculation

The emasculated panicles are covered with craft paper bags. Date of emasculation, name of seed parent and the pollen parent are marked on the envelopes. The emasculated panicles are inspected in the morning on the following day. Florets, which are open and stigma appears fresh and greenish are pollinated immediately. If, among the emasculated flowers, a floret appears with an entire or broken anther, the entire panicle should be rejected.

For pollination, pollen is collected in the pollinating paper bags from the designated variety. The panicles of plants selected as male are covered with pollinating craft paper bags the previous day. Pollens are collected around 9.30 to 10 am in the bag by slight tapping of the panicle covered by bag. The emasculated flowers are pollinated and then covered by a paper bag. An information tag is tied around the emasculated panicle. Pollination of emasculated flowers is carried out for 2 to 3 days. The developing F_1 seeds are inspected after every 3 to 4 days and Malathion 50 EC is sprayed on 5 to 10 days old seed to control ear head pests.

7. BREEDING METHODS

Breeding methods employed for improvement of sorghum are introduction, selection and hybridization followed by handling the segregating material by pedigree method. Back-cross method is also regularly used by sorghum breeders. Cytoplasmic-genetic male-sterility (CMS) and genetic male-sterility occur in sorghum hence methods advocated for cross-pollinated crops are also used in sorghum.

Germplasm collection: Germplasm are utilized as source of desired genes. At the International Crop Research Institute for Semi Arid Tropics (ICRISAT), Hyderabad, about 26,000 germplasm entries of sorghum from various parts of the world and its wild relatives are conserved, studied and catalogued (Rana and Rao, 1989).

Introduction: MSCK 60 A straight introduction to India in 1962 was used as a female parent to develop experimental hybrids. IS 84 and IS 3691 from germplasm collection were used as restorers of hybrids CSH 1 and CSH 2, respectively.

Selection: Selection from land races and introductions have been employed as a plant improvement technique in sorghum. Pure line selection was used to select land races and develop varieties like Vidisha 60-1, Ujjain 6, Maldandi 35-1, Aispuri, etc. Mass selection was seldom used in India for breeding new varieties of sorghum.

In the early years dwarf early maturing exotic lines were crossed with tall late maturing Indian varieties. From this program high yielding, mid-tall to tall, tan coloured and shoot fly and stem borer resistant varieties such as CSV 10, CSV 11, CSV 13, SPV 462 were evolved.

During nineties, SPV 462, CSV 13 (SPV 475), CSV 10 (SPV 346), CSV 11 (SPV 351) were used to develop high yielding varieties e.g., CSV 15, Jawahar Jowar 938 and Jawahar Jowar 1041.

Back-cross method: It has been used for transfer of simply inherited characters to an agronomically desirable variety. For example, resistance to head smut was transferred through back-cross method. In United States of America it was used for conversion of tropical germplasm (late maturing and tall types) to dwarf, early type suitable for tropical conditions.

Development of cystoplasmic-genetic male sterile lines (CMS), presently milo-kafir male-sterility, designated as A_1 type male sterility, results due to the interaction of recessive nuclear genes for male-sterility in Kafir with sterility causing factors in Milo cytoplasm. Male sterile lines are developed by two approaches (i) A number of lines from the germplasm, which are agronomically desirable, are used as pollen parents to pollinate a known male-sterile line (IMS 9A). Only those lines from germplasm which possess recessive genes in homozygous state maintain male sterility (i.e. ms ms). After pollinating individual heads of IMS 9A with different germplasm lines, seeds are collected separately. The seeds from plants used as pollen parents are numbered and harvested, separately. In the next season, paired rows of hybrid seed (IMS9A x germplasm line X) and Xth germplasm line are planted together. At flowering, all the plants are selfed. After completion of the flowering, study of the head indicates, which germplasm line carried the maintainer gene. All the plants in the hybrid row, i.e. (IMS 9A x germplasm line), will be sterile, whose pollen parent from germplasm carried ms ms genes in homozygous condition. Five to six back crosses of such germplasm line to IMS 9A will substitute most of the genotype of IMS 9A with the genotypes of specific germplasm line. Using this procedure several germplasm lines were converted to male-sterility e.g. IS 2219A, IS 2077A, IS 10246A.

The first step of male-sterile conversion program of the second approach is to cross the variety (to be converted to male-sterility) as a pollen parent to 'B' line, say IMS 9 B. Grow the F_2 generation and pickup plants which look similar to the variety to be converted into a male-sterile line. Select considerable number of plants and number them. Cross each plant separately to IMS 9A plants. Put the number of pollinator F_2 plants on envelops enclosing heads of IMS 9A.

In the subsequent year, grow paired rows of hybrid (i.e., IMS 9 A x 10th F_2 plant) and x 10th F_2 plant. Self all the plants. At flowering stage, find out hybrid rows, which are completely sterile. Pollinator F_2 plants of these hybrids carry ms ms genes. The F_2 pollinator parents are homozygous for sterility maintenance but may not be homozygous for other characters. Selection from F_2 pollinator row, say 10th row, plants which are similar in phenotype to variety, and pollinate the sterile plant from adjacent row (i.e., IMS 9A x 10th F_2 plant grown along with 10th plant row). Cross other selected plants from other rows with sterile plants from their corresponding rows. Sow paired rows in subsequent years and repeat the process. Five to six back crosses will result in a number of male-sterile plants. Their corresponding pollen parents are their B lines.

Hybrid varieties: Phenomenon of heterosis was observed in sorghum as early as 1927 (Conner and Karper) but its is commercially exploited with the discovery of cytoplasmic genetic male sterility system (Stephens and Holland, 1954). CSHI the frst commercial sorghum hybrid in India was released in 1964. CSH 5, CSH 6, CSH 9, were based on new male sterile lines evolved in viz., 2077A, 2219 A and 296 A respectively. Restorer IS 3541 (IS 3541, a tall zerazera) had excellent general combining ability used in three hybrids viz., CSH 5, CSH 6, and CSH 9. RS 29, a very good tall restorer was used for CSH 13 with a tall hybrid with 296.

Population improvement through recurrent selection: Recurrent selection permits recombination of genes among selected individuals and hence with each cycle genetic variability does not decrease. Random mating populations can be established in sorghum due to availability of male-sterile lines. A large number of superior genotypes are individually crossed to a male-sterile line. Equal quantity of hybrid seed from each cross is composited and planted in an isolated field. Male-sterile plants appearing in field are tagged. As a result of random mating seeds set on male-sterile plants are collected, bulked in equal quantity and the cycle is repeated. During this process new promising lines can also be infused by pollinating the male-sterile plants. Various recurrent selection procedures such as half-sib family selection, full sib family selection and S_1 progeny evaluation can be used.

Breeding for resistance to biotic stresses: Resistant parent is crossed with agronomically superior variety, in which resistance is to be incorporated. Segregating material is screened for its reaction under conditions which favour heavy incidence of pest. Pedigree method of handling the segregating material is usually adopted.

Shootfly – Atherigona Soccata It, is the most serious pest of sorghum. The adult fly lays eggs on the undersurface of leaves of 5 to 20 days old seedlings. After hatching eggs maggots move towards growing points and kill the plant by making dead hearts. Screening of sorghum for reaction to shootfly is carried out in the field. Interlards of highly susceptible cultivar such as CSH 1 or CK 60B, 3 weeks earlier and application of fishmeal are used for increase in population of shoot fly. Observations are recorded on eggs laid per plant and thereafter dead hearts. It has been found that simple recessive gene for glossy leaves contributes to shootfly resistance.

Stem borer (*Chilo partellus*) Artificial infestation is preferred by placing first instar larvae in leaf whorl of 15-20 days old plant for uniform infestation. Test plants are scored for different types of damage i.e., leaf feeding dead hearts, stem tunneling and peduncle damage

8. ORGANIZATIONS INVOLVED IN SORGHUM IMPROVEMENT

First effort under combined project to improve sorghum was made in India through the project initiated by Government entitled the Project for Intensification of Regional Research on Cotton, Oilseeds and Millets (PIRRCOM). It functioned with the cooperation of the Rockefeller Foundation of United States. PIRRCOM functioned upto 1963.

The All India Coordinated Sorghum Improvement Project (AICSIP) was established in India by the ICAR from 1969. This organization operated in all the sorghum growing states of the country.

The International Crops Research Institute for Semi Arid Tropics (ICRISAT), an International Organization for research on sorghum working since 1972.

9. LIST OF VARIETIES UNDER CULTIVATION IN INDIA

Varieties and hybrids recommended by the All India Coordinated Research Project and State Agricultural Universities are listed in Table 2 & 3.

Table 2. Hybrids recommended for cultivation in India

S. No.	Hybrid	Recommended for state
1	CSH 1 (CK 60A x JS 84) Release 1964	Maharashtra, Karnataka, Madhya Pradesh, Rajasthan, Tamilnadu, Uttar Pradesh
2	CSH 5 (2077 A x CS 3541) Release 1976	Andhra Pradesh, Madhya Pradesh, Gujarat, Rajasthan
3	CSH 6 (2219 A x cs 354) Release 1978	Maharashtra, Karnataka, Andhra Pradesh, Rajasthan, Tamilnadu
4	CSH 9 (296 A x CS 3541) Release 1982	Maharashtra, Andhra Pradesh, Madhya Pradesh, Gujarat, Rajasthan, Uttar Pradesh
5	CSH 14 (AKMS 14A x AKR 150) Release 1978	Maharashtra, Karnataka, Andhra Pradesh, Gujarat, Rajasthan, Tamilnadu, Uttar Pradesh
6	CSH 13 (296 A x RS 29) Release 1978	Entire Country
7	CSH 17 (AKMS 14 A x RS 673) Release 1999	Karnataka, Madhya Pradesh, Gujarat, Tamilnadu
8	CSH 18 (IMS 9 A x Indore 12) Release 1999	Entire country

Table 3. Varieties recommended for cultivation in India

S.N.	Varieties	Recommended for state
1	CSV 13 (SPV 475)	Entire country
2	CSV 15 (SPV 946) Release 1996	Entire country
3	PKV 400, SPV 699	Maharashtra
4	JJ 741, JJ 938, JJ 1041	Madhya Pradesh
5	SB 1066, SPV 462, ICSV 74	Karnataka
6	Moti, SPV 462 (PSV 1)	Andhra Pradesh
7	GJ 35, GJ 37, GJ 39, GJ 40, GJ 41	Gujarat
8	SPV 245, SPV 96	Rajasthan
9	BSr 1, APK 1 (SPV 881) Co 24, CO 25, CO 26	Tamilnadu

10. HYBRID SEED PRODUCTION

Sorghum does not tolerate water logging and hence selected land should be well drained with pH around 5.5 to 8.5. High pH and iron deficiency may cause female sterility. Sorghum seed production plot should be isolated at least 400m on all the sides from grassy sorghums, Sudan grass and Johnson grass. From fields of grain sorghums the seed plot should be atleast 100m away on all the sides, and from forage sorghum at least 200 m.

For successful hybrid seed production, on commercial basis, planting ratio of female to male line may be 6 : 2 or 4 : 2. Where the male line has shorter heads and shorter span of flowering as compared with the female line (as in case of CSH 6 and 14) it is desirable to plant 4 lines of A parent alternating with 2 rows of R parent in entire fields. However, when the panicles of female parents are longer (as in case of 2077 A, 296A) and male parent has a longer span of flowering the ratio of planting could be 6 rows of A alternating with 2 rows or R. Plant 2 to 3 seeds of sun-hemp at the start of male rows, so that male rows could be differentiated from beginning. At least 4 border rows of restorer should be planted around the seed plot to guard the seed plot from contamination from foreign pollen and to provide additional pollen to the female. Seed rate of female and male parent is 8 and 4 kg/ha, respectively.

The female and male parents of some hybrids flower at different times hence seed set is adversely affected. The female and male parents are photo- and thermo-sensitive and hence may react variably under different day lengths and temperatures. Several methods are employed to ensure nicking (Rana et. al., 1999). Staggered planting i.e. planting of female and male parents on different date for synchronized flowing. Foliar spray of 2 per cent urea to the late parent, 3 to 5 times on subsequent days at 1 to 2 days interval is recommended for correction of 2-4 days difference. Ratooning of early flowering parent may attain synchronized flowering. In case of failure to attain nicking, dusting of the pollen collected from male parent grown in vicinity may enable good seed setting. Generally stigma retains receptivity upto 4 to 5 days (2219 A, 296 A, AKMS 14 A) after flower opening. Stigma receptivity and pollen production is influenced negatively by higher temperatures. Plants of A line in rows of R line, plants of R line in rows of A line, pollen shedder from B line, any plant phenotypically dissimilar to A line and R line plants and plants infected with designated diseases such as grain smut (*sporisorium sorghi)* and head smut are removed from seed plot before flowering. Male rows are harvested before harvesting of seed parent.

REFERENCES

Celarier, R.P. (1959) Cytotaxonomy of the *Andropogoneae* III sub tribe *Sorgheae,* genus *Sorghum. Cytologia,* Tokyo, 23:395.

Clahyton, W.D. (1961) Proposal to conserve the generic name *Sorghum* Monech *(Gramineae)* versus *Sorgum* Adans *Gramineae).* Taxon, 10 (8):242.

Conner, A.B. and R.E. Karper (1927) Hybrid vigour in sorghum. Texas Agricultural Experiment Station Bulletin No. 359. College Station, Texas, USA, Texas A & M University.

Doggett, H. (1970) Sorghum. Publ. Longmans Tropical Agriculture Series.

Easthi, J.D. (1972) Photosynthesis and translocation in relation to plant development In "Sorghum in Seventies" Ed. Rao, N.G.P. and L.R. House. Oxford & IBH Publishing Co., New Delhi pp. 66.

Eastin, J.D., J.H. Hultquist and C.Y. Sullivan (1973) physiologic maturity in grain sorghum. *Crop Sci.,* 13: 175-178.

Endrizzi, J.E. (1957) Cytological studies of some species and hybrid in the *Eu.-Sorghums. Bot. Gaz. 119:1.*

Hadley, H.H. (1953) Cytological relationship between *Sorghum vulgare and sorghum halepense. Agron J., 45:139.*

Harlan, J.R. and J.M. J. Dewet (1972) A simplified classification of cultivated sorghum. *Crop Sci.,* 12 (2) :172-176.

House, L.R. (1982) A look ahead into the 1980: In "Sorghum in eighties". Proceedings of the International Symposium on Sorghum. ICRISAT, Patancheru. 2-7 Nov. 1981, p. 61-68.

House L.R. (1985) A guide to sorghum breeding. 2nd Edition. International Crops Research Institute for Semi-Arid Tropics (ICRISAT), Patancheru, Hyberabad p. 206.

Jayasekara, M. and J.L. Jinks (1976) Effect of gene dispersion on estimates of components of generation means and variances. *Heredity,* 36: 31-40.

Kambal, A.E. and O.J. Webster (1966) Estimates of general and specific combining ability in grain sorghum, *sorghum vulgare.* Pers. *Crop Sci., 5:521-523.*

Karper, R.E. and A.E. Conner (1916) Natural Cross- pollination in Milo. *J. Amer. soc. Agron.,* 11:257.

Karper, R.E. and A.T. Chisholm (1936) Chromosome numbers in Sorghum. *American*

Kottur, G.L. and R.K. Kulkarni (1922) Cross- fertilization in jowar. *Agri. J. India,* 17: 413.

Kottur, G.L. and V.M. Chavan (1928) Selections in jowars of the Bombay Karnatak. *Bom. Dept. Agri. Tech. Bull.* 151.

Maunder, A.B. and R.C. Pickett (1959) The genetic inheritance of cytoplasmic genetic-make sterility in grain sorghum. *Agron. J.* 51:47-49.

Maunder, A.B. and G.J. Sharp (1968) Localization of out-cross within the panicle of fertile sorghum. *Crop Sci.*, 3: 449.

Murdock, G.P. (1959) Staple subsistence crops of Africa. *Geographical Review,* 50 : 521-540.

Murty B.R., A. Arunachalam and M.B. L. Saxena (1967) Classification and catalogue of world collection of sorghum. *Indian J. Genet.,* 27 (Spl. No.) : 1-7.

Murty, D.S., R. Tabo and O. Ajaiji (1994) Sorghum Hybrid Seed Productions and Management. ICRISAT, Hyderabad Information Bulletin No. 41.

Patel, G.B. (1926) The extent of natural cross-fertilization in jowar at Surat. *Agri. J. Ind.,* 21:366.

Piggott, S. (1950) Prehistoric India. Penguin Books.

Poehlman, J.M. and D. Borthakur (1969) Breeding Asian Field Crops. *Oxford & IBH Publishing Co., New Delhi* pp. 182-202.

Pooni, H.S. and J. L. Jinks (1981) The true nature of the non-allelic interactions in *Nocotiana rustica* revealed by the association crosses. *Heredity,* 47 : 253-258.

Ramnathan, V. (1924) Some observations on Mendelian characters in sorghum. *Madras Agric. J.* 12: 1.

Rana, B.S., M.H. Rao, S. Indira, B.U. Singh, Chari Appaji and V. Tonapi (1999) Technology for increasing sorghum production and value addition. Publ. N. R.C.S., ICAR, Hyderabad.

Rana, B.S.D. and N.G.P. Rao (1989) Often cross-pollinated crops: Sorghum. In "Plant Breeding – Theory and Practice" Ed. V.L. Chopra, *Oxford & IBH Publishing Co.* Pvt. Ltd., New Delhi. pp. 225-235.

Rao, D.V.N. and K.O. Rachie (1965) Natural Crossing in sorghum as affected by locality and season. *Indian J. Agric. Sci., 35; 8-13.*

Rao, N.G.P. (1970) Genetic analysis of some exotic x Indian crosses in sorghum. I. Heterosis and its interaction with seasons. *Indian J. Genet.,* 30:347-351.

Rao, N.G.P. and J. Venkateswarlu (1971) Genetic analysis of some exotic x Indian crosses in sorghum. III. Heterosis in relation to dry matter production and nutrient uptake. *Indian J. Genet.,* 34:156-176.

Rao, N.G.P. (1982) Transforming traditional sorghums in India. In "Sorghum in eighties". Proceedings of the International Symposium on sorghum. ICRISAT, Patancheru Nov. 2-7, 1981, pp 39-59.

Schaffert, R.E. and L.M. Gourley (1982) Sorghum as an energy source. In "Sorghum in eighties". Proceedings of the International Symposium on sorghum. ICRISAT, Patancheru Nov. 2-7, 1981, pp 39-59.

Schertz, K.F. and D.R. Pring (1982) Cytoplasmic sterility systems in sorghum in eighties". Proceedings of the International Symposium on Sorghum. 2-7 Nov., 1981, ICRISAT, Hyderabad, pp. 373-383.

Sivakumar, M.V.K. and S. M. Virmani (1982) The physical environment. In "Sorghum in eighties". Proceedings of the International Symposium on Sorghum. ICRISAT, Patancheru, A.P. India, 2-7 November 1981, pp. 83-100.

Smith, H.H. (1952) Fixing transgressive vigor in *Nicotiana rustica.* In "Heterosis" Ed. G. W. Gowen, Iowa State College Press, Ames, Iowa pp. 161-174.

Snowden, J.C. (1936) Cultivated races of sorghum. Allard and Sons. London pp.274.

Stephens, J.C. and J.R. Quinby (1934) Anthesis, pollination and fertilization in sorghum. *J. Agric. Res.* 49: 123.

Stephens, J.C. and R.F. Holland (1954) Cytoplasmic male sterility for hybrid sorghum seed production. *Agron. J.* 46:20-23.

Tripathi, D.P., S.L. Mehta, B.S. Mehta and N.G.P. Rao (1980) Characterization of diverse cytoplasmic-genetic male-sterile in sorghum (*Sorghum bicolor* (L.) Moench). *Sorghum Newsletter, 23:107-108.*

Vanderlip, R.L. (1972) How a sorghum plant develops, Kansas State University, Manhattan, pp. 19.

Watt, G. (1983) dictionary of economic products of India, 6:291 London and Calcutta.

CHAPTER - 6

CHICKPEA

P.M. Gaur

1. NAMES

English – chickpea; Hindi – *chana*; Local names – *boot, hola, hora*; Other names – Bengal *gram*, Egyptian pea, *garbanzo* bean, *gram*, *grao-de-beco, hoos, kechererbse*, and *pois chicle*.

2. TAXONOMICAL CLASSIFICATION, CHROMOSOME NUMBER AND ORIGIN

2.1 Taxonomy

The word *Cicer* is a derivative of the Greek *kiros* referring to the well-known Roman family *Cicero*. The word *arietinum* is also derived from the Latin *aries*, meaning ram, referring to the shape of the chickpea seed, which resembles a ram's head. The shape of chickpea seed is similar to a chicken's head with a characteristic 'beak', hence the name 'chickpea'.

All grain legumes including chickpea (*Cicer arietinum* L.) are members of the family Leguminosae and the sub-family Papilionoideae. The genus *Cicer* was earlier classified in the tribe Vicieae Brown with the genera *Pisum, Lens, Vicia* and *Vavilovia* (Hutchinson 1964), but is now classified in its own monogenetic tribe Cicereae Alef (Kupicha 1977).

The genus *Cicer* includes 43 species, nine of which are annual, 33 are perennial and one with unspecified life cycle (van der Maesen 1987). The species *C. arietinum* is the only cultivated species of this genus. Four sections have been recognized in the genus *Cicer* on the basis of morphological characteristics, life cycle, and geographical distribution (reviewed by van der Maesen 1987). The cultigen (*C. arietinum*) and seven annual wild species (*C. reticulatum, C. echinospermum, C. pinnatifidum, C. judaicum, C. bijugum, C.*

yamashitae, C. cuneatum) were grouped in the section *Monocicer;* one annual (*C. chorassanicum* M. Pop.) and one perennial [*C. incisum* (Willd) K. Maly] species in the section *Chamaecicer;* 23 perennial species in the section *Polycicer*; and seven woody perennial species in the section *Acanthocicer*.

Annual species of *Cicer* have been subjected to numerous taxonomic studies. Ladizinsky and Adler (1975) studied cytogenetic relationships among seven annual *Cicer* species. Based on crossability and fertility of hybrids in interspecific crosses, these species were classified in three groups. *C. arietinum, C. reticulatum* and *C. echinospermum* were placed in the first group, *C. judaicum, C. pinnatifidum* and *C. bijugum* in the second group and *C. cuneatum* in the third group. Based on crossability and seed protein banding patterns, *C. reticulatum* has been proposed as wild progenitor of the cultivated chickpea (Ladizinsky 1975).

The phylogenetic relationships among nine annual species have also been studied through data on allozyme polymorphism (Kazan and Muehlbauer 1991, Ahmad *et al.* 1992, Labdi *et al.* 1996, Tayyar and Waines 1996) seed storage protein banding patterns (Ahmad and Slinkard 1992) and RAPD markers (Ahmad 1999). These studies placed the annual *Cicer* species in four phylogenetic groups. *C. arietinum, C. reticulatum* and *C. echinospermum* formed one group while *C. pinnatifidum, C. bijugum* and *C. judaicum* formed another group. *C. chorassanicum* was grouped with *C. yamashitae* whereas *C. cuneatum* formed an independent group and showed the largest distance from *C. arietinum*.

The phylogenetic relationships of three perennial species (*C. anatolicum, C. songaricum, C. microphyllum*) with annual *Cicer* species have also been reported (Tayyar and Waines 1996, Gargav and Gaur, 2001). All three perennial species have been placed in the third phylogenetic group which contained *C. chorassanicum* and *C. yamashitae*. These results suggest that *C. yamashitae* and/or *C. chorassanicum* may be the first annual species evolved from perennial species.

2.2 Chromosome Number

Almost all *Cicer* species have 2n=2x=16 chromosomes. The chromosomes have been numbered from 1 to 8 in order of decreasing size of the chromosomes and the size difference between pair one and pair eight has been found to be in the ratio of 3:1 (Ahmad and Godward 1980). Ahmad and Hymowitz (1993) recorded the total chromosome length at pachytene stage as 353.53 μm and also found that the chromosome size ranged from 30.53 to 58.05 μm.

The chickpea chromosomes are small which makes the karyotype analysis difficult. The chickpea karyotype that has emerged from various cytological investigations has the following features: a pair of very long chromosomes, distinctly satellited and sub-metacentric; six pairs of metacentric to sub-metacentric chromosomes; and a pair of very short metacentric chromosomes (reviewed by Gupta and Bahl 1983).

Both, spontaneous (Sen and Jana 1956) and induced (Ramanujam and

Joshi 1941, Akhtar 1954, Sharma and Gupta 1982, Pundir et al. 1983), autotetraploids have been reported in chickpea. Seed treatment with 0.1 to 0.25% colchicine for 4 hours has been found effective in inducing autotetraploidy. The autotetraploids showed predominantly bivalent pairing and normal disjunction at anaphase I (Sharma and Gupta 1983).

2.3 Origin

The chickpea most probably originated in an areas of present-day south-eastern Turkey and the adjoining northern region of Syria (van der Maesen 1987), because the proposed wild progenitor (*C. reticulatum*) of the chickpea and its other closely related wild species (*C. echinospermum, C. bijugum*) are found there. Since early botanists did not recognize these close relatives, various diffuse origins have been postulated for chickpea. De Candolle (1882) vaguely identified the region between Greece and the Himalayas as the place where chickpea originated. Vavilov (1926) suggested Southwest Asia and the Mediterranean as the primary centres of origin (now centres of diversity) with Ethiopia as the secondary centre. Harlan (1969) pointed out that maximum variability of chickpea was found in Ethiopia. Hence, he considered Ethiopia as a centre of diversity of chickpea.

Chickpea have diverged in two directions from its centre of origin. In the Western parts of its distribution where chickpea is grown in spring summer environments, cultivars had large, light coloured seeds with little wrinkling on the seed coat and tall plants with large leaflets and flowers. In the Eastern and Southern parts of its distribution where the crop is grown in cool dry seasons, cultivars had quick maturity, thin stem and small leaflets and flowers (Ramanujam 1976).

Within the cultivated chickpea two distinct groups of cultivars are found. Large seeded, cream-coloured chickpea called *kabuli* type and small seeded, dark-coloured chickpeas called *desi* type. Vavilov (1950) found that large seeded cultivars concentrated around the Mediterranean basin, whereas, small seeded cultivars predominated eastwards. These groups more or less coincide with the two races proposed by Moreno and Cubero (1978) *macrosperma* and *microsperma*. Race *macrosperma* was derived from race *microsperma* through selection during evolution of chickpea.

3. CLIMATIC REQUIREMENT, AREA AND DISTRIBUTION

3.1 Climatic requirement

Chickpea is a winter crop in the tropics and a summer or spring crop in the temperate region of the world. It requires cool, dry and bright weather. Temperature, day length, and availability of moisture are the three major factors affecting flowering in chickpea. Temperature is generally more important than day length. In general, flowering is delayed under low temperatures but more branching occurs. Progress towards flowering is rapid during long-days, while under short-days, flowering is delayed. However, some chickpea varieties are less sensitive to day length than others. Chickpea is susceptible to cold conditions, especially at flowering. At low temperature (mean daily temperatures below 15□C) pollen tubes grow slowly, fertilization is less likely and the flower often aborts.

3.2 Area and distribution

Chickpea is the third most important pulse crop in the world, after dry beans (*Phaseolus vulgaris* L.) and dry peas (*Pisum sativum* L.). It is mainly confined to Indian subcontinent where India alone contributes about 70% to the global chickpea production. The other important chickpea growing countries are Pakistan, Iran, Turkey, Canada, Australia and the United States of America. In India, during 1998-1999, chickpea was grown in the area of 8.41 million hectares and the total production was 6.68 million tones (Agricultural Statistics at a Glance 2000). The important chickpea growing states of India are Rajasthan, Madhya Pradesh, Uttar Pradesh, Maharashtra and Haryana. Rajasthan and Madhya Pradesh together contribute about 68% to the Indian and about 50% to the global chickpea production.

4. SALIENT FEATURES OF THE CROP CULTIVATION

In India, chickpea is grown in a number of production systems in *rabi* (post rainy season) following a *kharif* (rainy season) fallow or short-duration kharif crops. The sowing is usually done in the month of October or November. Chickpea is successfully grown on a variety of soil types including course textured-sands to fine-textured black soils with a pH ranging from 6.0 to 9.0. The seedbed should have loose tilth, be well drained and free from stubble and debris. Sowing is generally done on conserved soil moisture. Presowing irrigation is given where it is feasible and required.

The seeds should be treated with fungicides (2 g thirum + 1 g carbendazim per kg seed) before sowing. If chickpea has never been grown in a field before, the seeds should be inoculated with *Rhizobium* after their treatment with fungicides. The recommended doses of chemical fertilizers are 20 kg N, 50 kg P_2O_5, 20 kg S and 5 kg Zn. The fertilizer requirements should be adjusted according to the soil analysis report. A plant density of 30 plants m^{-2} is recommended. This stand can be obtained by spacing the rows 30 cm apart and placing the seeds every 10 cm within the row. Depending on the seed size, the quantity of seed required to show one hectare area may vary from 80 kg (small seeded varieties) to 180 kg (extra bold seeded varieties). The seeds should be sown deeply (5-8 cm) to make contact with moist soil. Pre-emergence herbicides (e.g. pendimethalin @ 1 kg/ha) should be used to control an early flush of weeds. Chickpea is generally grown as a rainfed crop, but two irrigations, one each at branching and pod filling stages, are recommended for higher yield.

Chickpea being rich source of protein is prone to insect pests and diseases. The integrated management of diseases and pest involves adoption of resistant varieties, seed treatment with appropriate fungicide/chemical/ bioagent, cultural practices and appropriate crop rotation. In general, foliar diseases (Ascochyta blight, Botrytis gray mold) are more prevalent in northern, northern-western, and eastern India, whereas root diseases (Fusarium wilt, collar rot, dry root rot) cause heavier losses in central and Peninsular India. Among insects, pod borer (*Helicoverpa* spp) is the most severe yield reducer throughout India. Spray of 0.07% endosulphon has been recommended. Three sprays of Nuclear Polyhedrosis Virus (NPV) @ 250 LE per ha on noticing first/second instar larvae or eggs of pod borer have also been found effective. The other pests (cut worm, aphids) are of minor importance and can be controlled by spray of endosulphon or quinalphos.

Drought occurs in most rainfed situations but is consistently a more severe constraint in peninsular India. Depending upon the total rainfall and its distribution, drought may occur at germination, and/or during the vegetative and reproductive stages of crop growth. Early varieties are used in central and peninsular India to escape from terminal drought.

5. BOTANICAL DESCRIPTION

5.1 Plant description

Chickpea is herbaceous annual. It has tap root system, which is usually deep and strong, and contributes to the chickpea's ability to withstand dry conditions. The roots develop symbiotic nodules with the Rhizobium bacteria, capable of fixing atmospheric nitrogen. The nodules are visible about one month after plant emergence, and eventually form slightly flattened, fan-like lobes. Practically all nodules are confined to the top 30cm of soil and 90% are within 15 cm of the surface.

The first two nodes of the main shoot invariably have scale leaves. The imparipinnate leaves with serrated leaflets arise alternatively from the third node. The first few leaves have five to seven leaflets. Fully formed leaves with 11 to 17 leaflets usually develop after the sixth node. Rachis is 25 to 30 cm long, ending in a terminal leaflet. Arrangement of leaflets may be opposite or alternate. They are sub-sessile, obovate-oblong to elliptic, 10mm long, and 4-12 mm wide. The base of leaflets is narrowly or broadly cuneate or rounded, top rounded or acuminate. Mutants with simple leaves, multipinnate leaves, narrow leaflets, broad leaflets and fewer leaflets have been reported. Stipules may be ovate or obliquely triangular.

The entire surface of the plant shoot, except the corolla, is densely covered with fine hairs known as trichomes. Many are glandular and secrete a highly acidic substance containing the malic, oxalic and citric acids. The acids play a role in protecting the plant against pests such as red-legged earth mite, Lucerne flea and aphids. These are also secreted through the root system and can solublize soil bound phosphate and other nutrients. A glabrous mutant has been reported which is highly susceptible to black aphids.

Stems may be straight or flexuous and ribbed. The final height of the plant generally fall in the range of 20-100 cm. Primary branches grow from the seed shoot as well as from buds at the lowest nodes and their number can very from one to eight depending upon the variety and growing conditions. These are thick, strong and woody, and determine the general appearance of the plant. Secondary branches are produced on the primary branches. They are less vigorous but contribute to a major proportion of the plant yield. Tertiary branches growing from the buds located on secondary branches are leafier and carry fewer pods. Five growth habits based on angle of branches from the vertical are classified: erect, semi-erect, semi-spreading, spreading and prostrate. The erect or semi-erect varieties enable mechanical harvesting.

The pod is typically inflated ending in a mucro that sometimes looks almost like a thorn. Pod size is largely unaffected by the environment but vary greatly between varieties. It is not an easy task to describe pod shape in chickpea. However, the majority of pod shapes fall in the category of rhomboid, oblong and ovate. The number of pods per plant is highly influenced by the environment and generally falls in the range of 20 to 150. The number of seeds per plant varies from one to three.

Seeds are characteristically beaked and very frequently ram head shaped and strongly wrinkled or ribbed. The coat can be smooth or tuberculate. The post-chalazal bundles vary in degree of forking and in depth. Except with spherical and nearly spherical seeds, the two cotyledons are separated by a groove, indented in the wrinkled seeds. The beak colour is sometimes different from coat colour. The hilum is small, indented, elliptical or sub-orbicular, whitish or grayish with a colour rim or corona (halo). Roughness and shape depend on cotyledons, which are part of the embryo, but colour is mainly in the testa, which is pure maternal tissue. Seed coat may be white, creamy, yellowish with an orange tinge, most often brown in various shades, black or dull green. Length and with of seed can very between 4-12 and 4-8 mm, respectively. The weight of the seed is a very stable characteristic, almost as stable as the size of the pod, and generally falls in the range of 0.1 to 0.6 g.

5.2. Stages of development

Chickpea seedlings have hypogeal emergence, i.e. their cotyledons remain underground inside the seed coat while providing energy to the rapidly growing roots and shoots. Emergence occurs 7-15 days after sowing, depending on temperature conditions and depth of sowing. Root growth is more rapid before flowering but will continue until maturity under favourable conditions. In deep well-structured soils, roots can penetrate more than three meters deep.

The growth of chickpea plant is indeterminate, i.e. vegetative growth continues even after the start of flowering. Hence, there is often a sequence of leaf, flower bud, flower and pod development along each branch. The duration of vegetative growth before flowering generally ranges from 40 to 80 days depending on the variety, location, availability of soil moisture, and weather conditions. A number of pseudo–flowers or false flower buds develop during the change over from leaf buds to flower buds on the stem.

Excessive vegetative growth is a problem of long growing season environments, as in northern and eastern India when soil moisture and temperature conditions remain favourable during chickpea's early reproductive growth stages.

Under favourable conditions, the time taken from fertilization of the ovule to the first appearance of a pod is about six days. The seed then fills over the next three to four weeks. After pod set, the pod wall grows rapidly for the first 10 to 15 days while seed growth mainly occurs later. Any damage to the pod that punctures the pod wall during the first 20-30 days will result in seed abortion. Seed filling and subsequent seed size is highly dependent on weather conditions.

Soon after the development of pods and seed filling, senescence of subtending leaves begins. If there is plenty of soil moisture, flowering and podding will continue on the upper nodes. However, as soil moisture is depleted, flowering ceases and eventually the whole plant matures.

5.3 Floral biology

Chickpeas have typically papilionaceous flowers. Solitary flowers are borne in axillary racemes, although cultivars with two or rarely with three flowers racemes are also found. Peduncles are 13-17 mm long, ending in a small arista. Bracts are triangular and small. The pedicels (6 to 13 mm) are straight when flowering and curved when bearing fruit. The calyx (6 to 13 mm) is glabrous at the base and has five partly joined sepals. The corolla (8 to 11 mm) is veined and may be pink, purple, red, white, blue, greenish white or pinkish white in colour. The standard petal is broad and clawed, the wings are free and the keel incurved. The diadelfous (9 fused and 1 free) stamens have 6 to 8 mm long filaments and the anthers are dorsifixed. Ovary is ovate, 2 to 3 mm long and 1 to 1.5 mm wide, generally with 2, sometimes 4, ovules. The style (3 to 4 mm) is upturned with slightly broadened stigma.

6. SELFING AND CROSSING PROCEDURE

Chickpea is a self-pollinated crop having an out crossing less than 1%. Self-pollination takes place 1 or 2 days before the flower opens. Eshel (1968) proposed the following categories for the developmental stages of bud and flower in chickpea —

A. *Closed bud*: in this stage, the stigma is immature and the anthers are still at the base of the bud.

B. *Hooded bud*: the corolla has elongated, and the anthers are about half the height of the style. The stigma is receptive and remains so until stage D. Emasculation is done at this stage.

C. *Half open flower:* the anthers have now reached the same height as the stigma and the anthers and the pollen is mature just before the anthers dehisce. Pollination now takes place and the keel remains closed preventing any foreign pollen from reaching the stigma. Pollen for crossing is collected at this stage.

D. *Fully open flower:* The anthers are shriveled, while the standard and wings are fully expanded. Fertilization takes place 24 hours after pollination.

E. *Fading flower:* the post fertilization stage, in which the ovary starts to elongate.

Hybridization in chickpea is generally not begun until the appearance of first pod on the female plant. Variables results have been found with regard to the time of emasculation. In some places In northern India, good pod setting is obtained by emasculation between 1400 and 1800 hrs and pollination the next morning between 0800 and 1300 hrs. As the season advances and days become warmer, morning emasculation followed by immediate pollination is more effective. Simultaneous emasculation and pollination was best in southern India, Chile, Israel, Spain and Turkey, but evening emasculation and next day pollination was more effective in Mexico.

A healthy large flower bud (hooded bud or Stage **D** of Eshel 1968) where petals are just showing should be selected for emasculation. The bud should be held tightly at its base, between the thumb and first finger. The front sepal may be drawn back or snipped off. The keel petal should be gently pushed downwards or open with a sharp- pointed forceps. This exposes the anthers, which can be removed from the filaments. The peduncle, stigma, and style are fragile and should not be touched during the operation.

Half-open flowers should be selected for pollen collection. Pollen at this stage is yellow and slightly sticky. Pollen viability remains high for 24 hrs, and then rapidly declines and the pollen becomes grayish or cream-coloured and powdery. The flower should be held tightly at its base and all petals should be gently removed without disturbing the pollen grains. This flower should then be used as a brush to apply pollen to the stigma of emasculated bud. The pollen should be applied liberally, but lightly, avoiding damage to the stigma, style or pedicels.

Small tags, labeled with pencil, should be used for tagging the pollinated bud. The tags should be tied on the stem below the pollinated bud. In normal circumstances, there is no need to protect the pollinated flower.

The rate of pod setting varies considerably depending on the cross-compatibility of parental lines, stage of bud selected for emasculation, time of emasculation and pollination, weather condition, and more importantly the skill of the person making crosses. In majority of cases, the pod setting falls in the range of 5 to 25%.

7. BREEDING OBJECTIVES

7.1 Enhancement of yield potential

The average productivity of chickpea in India is still below one ton per hectare. Development of high yielding varieties is the major objective in chickpea breeding.

7.2 Resistance to biotic stresses

Several diseases and pests cause severe damage to chickpea and largely responsible for instability in the chickpea production in the country.

The major diseases and pests are as follows —

Root diseases: Important root diseases of chickpea caused by fungi are fusarium wilt, dry root rot, collar rot and black root rot. The pathogens of these diseases are competitive facultative saprophytes, which can survive indefinitely in the soil. The root diseases of chickpea are important in areas between latitude 0° to 25° where the chickpea-growing season is dry and warm.

Fusarium wilt (*Fusarium oxysporum* f. sp. *ciceri*) is the most important root disease of chickpea. It is a typical vascular disease that causes browning and blackening of xylem. Affected seedlings first show drooping of the leaves and then collapse. The adult plants show typical wilt symptoms. The roots do not show external rotting but when split vertically the vascular tissues show brown to black discolouration. The fungus is seed and soil borne and can survive in the soil in the absence of the host for at least 6 years.

Dry root rot (*Rhizoctonia bataticola*) is a serious disease whenever the crop is exposed to temperature above 30°C. The disease generally appears around flowering and podding time. Dry soil conditions promote the disease. The whole plants dry up and turn straw–coloured. Roots become brittle and have only a few lateral roots or none at all.

Collar rot (*Sclerotium rolfsii*) is becoming a serious problem in Madhya Pradesh in areas where chickpea is grown following rice or soybean. High soil moisture, the presence of undercomposed organic matter on the soil surface, low soil pH, and high temperature between 25° to 30°C favour the disease. The affected seedlings turn yellow and wilt. The collar region is constricted that begins to rot. White strands are seen growing over the affected tissue.

Black root rot (*Fusarium solani*) is a minor disease that occurs in soils with relatively high moisture content. Dark brown lesions are seen on stems above the collar region and the roots turn black and rot.

Root-knot nematodes (*Meloidegyne* spp usually *M. incognita* and *M. javanica*) may cause negligible to very high yield loses in chickpea. They damage roots causing vascular disorders, and extensive necrosis, suppress rhizobium nodulation, reduce grain quality, and may increase severity of many soil-borne diseases. Infected plants are stunted, early senescing and poor yielding.

Foliar diseases: Ascochyta blight and botrytis gray mold are the major and alternaria blight and Stemphylium blight are the minor foliar diseases of chickpea. The incidence of these diseases is confined to the areas between latitudes 25° and 45° where weather is cool and humid during the crop growing season.

Ascochyta blight (*Ascochyta rabiei*) is a potentially devastating disease and prevalent only in areas between the latitudes 30° and 45°, where relatively low temperatures prevail during the crop season. Appearance of this disease is not regular and sometimes depends on winter rains. The severe symptoms of disease usually appear around flowering and podding time as patches of blighted plants in the field. Circular spots typically appear on leaves and pods, elongated lesions on stem, and deep cankerous lesions on seeds.

Botrytis gray mold (*Botrytis cinerea*) causes concealed damage in chickpea and its importance has only recently been realized. It is an important disease in *Terai* and northeastern regions of the country. Without visible symptoms on foliage, the disease can cause flower drop resulting in poor pod setting, and extension of the crop duration.

Pests: Pod borer (*Helicoverpa* spp, usually *H. armigera*) is the most important pest of chickpea in all chickpea growing areas. A survey indicated a range of 0 to 84% pod damage at different states, with an overall average of 8% (Sithanantham et al. 1983). Larvae feed on leaves during the vegetative stage and on flowers and pods as soon as they have developed.

Other pests are of minor importance in chickpea. Cutworm (*Agrotis ipsilon*) may reduce plant stands, particularly when high levels of moisture and organic matter are present in the soil at sowing time. Gray-black larvae hide beneath the soil surface during the day, and become active at night. They cut seedlings at or below ground level.

In stores, chickpea can be severely damaged by bruchids (*Callosobruchus* spp) that can destroy the seeds, completely.

7.3 Tolerance to abiotic stresses

Chickpea is sensitive to a number of abiotic stresses. The importance of different abiotic stresses varies depending on growing area. The commonly experienced abiotic stresses are as follows —

Drought : Drought is the most important abiotic constraint to yield in chickpea as the crop is generally raised on conserved soil moisture during winter season. It occurs in most rainfed situations but is consistently a more severe problem in peninsular and central India than in northern India. Depending upon the total rainfall and its distribution, drought may occur at germination, and/or during the vegetative and reproductive stages of crop growth. The drought at the germination reduces plant stands, whereas at the later stages it reduces plant growth, causes early senescence and reduces pod and seed development.

Temperature constraints : Optimal growth and development of chickpea takes place under temperature ranges from 15°-21° to 23°-29°C. The low and the high temperature regimes are harmful to chickpea. The temperature below 0°C causes freezing injury to the plant. The low temperature (0° to 5°C) causes chilling of the plant and consequently flower drop and pod abortion. The high temperature (30°C) reduces growth and development of the plants.

Salinity: The salinity is a common problem in several parts of India. The area of saline soils is estimated to be 6.1 million hectares. Saline-soils mainly constitute soluble salts and their electrical conductivity in soil saturation extract (ECe) is more than 4.0 dS/m. Chickpea is highly sensitive to salinity and only salt tolerant lines can be grown successfully in soils having ECe higher than 4.0 dS/m.

7.4 Shortening of maturity duration for greater adaptation

Early maturing varieties of chickpea are needed for fitting the crop under different cropping systems and situations. In peninsular India, chickpea is normally sown in October and forced to mature by mid-February owing to raising temperatures and drought stress. The growing season here is very short (90-120 days). In many parts of India, where chickpea is grown following a rainy season crop (e.g. rice, soybean) the sowing of chickpea is done late (November-December) due to delay in the harvesting of rainy season crop and field preparation after that. It reduces growing season for chickpea. Growing early maturing varieties of chickpea is the only viable solution for these situations. There is still scope for expansion of the chickpea area in the country. Using satellite image analysis, ICRISAT has identified about 13 million ha rice fallow land in India. A large proportion of this area can be utilized for growing chickpea if suitable varieties are available.

7.5 Development of input responsive cultivars for irrigated areas

Chickpea has been traditionally grown on marginal soils as rainfed crop from time immemorial. The natural selection has favoured genotypes, which could survive under adverse conditions, and the genotypes, which could have yielded more under high input conditions, have been neglected and probably lost. The application of more than two irrigations and higher levels of fertilization to the presently available chickpea cultivars generally lead to excessive vegetative growth, poor pod setting and lodging. The area of irrigated chickpea is gradually increasing in some states. About one-third of area under chickpea in Madhya Pradesh is now irrigated. There is a need to develop varieties which response favourably to high input conditions.

8. BREEDING METHODS

Chickpea is a self-pollinated crop and thus the breeding methods applicable for autogamous crops are employed. These include —

Introduction: When chickpea is introduced in any new area, introductions may be directly released as varieties after their evaluation for yield and other traits.

Pure line and mass selection: These were the widely used methods in the past when sufficient variability was available in land races. The prominent varieties developed through these methods include Chaffa, Annegeri, CO 2, JG 315, CSJ 8962, JG 74, DCP 92-3, Pragati, KGD 1168, BG 267 and Ujjain 21.

Hybridization: Hybridization is now invariably being used to create genetic variability in chickpea. The main objective of hybridization is to transfer one or more characters from one variety to other or to recover transgressive segregants for yield. Selection in segregating generations is done by pedigree, bulk or various modifications of these methods. Chickpea shows high genotype x environmental interactions and thus making visual selection for yield ineffective. It makes bulk method or modified bulk method as preferred method of selection over pedigree selection. However, pedigree selection is expected to be the most effective method for resistance to diseases and pests. The varieties developed through hybridization include C 235, L 144, L 550, H 82-2, Gaurav, BG 261, BG 256, KPG 59, Radhey, RSG 44, Phule G 5, K 850, Avrodhi, PBG 1, ICCV 10, Pusa 372, Vijay, GNG 469, GNG 663, Vishal, Pusa 391, GPF 2, BG 1003, CO3, JG 11, JG 130, BGD 72 and GCP 101.

Mutation: This is a less commonly used method and has been used to improve varieties for specific characters. The varieties release through mutation breeding include BGM 408, BGM 413, BGM 417, RSG 2 and RS 11.

Ideotype breeding: It has been suggested that an ideotype approach may be of particular help to chickpea breeding programme as both environmental and genotype x environment effects are large in this legume. Attempts have been made to describe chickpea ideotypes for different growing conditions. The emphasis has been laid on reducing internodal length, secondary branches and plant spread.

Development of transgenics: The advances in gene cloning and plant transformation technologies have made it possible to transfer genes across organisms. Transgenic plants are now available in many crops for some focused traits. Despite extensive screening efforts in cultivated and wild *Cicer* species, a good source for resistance to *Helicoverpa* pod borer has not been identified. Similarly, good source of resistance is not available for collar rot. Biotechnological approaches can be applied in chickpea for introduction of novel genes for resistance to biotic and abiotic stresses and for protein quality from other plant genera or organisms. Efficient regeneration protocols for chickpea have already been established and efforts are being made to introduce Bt and CpTi genes for resistance to *Helicoverpa* and PGIP gene of raspberry for resistance to botrytis gray mold.

Marker assisted selection: Molecular markers can greatly facilitate indirect selection for traits that are difficult or inconvenient to score directly. Attempts have been made to develop basis for markers assisted selection. The first linkage map of chickpea was published in 1990 and consisted of 26 isozyme and three morphological trait loci (Gaur and Slinkard 1990a,b). Several additional isozyme loci and morphological trait loci were mapped in the subsequent studies (Kazan et al. 1993, Simon and Muehlbauer 1997, Idnani 1998). The use of DNA markers in gene mapping has greatly accelerated progress in development of a detailed gene map of chickpea. Only RFLP (restriction fragment length polymorphism) markers were available in the beginning but now the simple and flexible application of polymerase chain reaction (PCR) has given a whole series of new DNA markers, e.g. RAPD (randomly amplified polymorphic DNA), AFLP (amplified fragment length polymorphism), STMS (sequence tagged microsatellite site), ISSR (inter-simple-sequence-repeat), SCAR (sequence-characterized amplified region), DAF (DNA amplification fingerprints), etc. A linkage map of DNA markers was first published in 1997 which contained 10 RFLP and 45 RAPD markers (Simon and Muehlbauer 1987). A molecular map based on 120 STMS markers was published by Winter et al. (1999). More recently, recombinant inbred lines (RIL) have been used to develop a molecular map of chickpea, which consists of 118 STMS, 96 DAF, 70 AFLP, 37 ISSR, 17 RAPD, 2 SCAR, 3 cDNA and 8 isozyme markers (Winter *et al.,* 2000). Interspecific segregations have been used in these linkage studies, because of the low level of polymorphism in the cultivated chickpea. A molecular map based on intra-specific cross (*kabuli-desi* cross) has also been developed. This map consists of 114 markers (STMS and RAPD) and its length is 550 cM (Jagdish Kumar, pers. communications).

DNA markers have been successfully used to tag genes for resistance to Fusarium wilt in chickpea. Two SCAR markers were found associated with resistance to race 1 (Mayer *et al.,* 1987) and one ISSR marker with resistance to race 4 (Ratnaparkhe *et al.,* 1998). The genes for resistance to races 4 and 5 have been found to be linked and located close to one STMS and one SCAR marker (Winter *et al.* 2000).

9. INTERNATIONAL AND NATIONAL ORGANIZATIONS INVOLVED IN THE CROP IMPROVEMENT

International Crop Research Institute for the Semi-Arid Tropics (ICRISAT): The ICRISAT was established in 1972 through the Consultative Group on International Agricultural Research (CGIAR) and has its main research station at Patancheru, India. The chickpea genetic resources available at ICRISAT include over 17,000 accessions of the cultivated species, 158 accessions of the annual wild species and 77 accessions of the annual perennial species. These resources are maintained in the genebank in long term and/or medium term-storage facilities. In the last 25 years (1974-1999), ICRISAT has distributed over 110,000 samples to the scientists in 84 countries. Fifteen germplasm lines have been released as cultivars in 13 countries. The most popular varieties of chickpea developed by the ICRISAT include Sweta (ICCV 2), Kranti (ICCC 37), Bharti (ICCV 10) and KAK 2 (ICCV 92331).

The International Center for Agricultural Research in the Dry Areas (ICARDA): The ICARDA is also one of the centers of CGIAR. It was established in 1977. With its main research station and offices based in Aleppo, Syria, the ICARDA is responsible for the improvement of chickpea in the Central and West Asia and North Africa (CWANA) region.

National Research Network: In India, systematic research work on chickpea was started under the ambit of the All India Co-ordinated Pulse Improvement Project (ACPIP) in 1967. Initially, the programme was launched at 12 centres located in SAUs with headquarter of the project at IARI, New Delhi. In 1977, the project was upgraded to Project Directorate and its head quarter was shifted to Regional Research Station of IARI at Kanpur. In 1993, the Directorate of Pulses Research was upgraded to Indian Institute of Pulses Research (IIPR). Realizing the importance of chickpea, an independent AICRP on chickpea was launched in the same year with 9 mandatory and 12 verifying centres.

10. MAJOR ACHIEVEMENTS

Enhancement of yield potential: Considerable progress has been made in development of varieties with enhanced yield potential. More than hundred improved varieties have been released in chickpea during the last 50 years. Despite the progress made in cultivar development, the average productivity of chickpea has remained low. There are production gaps between yields obtained from different management practices as evident from the yields obtained at research stations, those obtained at farmers' fields with improved cultivars and improved management practices, and those recorded at state level. The average productivity can be increased appreciably if seeds of the improved varieties are popularized and improved technology is transferred to the farmers.

Resistance to diseases and pests: A major emphasis in chickpea breeding has been on development of varieties resistant to diseases. The efforts have been highly remunerating, particularly in development of Fusarium wilt resistant varieties.

Four races (1 to 4) of Fusarium wilt have been identified from India (Haware and Nene 1982). Three independent loci, designated h_1, h_2 and h_3 have been reported for resistance to race 1 (Singh et al. 1987). The wilt resistant variety JG 315 was defined as resistant with homozygous recessive genes at all three loci (h_1h_1, h_2h_2, h_3h_3). The availability of good wilt sick fields at several research centres has helped screening of a large number of germplasm lines and segregating populations for resistance to Fusarium wilt. A number of germplasm lines have been identified as stable sources of resistance. The lines GL 87078 and GL 87079 showed resistance to all the four races of wilt reported to be prevalent in India. Development of wilt resistant varieties in chickpea in India, is a significant success story. Now almost all varieties being released in India are resistant to Fusarium wilt.

Ascochyta blight: Inheritance of Ascochyta blight has been found complex. Three recessive and complementary major genes with several modifiers have been found to confer resistance. Absence of one or two of the major genes confers susceptibility, whereas the presence of the modifiers determines the degree of resistance (Tekeoglu et al. 2000). A simple and reliable field screening technique to evaluate a large number of germplasm and breeding lines was developed at ICARDA in the late 1970s (Singh et al. 1981). Using this technique, about 20,000 accessions of chickpea from ICARDA and ICRISAT have been screened. A large number of resistant lines have been identified including five lines (ILC 200, ILC 6482, ICC 4475, ICC 6328, and ICC 12004) which were found to be resistant under field conditions over number of years and under greenhouse conditions. Attempts have also been done to identify lines with multiple-race resistance. Some lines, including ILC 72, ILC 201, ILC 202, ILC 2506, ILC 2956, ILC 3279 and ILC 3996, were resistant to three of the six races tested.

More than 14,800 germplasm accessions were screened for tolerance to pod borer in unsprayed areas at ICRISAT during 1984 to 1990. Over 6 years, ICC 506 showed a mean of 8.6% pod damage as against 29.9% in the popular cultivar Annigeri (Lateef and Pimbert, 1990). The national chickpea research programme in India identified many tolerant lines and recommended some for use as parents in crossing programmes.

Tolerance to abiotic stresses: Methods have been developed to screen for drought tolerance in chickpea using line-source sprinkler system (Saxena et al. 1993). Two drought tolerant lines, ICC 4958 and ICC 10448, were identified at the ICRISAT. ICC 4958 gave 30% more root volume than the popular variety Annigeri. ICC 4958 has been widely used as a parent in chickpea hybridization aimed at developing drought tolerant varieties. The ICAR and the ACIAR (Australian Centre for International Agricultural Research) launched a collaborative project in July 1998 to identify traits for yield improvement of chickpea in drought prone environments of India and Australia. Experiments of this project are being conducted at five locations in Australia and seven locations in India. The traits identified for drought tolerance will be tagged using molecular markers to facilitate breeding for drought tolerance.

Easy and reliable screening techniques have been developed at ICARDA to evaluate chickpea germplasm for cold tolerance. More than 10,000 germplasm lines were evaluated and several cold tolerant lines were identified, including ILC 8262, ILC 8617 and FLIP 87-82C (Singh et al. 1989). The cold tolerant sources have been used in breeding programmes and several cold tolerant lines have been developed. These breeding lines can tolerate cold up to −10□C. Inheritance studies revealed that resistance to freezing was controlled by at least five genes with both additive and non-additive effects and high heritability estimates (Singh et al. 1989).

Genetic variation for salinity tolerance has been found in chickpea. The genotypes ICCC 32, ICCL 86446, CSG 8977, CSG 8962 and CSG 8943 of *desi* type and the genotype CSG 88101 of *kabuli* type have shown high tolerance to salinity. A variety, Karnal Chana 1 (CSG 8962) has been released for salt affected soils of the north-west plain zone. This variety can suitably be grown in sandy loam soils with ECe ranging from 4 to 6 dS/m.

Shortening of maturity duration for greater adaptation: The development of early maturing *kabuli* varieties, like ICCV 2 and KAK 2, have made it possible to grow *kabuli* chickpea in southern and central India where the chickpea growing season is short. Varieties have been developed for late sowing conditions, e.g. BG 372, KPG 59 and PBG 1. The genetic yield potential of these varieties is 20 to 25 q/ha. Thus, it has been possible to reduce the maturity duration without sacrificing the yield potential.

Development of input responsive cultivars for irrigated areas: Under AICRP on chickpea, there are separate trials for development of varieties for high input conditions. A non-lodging variety DCP 92-3 has been released for cultivation in high input management areas of northwest plain zone.

Interspecific gene transfer: Efforts have been made to utilize wild species in chickpea improvement. The cultivated chickpea has been successfully crossed to five of the eight annual wild species. Ladizinsky and Adler (1976a,b) were the first to make interspecific crosses of *C. arietinum* with *C. reticulatum* and *C. echinospermum*. Verma *et al.,* (1990) successfully crossed *C. arietinum* with *C. echinospermum, C. judaicum, C. bijugum and C. pinnatifidum* and obtained fertile hybrids in all crosses. A hybrid was also obtained from *C. arietinum* x *C. cuneatum* cross through ovary culture but it was sterile (Singh and Singh 1989).

Table 1. Useful genes identified in wild *Cicer* species

A. Genes for resistance/tolerance to biotic and abiotic stresses	
Fusarium wilt	*C. bijugum, C. echinospermum, C. pinnatifidum, C. reticulatum*
Ascochyta blight	*C. bijugum, C. judaicum, C. echinospermum, C. pinnatifidum*
Botrytis gray mold	*C. judaicum, C. echinospermum, C. pinnatifidum*
Bruchid	*C. bijugum, C. echinospermum, C. pinnatifidum*
Leaf miner	*C. bijugum, C. cuneatum, C. echinospermum, C. judaicum, C. pinnatifidum*
Cyst nematode	*C. bijugum, C. echinospermum, C. pinnatifidum, C. reticulatum*
Root-knot nematodes	*C. bijugum, C. pinnatifidum*
cold	*C. bijugum, C. echinospermum, C. pinnatifidum, C. reticulatum*
B. Genes for other traits	
High yield	*C. reticulatum, C. echinospermum, C. judaicum*
High nodulation	*C. reticulatum*
High protein	*C. bijugum, C. pinnatifidum, C. reticulatum, C. yamashitae*

High yielding lines have been obtained from some interspecific crosses. Jaiswal et al. (1986) reported high yielding F_5 lines from *C. arietinum* x *C. reticulatum* crosses. In addition to high yield, high nodulation was observed by Jaiswal and Singh (1990) in a cross of *C. arietinum* with *C. reticulatum.* Singh and Ocampo (1997) developed high yielding lines from *C. arietinum* x *C. reticulatum* and *C. arietinum* x *C. echinospermum* crosses. Nine F_7 lines outyielded the cultigen parent by up to 39%. These lines were not only higher yielding but also free from any known undesirable traits from the wild species. In addition to the above two wild species, *C. bijugum* has also been found promising in yield improvement of chickpea. Verma *et al.* (1995) recovered F_3 recombinants from *C. arietinum* x *C. bijugum* crosses with a very large number of secondary branches, a high pod number and high yield.

The wild species of chickpea have already shown promise in gene mapping and developing high yielding lines. The much of the treasure of useful genes present in these species remains unexploited. Thus, emphasis should be given on strengthening interspecific hybridization in chickpea breeding.

Table 2. Recently developed promising varieties of chickpea.

Year	Variety	Yield (q/ha)	Important features	Area of adoption (states)
1981	JG 315	15.0	Resistant to wilt	M.P., Maharashtra, Gujarat
1983	JG 74	18.0	Resistant to wilt	NEH, West Bengal, Assam, M.P.
1985	Gaurav	18.0	Resistant to wilt	Punjab, Haryana, Delhi, North Rajasthan, West U.P.
1986	Vishwash (Phule G5)	20.0		M.P., Maharashtra, Gujarat
1987	Avrodhi	22.0	Resistant to wilt	U.P.
1988	Pusa 267 (BG 267)	21.0	Kabuli	Punjab, Haryana, Delhi, North Rajasthan, West U.P.
1988	PBG 1	18.0	Tolerant to Ascochyta blight	-do-
1990	Haryana Chana No. 1	22.0	Resistant to wilt	-do-
1990	Phule G 12	21.0	Resistant to wilt	Maharashtra
1991	RSG 44	23.0		Rajasthan
1992	Bharti (ICCV 10)	20.0	Resistant to wilt, Ascochyta blight, dry root rot, suitable for late sowing	M.P., Maharashtra, Gujarat, A.P., Karnataka, Orissa, and T.N.
1992	Uday (KPG 59)	20.0	Tolerant to dry root rot, wilt and stunt	Punjab, Haryana, Delhi, North Rajasthan, West U.P. And NEPZ
1993	ICCV 2		Kabuli, Early maturing	Maharashtra
1993	Pusa 372 (BG 372)	22.0	Mod. resistant to wilt, ascochyta blight, dry root rot, suitable for late sowing	Punjab, Haryana, Delhi, Rajasthan, U.P., Bihar, W.B., M.P., Maharashtra, Gujarat
1995	Vijay (Phule G 81-1-1)	20.0	Resistant to wilt	M.P., Maharashtra, Gujarat
1995	GNG 663 (Vardan)	24.0	Resistant to wilt	Punjab, Haryana, Delhi, North Rajasthan, West U.P.
1995	CPF 2	21.0	Resistant to wilt, tolerant to ascochyta blight	-do-
1995	KWR 108	20.0	Resistant to wilt	East U.P., Bihar, W.B.
1996	JG 218	18.0	Resistant to wilt	M.P.
1996	GNG 469 (Smart)	21.0	Tolerant to wilt, Ascochyta blight and root rot	Punjab, Haryana, Delhi, North Rajasthan, West U.P.
1996	Pusa 391 (BG 391)	18.0	Moderately resistant to wit and dry root rot	M.P., Maharashtra, Gujarat
1996	Vishal (Phule G 87207)	20.0	Resistant to wilt	-do-
1997	JG 322	18.0	Resistant to wilt	M.P.
1997	JGG 1	13.0	*Gulabi* type, good parching quality, Resistant to wilt	M.P.
1997	DCP 92-3	20.0	Resistant to wilt and lodging	Punjab, Haryana, Delhi, North Rajasthan, West U.P.

contd....

Year	Variety	Yield (q/ha)	Important features	Area of adoption (states)
1998	Karnal Chana 1 (CSG 8962)	20.0	Resistant to wilt and tolerant to salinity	Punjab, Haryana, Delhi, North Rajasthan, West U.P.
1999	BG 1053	18.0	Kabuli, moderately resistant to wilt & root rot, bold seeds	U.P, Bihar, Orissa, West Bengal
1999	BGD 72	20.0	Moderately resistant to wilt and dry root rot, bold seeds	M.P., Maharashtra, Gujarat
1999	JG 11	18.0	Resistant to wilt, moderately resistant to dry root rot, early maturity	A.P., Karnataka, Tamil Nadu
1999	KAK 2	18.0	Bold kabuli, early maturity	M.P., Maharashtra, Gujarat
1999	CO 3	18.0	Tolerant to root rot, bold seeds	Tamil Nadu
1999	WCG 2 (Surya)	18.0	Moderately resistant to wilt, dry root rot	U.P.
1999	GCP 101	20.0	Tolerant to wilt	M.P., Maharashtra, Gujarat
2000	JG 130	20.0	Resistant to wilt, bold seed size	M.P.
2000	JG 16 (SAKI 9516)	20.0	Resistant to wilt, tolerant to collar rot, botrytis gray mold and stunt.	M.P., Maharashtra, Gujarat

11. REQUIREMENTS FOR SEED PRODUCTION

Chickpea is strictly an autogamous crop. Autogamy facilitates seed production as well as maintenance of genetic purity. An isolation distance of 3m from other chickpea field is sufficient for seed production.

REFERENCES

Ahmad, F. (1999). Random amplified polymorphic DNA (RAPD) analysis reveals genetic relationships among the annual *Cicer* species. *Theor. Appl. Genet.* 98:657-663.

Ahmad, F., Gaur, P.M. and Slinkard, A.E. (1992). Isozyme polymorphism and phylogenetic interpretations in the genus *Cicer* L. *Theor. Appl. Genet.,* 83:620-627.

Ahmad, F. and Hymowitz, T. (1993). The fine structure of chickpea (*Cicer arietinum* L.) chromosomes as revealed by pachytene analysis. *Theor. Appl. Genet.* 86 : 637-641.

Ahmad, F. and Slinkard, A.E. (1992). Genetic relationships in the genus *Cicer* L. as revealed by polyacrylamide gel electrophoresis of seed storage proteins. *Theor. Appl. Genet.* 84 : 688-692.

Ahmad, S. and Godward, M.B.E. (1980). Cytological studies on the cultivars of *Cicer arietinum* L. from Pakistan. *Caryologia.* 33 : 55-68.

Akhtar, M. (1954). A study on the introduction of polyploidy in gram. *Proc. Sci. Conf.* 6:124-125.

de Candolle, A (1883). Origine des Plantes Cultivees. Paris :258-260.

Gargav, Shweta and Gaur, P.M. (2001). Phylogenetic relationships of perennial chickpea species *Cicer microphyllum* (Benth.) with its annual relatives as revealed by allozyme polymorphism. *Indian J. Genet.,* 61 : 218-222.

Gaur, P. M. and Slinkard, A. E.. (1990a). Inheritance and linkage of isozyme coding genes in chickpea. *J. Hered.*, 81: 455-461.

Gaur, P. M. and Slinkard, A. E.. (1990b). Genetic control and linkage relations of additional isozyme markers in chickpea. *Theor. Appl. Genet.*, 80:648-656.

Gupta, P.K. and Bahl, J.R. (1983). Cytogenetics and origin of some pulse crops. In *Cytogenetics of crop plants*. M.S. Swaminathan, P.K. Gupta and U. Sinha, eds. Macmillan India Ltd., Delhi, Bombay, Calcutta, Madras, pp. 405-440.

Harlan, Jack, R. (1969). Ethiopia : a center of diversity. *Econ. Bot.* 23 : 309-314.

Hutchinson, J. (1964). The genera of flowering plants. Dicotyledons. Vol. 1. Oxford Univ. Press. Oxford, UK.

Haware, M.P. and Nene, Y.L. (1982). Races of *Fusarium oxysporum* f. sp. *ciceris. Plant Dis.* 66:809-810.

Eshel, Y. (1968). Flower development and pollen viability of chickpea (*Cicer arietinum* L.). *Israel J. Agric. Res.* 18:31-33.

Idnani, Namrata (1998). Inheritance and linkage studies with some isozyme and morphological loci in chickpea. M.Sc. Thesis, JNKVV, Jabalpur.

Jaiswal, H.K. and Singh, R.K. (1990). Breeding for increased nitrogen fixation ability among wild and cultivated species of chickpea. *Ann.Appl. Biol.* 117 : 415-419.

Jaiswal, H.K., Singh, B.D., Singh, A.K. and Singh, R.M. (1986). Introgression of genes for yield and yield traits from *C. reticulatum* into *C. arietinum. Int. Chickpea Newsl.* 14 : 5-8.

Kazan, K. and Muehlbauer, F. J.. (1991). Allozyme variation and phylogeny in annual species of *Cicer* (Leguminosae). *Pl. Syst. Evol.,* 175:11-21.

Kazan, K., Muehlbauer, F. J., Weeden, N. F. and Ladizinsky, G. (1993). Inheritance and linkage relationships of morphological and isozyme loci in chickpea (*Cicer arietinum* L.). *Theor. Appl. Genet.,* 86:417-426.

Kupicha, F.K. (1977). The delimitation of the tribe Viceae (Leguminosae) and the relationship of *Cicer* L. *Bot. J. Linn. Soc.* 74 : 131-162.

Labdi M., Robertson L. D., Singh K. B. and Charrier A. (1996). Genetic diversity and phylogenetic relationships among the annual *Cicer* species as revealed by isozyme polymorphism. *Euphytica,* 88:181-188.

Ladizinsky, G. (1975). A new *Cicer* from Turkey. Notes *R. Bot. Gard.* Edinburgh. 34 : 201-202.

Ladizinsky, G. and Adler, A. (1975). The origin of chickpea as indicated by seed protein electrophoresis. *Indian J. Bot.* 24 : 183-189.

Ladizinsky, G. and Adler, A. (1976a). Genetic relationships among the annual species of *Cicer* L. *Theor. Appl. Genet.* 48 : 197-293.

Ladizinsky, G. and Adler, A. (1976b). The origin of chickpea (*Cicer arietinum* L.). *Euphytica.* 25 : 211-217.

Lateef, S.S. and Pimbert, M.P. (1990) The search for host plant resistance of

Heliconerpa armigera in chickpea and and pigeonpea at ICRISAT. In: Summary Proceddings of the consultative group meeting on the host selection behaviour of *Heliconerpa armigera* 5-7 March 1990. ICRISAT, Patancheru. pp 14-18.

Mayer, M. S., Tullu, A., Simon, C. J, Kumar, J., Kaiser, W. J., Kraft, J. M. and. Muehlbauer, F. J. (1997). Development of a DNA marker for Fusarium wilt resistance in chickpea. *Crop Sci.,* 37:1625-1629.

Moreno, M.T. and Cubero, J.I. (1978). Variation in *Cicer arietinum* L. *Euphytica.* 27 : 465-468.

Pundir, R.P.S. and van der Maesen, L.J.G. (1983). Interspecific hybridization in *Cicer. Int. Chickpea Newsl.* 23 : 5-8.

Ramanujam, S. (1976). Chickpea : *Cicer arietinum* (Leguminosae-Papilionoideae). *In : Evolution of Crop Plants* (N.W. Simmonds, ed.) pp. 157-159.

Ramanujam, S. and Joshi, A.B. (1941). Colchicine – induced polyploidy in crop plants. Gram (*Cicer arietinum* L.). *Indian J. Agric. Sci.* 11 : 835-849.

Ratnaparkhe, M, Santra, D.K., Tullu, A., and Muehlbauer, F.J. (1998). Inheritance of inter-simple-sequence-repeat polymorphism and linkage with a Fusarium wilt resistance gene in chickpea. *Theor. Appl. Genet.* 96:348-353.

Saxena, N.P., Johansen, C., Saxena, M.C. and Silim, S.N. (1993). Selection for drought and salinity tolerance in cool season food legumes. *In:* K.B. Saxena and M.C. Saxena (ed.) Breeding for stress tolerance in cool season food legumes. ICARDA, Willey-Sayce, New York. Pp 240-270.

Sen, N.K. and Jana, M.V. (1956). A spontaneous autotetraploid gram (*Cicer arietinum* L.). *Curr. Sci.* 25:231-252.

Sharma, P.C. and Gupta, P.K. (1982). Chromosome associations in autotetraploid chickpea. *Int. Chickpea Newsl.* 6:3-4.

Sharma, P.C. and Gupta, P.K. (1983). Cytological studies in the genus *Cicer* L. *Proc. of XV Inter. Cong. Genet.*, New Delhi, 12-21 Dec. 1983. Abstract No. 1257. Oxford & IBH Pub. Co., New Delhi, India.

Simon C. J. and Muehlbauer, F. J. (1997). Construction of a chickpea linkage map and its comparison with maps of pea and lentil. *J. Hered.*, 88:115-119.

Singh, H., Kumar, J., Haware, M.P. and Smithson, J.B. (1987). Genetics of resistance to Fusarium wilt in chickpeas. *In:* R.P. Day and G.J. Jellis (ed.) Genetics and plant pathogenesis. Blackwell Scientific Publ., Oxford, U.K. pp 339-341.

Singh, K.B., Hawtin, G.C., Nene, Y.L. and Reddy, M.V. (1981). Resistance in chickpeas to *Ascochyta rabiei. Plant Dis.* 65:586-587.

Singh, K.B. and Ocampo, B. (1997). Exploitation of wild *Cicer* species for yield improvement in chickpea. *Theor. Appl. Genet.* 95 : 418-423.

Singh, K.B., Vito, M. di, Greco, N. and Saxena, M.C. (1989). Reaction of wild *Cicer* spp. lines to *Heterodera ciceri, Nematologia Mediterranea.* 17 : 113-114.

Singh, R.P. and Singh, B.D. (1989). Recovery of rare interspecific hybrids of gram (*Cicer arietinum* X *C. cuneatum* L.) through tissue culture. *Curr. Sci.* 58 : 874-876.

Sithanatham, S., Rao, V.R. and Ghaffar, M.A. (1983) International review of crop losses caused by insects on chickpea. In: B.H. K. Rao and K.S. R. K. Murthy (eds.) Proceeding of the National Seminar on Crop Losses due to Insect pests.

7-9 Jan 1983, Hyderabad. Indian J. Entomology 11 (Special issue): 269-283.

Tayyar, R.I. and Waines, J.G. (1996). Genetic relationships among annual species of *Cicer* (Fabaceae) using isozyme markers. *Theor. Appl. Genet.,* 92:245-254.

Tekeoglu, M. Tullu, A., Kaiser, W.J. and Muehlbauer, F.J. (2000). Inheritance and linkage of two genes that confer resistance to Fusarium wilt in chickpea. *Crop Sci.* 40: 1247-1251.

van der Maesen, L.J.G. (1987). Origin, history and taxonomy of chickpea. In : Chickpea, M.C. Saxena and K.B. Singh (Eds.), pp. 11-34. *CAB International*, Wallingford, U.K.

Vavilov, N.I. (1926). Studies on the origin of cultivated plant. Leningrad, pp. 129, 238.

Vavilov, N.I. (1950). The origin, variation, immunity and breeding of cultivated plants. Chronica Botanica, New York 13-1/6:26-38,75-78, 151.

Verma, M.M., Ravi, Sandhu, J.S. (1995). Characterization of the inter specific cross *Cicer arietinum* L. x *C. judaicum* (Boiss.). *Plant Breeding.* 114 : 549-551.

Verma, M.M., Sandhu, J.S., Brar, H.S. and Brar, J.S. (1990). Crossability studies in different species of *Cicer* (L.). *Crop Improvement.* 17 : 179-181.

Winter, P., Benko-Iseppon, A.M., Huttel, B., Ratnaparkhe, M., Tullu, A., Sonnante, G., Pfaff, T., Tekeoglu, M., Santra, D., Sant, V.J., Rajesh, P.N., Kahl, G., and Muehlbauer, F.J. (2000). A linkage map of chickpea (*Cicer arietinum* L.) genome based on recombinant inbred lines from a *C. arietinum* x *C. reticulatum* cross: localization of resistance genes for Fusarium wilt races 4 and 5. *Theor. Appl. Genet.* 101 : 1155-1163.

Winter, P., Pfaff, T., Udupa, S.M.. Huttel, B., Sharma, P.C., Sahi, S., Arreguin-Espinoza, R., Weigand, F., Muehlbauer, F.J., and Kahl, G. (1999). Characterization and mapping f sequance-tagged microsatellite sites in the chickpea (*Cicer arietinum* L.). *Genome.* 262:90-101.

CHAPTER - 7

PIGEONPEA

R.G. Satpute

1. TAXONOMICAL CLASSIFICATION, ORIGIN AND CHROMOSOME NUMBER

Pigeonpea [*Cajanus cajan* (L.) Millsp.] belongs to the genus *Cajanus,* sub-tribe *Cajaninae,* tribe Phaseoleae and family Fabaceae. Word *Cajanus* is derived from a malaya word *Katjang* or *'Katschang'* meaning pod or bean.

Based on the range of genetic diversity of the crop in India, Vavilov (1951) concluded that pigeonpea is originated in India. Peninsular India is the centre of diversity of *Cajanus* and its parent genus *Atylosia.* It has a chromosome number of x = 11; 2n = 22, 44, 66. It is also known as red gram, tur, arhar, gandul (spanish) and cango bean (English).

2. CLIMATIC REQUIREMENT, AREA AND DISTRIBUTION

Cajanus is mainly grown as *kharif* pulse-crop, which overlaps *kharif* and *rabi* seasons. It is predominantly a crop of tropical area mainly cultivated in semi-arid regions of India and Kenya. It is also cultivated in sub-humid regions of Uganda, the West Indies, Myanmar and the Caribbean regions. Pigeonpea can be grown between 14°N and 28°N latitude with the temperature ranging from 26°C to 30°C in the rainy season (June to October) and 17°C to 22°C in the post rainy (November to March) season. Annual mean rainfall of 600 to 1400mm with 80%-90% part of the rain received in the rainy season is also a prerequisite (Reddy and Virmani, 1981). The growing season extends from 100 to 280 days. Isoclimes similar to India exist in western Africa and Southern Sudan, a suitable environment for growing pigeonpea. It is very sensitive to low radiation at pod development therefore flowering during the monsoon and cloudy weather, leads to poor pod formation.

Total area under pigeonpea in the country is spread over more than thirty-three lakh hectares. India contributes about 90% of the total world production. Uttar Pradesh, Maharashtra, Madhya Pradesh, Gujrat, Karnatak, Tamil Nadu and Bihar are the principal pigeonpea producing states.

3. SALIENT FEATURES OF CROP CULTIVATION

The crop is sown during June and July at the onset of mansoon in a well-prepared and drained soil. Sowing of seed is done at the depth of 4-5 cm, preferably on ridges if the ridge-and-furrow system is adapted. Plant spacing depends on maturity group and cropping systems. Extra-early varieties are grown at 45 cm X 15 cm, medium maturing at 20 cm X 60 cm and late maturing at 25 cm X 75 cm spacing between plants and rows, respectively.

The seed rate depends on the desired plant density for a genotype (early, medium and late), cropping system (pure crop, mixed crop or inter crop), germinability and mass of seed. In general, 35kg seed/ha for extra early pigeonpea and 20kg seed/ha for medium and late maturing varieties are recommended.

Before sowing seed of the selected variety should be treated with thirum @ 3.0 g/kg + penta-chloro-nitro benzene (1.5 g/kg) to avoid the incidence of seed rot and seedling blight diseases.

It is necessary to keep the crop weed-free during the early growth period (4-6 weeks). Pre-emergence application of Prometryn (Gesgard or Caparol a.i. 1.25 kg/ha) controls the initial weeds. Pendimethalin (Stomp a.i. 1.0-1.5 kg/ha) or Metachlore (Dual a.i. 1.0 kg/ha) are the other pre emergence herbicides recommended for pigeonpea. A hand weeding after 3-4 week sowing is required to remove the late emerging weeds. Two to three hand weddings are recommended between 1st to 6th week of crop growth in case of no use of herbicide.

Pigeonpea is largely grown as a rainfed crop, however it is well established that flower initiation and pod setting stages are sensitive to drought stress. Excessive moisture is detrimental to pigeonpea. It promotes vegetative growth and enhances the incidence of *Phytophthora* and *Alternaria* blights.

Pigeonpea should be harvested when 75-80% of the pods attains physiological maturity. Delayed harvesting during bad weather may damage the seed. The harvested plants are sun dried and threshed with a small plot thresher or with sticks. The seeds are cleaned and dried either in a drier at 35-43C or in the sun to reduce the moisture content upto 6-8%.

4. BOTANICAL DESCRIPTION

The first two leaves in the seedling called primary leaves are simple, opposite and caduceus (Fig 1). The later leaves are pinnately trifoliate; leaflets are lanceolate to elliptical with acute ends. Pubescent leaves are arranged spirally.

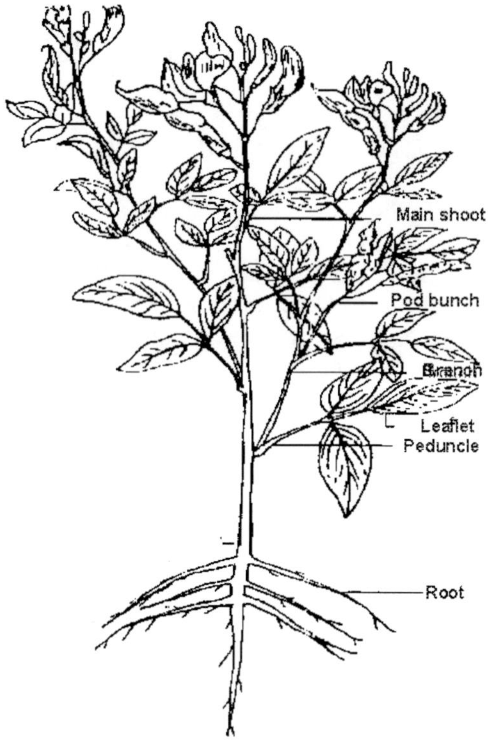

Fig. 1. A stylized pigeonpea plant

The plant is erect and branched (Fig. 2). The stem is woody, leaves are trifoliate and compound. It possesses a taproot system. It grows into a woody shrub of, 1-2 m length, which is harvested annually. As a perennial plant in can be grown in fence and agro-forestry where it may attain a height of 3-4 m. Roots are deep and wide spreading in the soil with well-developed lateral roots. Short-duration genotypes have smaller root system than long-duration genotypes. Spherical, oval, elongate, or branched nodules of 2-4 mm size are formed by the cowpea group of *rhizobia*, mainly on the upper 30cm of the root system. Nodulation declines towards pod filling (Kumar Rao 1990).

Strong woody stem of pigeonpea has colour varies from green, sun-red or purple. For agronomic purpose pigeonpea plants can be grouped as compact (erect) semi-spreading (semi-erect) and spreading types. Based on the flowering pattern it may be determinate or indeterminate (Fig 2). The determinate type first completes the vegetative phase and then enters the reproductive phase. In this type, the apical bud of the main shoot develops into an inflorescence and the sequence of inflorescence production is basipetal. The non-determinate type shows continuous vegetative and reproductive phases. In this type the flowering starts at nodes behind the apex and proceeds both acropetally and basipetally. Another group is semi-determinate which behaves between the determinate and non-determinate habit.

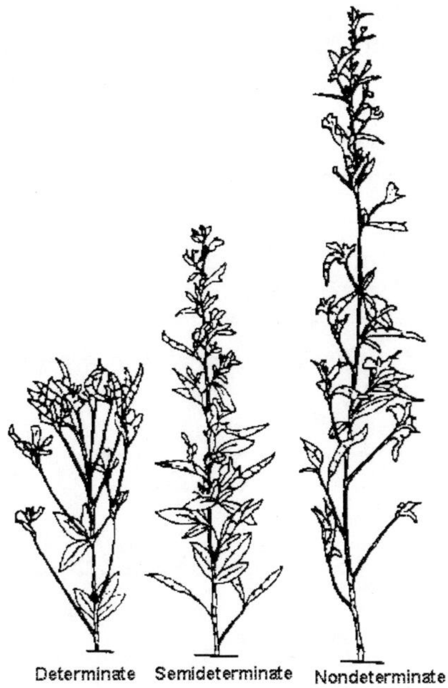

Fig. 2. Branching pattern in pigeonpea

Floral biology Flowers with long peduncle (Fig. 3) are borne on terminal or auxiliary racemes (4-12 cm). The raceme inflorescence forms a terminal in non-determinate type and a somewhat corymb-shape bunch in the determinate types. Flowering proceeds acropetally (in the direction of apex) in the raceme and on the branches too.

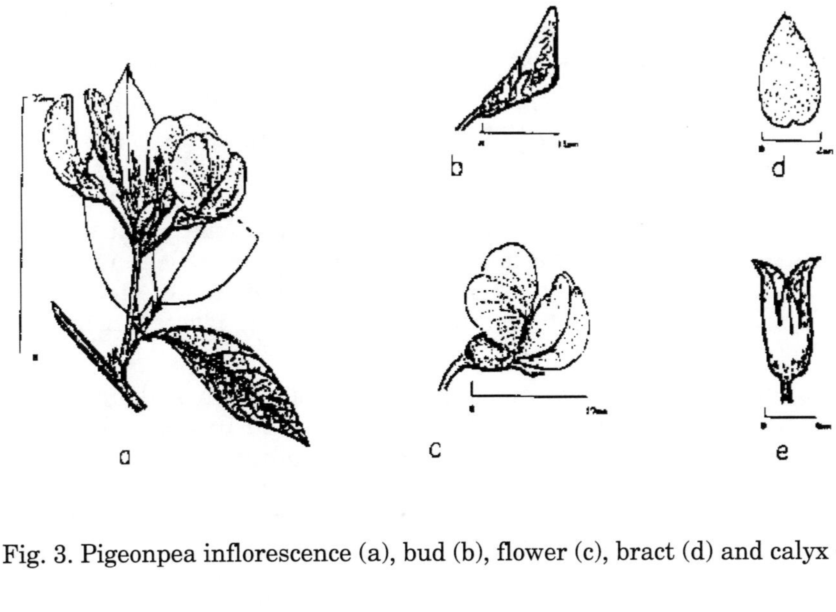

Fig. 3. Pigeonpea inflorescence (a), bud (b), flower (c), bract (d) and calyx (e)

Cluster of flower is formed at the top of the 1-8 cm long peduncle (Fig 3). Thin and long (5-15 mm) pedicel is covered with hairs. Colour of papileonaceous, bisexual and zygomorphic flower (Fig. 3) varies from yellow to yellow with purple veins and from yellow to diffused red. Calyx is five lobed, gamosepalus and companulate (bell shaped) with nerved teeth. The upper lobes are paired, free or partly free whereas the lower one is the longest (Fig.3).

The corolla is zygomorphic and bright yellow. The imbricate petal are of three prominent types i.e., the standard (broad, large, auricled and erect; wings (obliquely obovate with an incurred claw) and the keel (obtuse, round, incurred and boat shaped). The keel covers the androecium and gynaeceum (Fig. 4).

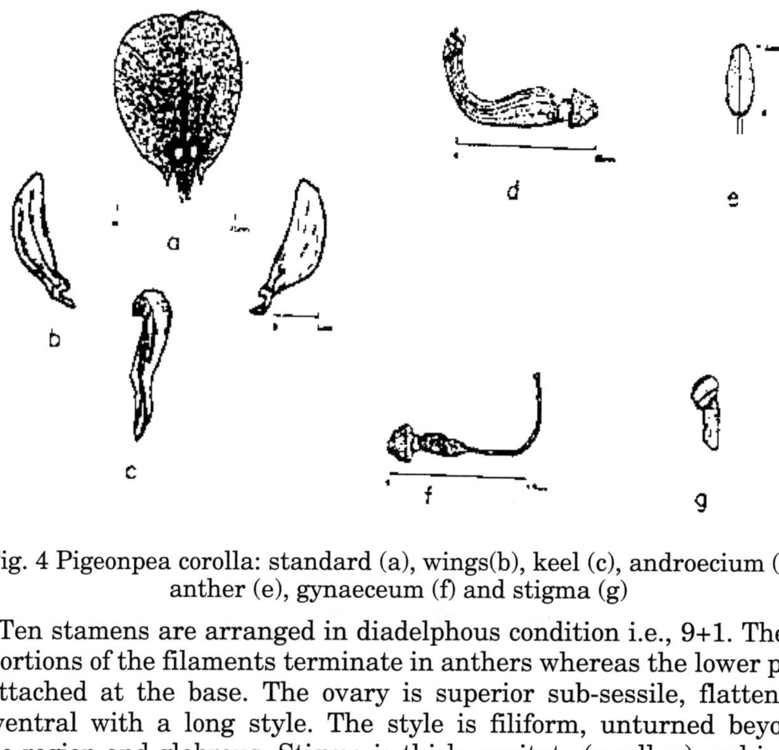

Fig. 4 Pigeonpea corolla: standard (a), wings(b), keel (c), androecium (d), anther (e), gynaeceum (f) and stigma (g)

Ten stamens are arranged in diadelphous condition i.e., 9+1. The upper free portions of the filaments terminate in anthers whereas the lower portions are attached at the base. The ovary is superior sub-sessile, flattened and dorsiventral with a long style. The style is filiform, unturned beyond the middle region and glabrous. Stigma is thick, capitate (swollen) and incurred. Stalk is short densely pubescent and glandular punctuate (dotted or pitted), ovary is monocorpellary and unilocular with marginal placentation of two to nine ovules.

Pigeonpea is a short day plant, though a combination of photo-period and temperature can affect the time of flowering and maturity (Green *et al.,* 1979). In a fully developed bud, anthers surround the stigma and dehisce a day before the flower opening. Anthesis in pigeonpea starts from 0600 and continues till 1600 with peak period between 0900 and 1000 (Sharma and Green 1980). The duration of flower opening depends on the climate and environment.

Fertilization occurs on the day of pollination. It is an often cross-pollinated crop with the extent of cross-pollination ranges from 3 to 40% (Khan, 1973). A plant produces numerous flowers of which only 10% pods are set (Pathak 1970).

The endosperm undergoes rapid development during the 1st week of anthesis, Large amounts of endospermic tissues with distinct embryo are visible by the end of the 2nd week. Later, due to rapid growth of cotyledons, endosperm is completely degenerated. In the cotyledons, synthesis of starch and protein starts about 17 days after pollination and continues for 14 days (Sehgal *et al.,* 1987). Colour of pod varies from green, purple, dark purple or combination of these colours. Two to seven seeds per pod are present in separate locules. Locules are formed by the cross wall formation during the first week after fertilization. Development of seed is visible after seven days of pollination while pod is formed 15-20 days after fertilization. Seeds reach at physiological maturity in 30 days and are ready for harvest in 40 days (Rao and Rao 1974).

Colour of oval, square, or elongate shaped seed ranges from white, cream, brown, black, pink or red to purple with 100 seed mass from 2.8 to 22.4g. Cultivated varieties possess 3-4 seeds per pod. Seed with no known dormancy germinates as hypogeal.

5. SELFING AND CROSSING PROCEDURE

Selfing is done by covering an individual inflorescence, a branch of a plant or whole plant. When a large amount of seed is required, groups of plants are covered with bee-proof cages. Depending on seed requirements, a large or small muslin bag is used for selfing of an individual plant. Bee-proof cages are preferred over cloth bags for large-scale seed production and for handling the segregating populations.

Emasculation: The tightly closed buds of approximately 66% the size of a mature bud (2 cm) are selected for emasculation (Fig 5). It is desirable to select two buds on one inflorescence. In this way, two to ten buds can be emasculated on a branch. All other buds are removed. For emasculation hold the bud firmly and remove the sepal covering the keel. Then open the corolla by inserting one tip of the forceps at the base of the keel. Now move the forceps upward to the tip of the standard and press the bud with a slight pressure until it opens. This makes the anthers visible. With the forceps remove the anthers from the staminal column, count the anthers and examine the bud with a magnifying glass to ensure that all the anthers or their broken parts are completely removed.

Pollination: The stigma became receptive before anthesis. Therefore, pollination can be done immediately after emasculation. Remove the corolla of the pollen-source bud to uncover the staminal columns and anthers. Hold the emasculated bud in one hand and transfer the pollen by touching the anthers to the stigma.

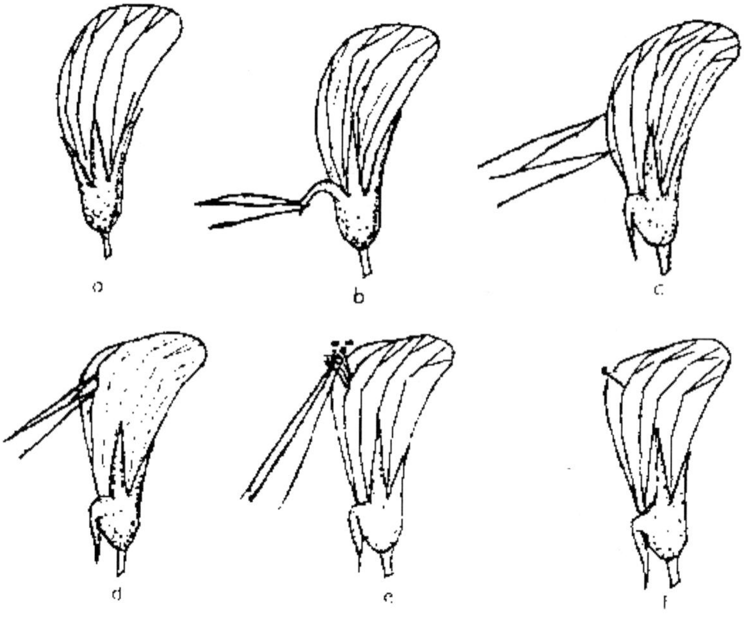

Fig 5 Steps of emasculation

Labelling: Individual row of genotype selected as female is sown for each cross. Therefore, labelling of individual bud is not necessary. An identification label for the male parent is written on each plot. To identify the artificially pollinated buds, a small bright coloured nylon thread is tied to the peduncle of the bud. Seeds are harvested only from such pods.

6. BREEDING OBJECTIVE

Increased yield with acceptable grain quality and stability is the major breeding objectives across phenological groups. Stability in production is sought by incorporating resistance or tolerance to biotic stresses, such as diseases (sterility mosaic, wilt and *Phytophthora* blight) and pests (pod borer and pod fly) and abiotic stresses such as water-logging, drought, acidity and salinity. The adaptation of improved genotypes of different phenological types to such systems as inter cropping, alley cropping, multiple cropping and multiple harvests (ratoonability) is an important objective in pigeonpea improvement.

7. BREEDING METHODS FOR CROP IMPROVEMENT

Adaptation: Most of the early improvement work was confined to selection and pedigree evaluation from land-races adapted to the region in which such selection was exercised. Pigeonpea in India is grown in widely diverse environmental conditions and cropping systems. By centralized pigeonpea improvement programmes, valuable contributions can be made viz., development of breeding population and lines specifically adapted to particular system. Introgression of desirable genes among the gene base of these systems. And the improvement of characters that are transferable across production systems, such as diseases and pest resistance. Studies on the expression of genetic variability and heritability of particular characters in different environment are needed. However, shuttle breeding or alternate selection cycles in diverse systems and seasons could lead to widely adapted genotypes.

Selection and hybridization: combining ability of the parents are determined by the diallel or line x tester mating schemes. Single crosses are commonly used by breeders but Hartwig (1972) suggested one or more back-crosses to the adapted parent to retain productivity and adaptation to a specific environment.

The commonly used breeding methods in pigeonpea are applicable to any self-pollinated crop, even in the presence of a considerable amount of out crossing. Consequently, controlled production of self-pollinated seed is necessary for genetic testing and maintenance. Green *et al.,* (1981) observed that pedigree selection has been useful in breeding for highly heritable traits such as sterility mosaic disease resistance, seed size, seed colour, growth habit and seed number per pod. Selections based on single-plant yield in early segregating generations have been found to be ineffective (Byth *et al.,* 1980). In view of the minimum efforts of selfing required and the apparent ineffectiveness of pedigree selection for yield *per se*, bulk hybrid advanced by single pod descent appears to be a better procedure for breeding high yielding lines (Green *et al.,* 1979).

Stratified mass selection (SMS) and mass selection with progeny testing (MSPT) for yield is useful in pigeonpea. Handling of segregating populations of biparental or multiple crosses as bulk population before extraction of pure lines has been considered a favourable practice.

8. ORGANIZATIONS INVOLVED IN THE CROP IMPROVEMENT

Pigeonpea is one of the mandate crops of International crops Research Institute for Semi-Arid Tropics (ICRISAT) where multidisciplinary research is undertaken. Its headquarter is at Patancheru, Andhra Pradesh, India, whereas at national level, Indian Institute of Pulses Research, Kanpur (IIPR) coordinates the research programme among various research organizations including State Agricultural Universities of the country.

9. VARIETIES UNDER CULTIVATION IN INDIA

Number of varieties for different zones as per the requirement have been released. The main factor for the adoption of these varieties is the place in crop rotation, therefore the varieties are classified based on the maturity group (Table 1). The pedigree and salient characters of the important varieties are described in table 2.

Table 1 Varieties of pigeonpea released in India for the different maturity groups

Extra-Short-duration	Short duration	Medium duration	Late-duration
140 days	140-160 days	160-200 days	200 days
UPAS 120, PCPL 87 (Pragati), ICPL 151 (Jagriti), CO2, AL 15, Pusa 33	TAT 10, T 21, Pusa Ageti, HY 2, CO 1, Pusa 84	NO 148, JA 4, Asha (ICPL 87119), BDN 1, BDN 2, C 11, Maruthi (ICPL 8863), HY 1, HY 3A, HY 5, Mukta (R 60), BSMR 736.	Laxmi (Kanke3), Gwalior 3, Bahar, MA3, NP (WR) 15

Table 2 Salient features of important pigeonpea varieties

Variety	Parentage	Maturity (days)	Salient features
Extra Short-duration Varieties			
ICPL 87 (Pragati)	T21 x JA 277	130-135	Short statured, determinate, brown streaks in pods; seed brown, 9.5 to 10.0 g/100 seed; yield potential 2.0 t ha^{-1}
ICPL 151 (Jagriti)	ICPL 6997 x Prabhat	120-130	Short statured, determinate, purple streaks on pods; 10.5g/100 seed, seed of cream colour; yield potential 2.0 t ha^{-1}
Pusa 33		130-135	Short statured, indeterminate; 9.0 g/ 100 seed, brown seed; yield potential 1.7 t /ha
Medium duration varieties			
JA 4	No 148xC 11	165-175	Medium tall, semi-spreading, indeterminate; seeds brown-red, 9.5 g/100 seed; yield potential 1.8 t/ha
No. 148	Local selection	180-185	Medium tall, semi-spreading, indeterminate, flowering seated deep into the branches; seed oval, brown red, 9.0-9.5 g/100 seed; yield potential 1.7 t/ha
JKM 7	ICPL 8863 (Maruthi) X LRG 30	195-200	Tall, semi-spreading, indeterminate, long branches; seed red brown, 8.5 g/ 100 seeds; yield potential 1.8 to 2.0 t/ha
Asha (ICPL 87119)	C11 x ICPL6	195-200	Plants medium tall, semi-compact, semi determinate, profuse primary and secondary branches, resistant to wilt and sterility mosaic diseases; seeds brown, 9.5 g/100 seed; yield potential 2.0-2.1 t/ha.
Late maturing varieties			
Gwalior 3	Local selection	250-260	Tall, semi-spreading, indeterminate, bushy plants; seeds oval, light brown, 9.0 g/100 seeds; yield potential 1.5 – 1.8 t/ha.
MA 3	-	250	Tall, indeterminate; seeds oval, brown, 9.5 g/100; yield potential 2.2 t/ha

10. MAJOR ACHIEVEMENTS IN THE CROP THROUGH BREEDING

Traditionally, medium and long-term maturity pigeonpea has been cultivated as mixed crop. However, of late shift towards mono-cropping with sufficient inputs in certain agro-ecological niches has been accepted. This could largely became possible mainly due to development of resistant varieties against major diseases like sterility mosaic disease (SMD), wilt and *Alternaria* blight. The early and extra-early maturing varieties of 120-150 days are becoming popular in MP and Punjab for pigeonpea wheat rotation.

Fusarium wilt continued to be the major problem but losses have been mitigated to some extent by adoption of resistant varieties (ICP 8863, ICPL 87119, BSMR 736 etc.) in endemic regions of Karnataka, Andhra Pradesh, Madhya Pradesh and Maharashtra.

Hybrid pigeonpea : Harnessing hybrid vigour as an additional tool to augment productivity in pigeonpea has already taken a firm shape. By the exploitation of genetic male sterility (GMS), Four early duration hybrids viz. ICPH 8 in central zone, PPH 1 in Punjab and COH 1 and COH 2 in Tamil Nadu have been released. Hybrids AKPH 4101 in early and AKPH 2022 in medium duration are the latest release for the central zone. Cytoplasmic genetic male sterility (CMS) which will usher a new era of dynamism in commercial exploitation of hybrid vigour in pigeonpea has been identified. Stabilization of CMS lines and search for suitable restorer has successfully been achieved. Evaluation of 5 cytoplasmic male sterile (CMS) lines viz., 1024 A, GT 288 A, 67 A, AKCMS 1 and ICRISAT - CMS confirmed the stability of male sterility in GT 288 A across the locations and environment. ICRISAT –CMS has shown 95% sterility. Few fertility restorers have been found against GT 288 A, which are under evaluation. GT 288 A is being converted into early, medium and late duration agronomic bases with resistant to major diseases for different agro-ecological regions.

To produce hybrid seed in bulk, male sterile lines are planted in the ratio of six male sterile rows (Female): one pollinator row (Male). The hybrid seed plot is surrounded by four pollinator rows to provide sufficient pollen load. In genetic male sterility (GMS) system 50% plants appears male fertile in the female (MS) rows. Therefore, these fertile sibs needs to be uprooted immediately as the first bud appear on the plant. The male sterile sibs those remain are to be tagged in the female rows. Periodic picking of immature pods from the pollinator rows may prolong their flowering time. It is possible to produce several hybrids in one isolation block using a common male parent and several male sterile, if their flowering can be synchronized. Appropriate isolation distance of 250m between two seed blocks should be maintained to avoid contamination.

REFERENCES

Byth, D.E., Green, J.M. and Hawtin, G.C. 1980. ICRISAT/ICARDA chickpea breeding strategies. In: proceedings of the Intl. Workshop in chickpea improvement. 28 February – 2 March 1979. ICRISAT Centre, India, Patancheru, A.P. pp, 11-27.

Chandra, S., Asthana, A.N., Ali, M., Sachan, J.N., Lal, S.S., Singh, R.A. and Gangal, L.K. 1983. Pigeonpea cultivation in India – a package of practices. Kanpur, U.P. India Directorate of Pulses, 44 pp.

Green, J.M., Sharma, D., Saxena, K.B., Reddy, L.J. and Gupta, S.C., 1979. Pigeonpea breeding at ICRISAT. Presented at the Regional Workshop on Tropical Grain Legumes, 18-22 June 1979, Univ. of West Indies St. Angustin, Trinidad.

Green, J.M., Sharma, D., Reddy, L.J. Saxena, K.B., Gupta, S.C., Jain, K.C., Reddy, B.V.S. and Rao, M.R. 1981. Methodologies and progress in the ICRISAT pigeonpea breeding program. In: Proceedings of the Intl. Workshop on pigeonpea, Vol. – 1, 15-19 Dec., 1980, ICRISAT Centre, India Patancheru, A.P. pp. 437-449.

Hartwig, E.E. 1972. Utilization of soybean germplasm strains in a soybean improvement program. Crop Science 12, 456-859.

Khan, T.N. 1973. An approach to the breeding of pigeonpea (*Cajanus cajan* (L.) Millsp.) formation of composites. Euphytica 22: 373-377.

Kumar Rao, J.V.D.K. 1990. Pigeonpea nitrogen fixation. Pages 233-254. In: the pigeonpea (Nene, Y.L., Hall, S.D. and Sheila, V.K. eds.) Walling Ford, U.K. CAB International.

Pathak, G.N. 1970. Redgram. Pages 14-53 In: Pulses Crops of India, new Delhi. India, ICAR.

Rao, M.K.V. and Rao, R.G. 1974. Gibberellin like substance in developing and germinating seeds of pigeonpea. Indian J. Plant Physiology. 17 (1,2): 65-72.

Sehgal, C.B., Gandhi, V. and Varma, B. 1987. Histological and histochemincal studies on the cotyledons of some legumes. II Resource Metabolites during seed development. Cytologia 52: 847-858.

Sharma, D. and Green, J.M. 1980. Pigeonpea pages 471-481 In. Hybridization of crop plants (Fehr, Walter R. and Hadley Henry, H. eds.) Madison WI, USA. American Soc. of America.

Singh, L. Gupta, S.C. and Faris, D.G. 1990. Pigeonpea breeding. Pages. 375-399. In. The pigeonpea (Nene, Y.L., Hall, S.D. and Sheila, V.K. eds.) Walling Ford, U.K. CAB International.

Vavilov, N.I. 1951. The origin, variation immunity and breeding of cultivated plants. Chronica Botanica 13 (1.6): 1-366.

CHAPTER - 8

MUNGBEAN AND URDBEAN

B. M. Moghe and Mridula Billore

1. NAME OF THE CROPS

Hindi	— Urid or Urd, uridbean
English	— Blackgram
Botanical name	— *Vigna mungo*
Hindi	— Moong or Mung, mungbean
English	— Green gram
Botanical name	— *Vigna radiata*
Family	— Leguminoceae
Sub-family	— Papilionaceae

2. TAXONOMICAL CLASSIFICATION AND CHROMOSOME NUMBER

Vigna radiata var. sublobata is considered to be the putative progenitor of moongbean. There exists a range of opinions as regards the exact Taxonomical position and origin of the cultivated urid at present. Verdcourt (1971) reported that one wild species *Vigna radiata* var. sublobata has given rise to two cultigens namely *Vigna mungo* (black gram) and *Vigna radiata* (green gram). Chandel (1984) reports that *Vigna mungo* domesticated from *Vigna mungo* var. silvistris instead of *V. radiata* var. sublobata. Karmakar and Dana (1987) report that var. sublobata was not separated from mungbean (*V. radiata*) by any isolating barriers while it was separated from urdbean (*V.mungo*) by multiple type of isolating mechanism.

Earlier called the genus "Phaseolus" is now newly called as genus "Vigna" either for urd and mung. Black gram (urd) belongs to the species "mungo" while the green gram to the species "radiata". Both green gram (*V. radiata* L. Wilzek) and black gram (*V. mungo* L. Hepper) have x = 11, 12; 2n = 22, 24.

3. ORIGIN AND DISTRIBUTION

South west Asia and the Mediterranean region are rich in primitive and wild relatives of pulses. India happens to be the origin of gene centre of urdbean and moong-bean both (Zukovskiji, 1962). Moongbean is the third important pulse crop in India only after chickpea and pigeonpea. It is widely spread over South Asian countries viz., India, Pakistan, Bangladesh, Srilanka, Philippines, Taiwan, Thailand and Nepal etc. Although introduced in Southern United States, the Westindies and other tropical areas, the uridbean is extensively cultivated in India since ancient times. The major uridbean growing countries are Pakistan, Bangladesh, Srilanka, Burma, and Nepal etc.

4. CLIMATIC REQUIREMENT, AREA AND DISTRIBUTION

V. radiata (green gram) ranks third important pulse crop in India. The states accounting for the major area under cultivation are Orissa, A.P., M.P., Maharashtra and Rajasthan. It is a warm and humid climate crop mostly kharif. This is also cultivated in summer with irrigation.

Vigna mungo (black gram) is the fourth important pulse crop in India. It is mostly grown in Madhya Pradesh, Maharashtra, TamilNadu, Uttar Pradesh and Rajasthan during kharif and in Andhra Pradesh and West Bengal in Rabi. Commonly warm and humid climate is suitable for black gram cultivation. The area, production, productivity and their overall status of total pulses, mungbean and uridbean in India from 1994-1999 have been presented in Table 1.

5. SALIENT FEATURES OF CROP CULTIVATION

1 Soils — Light to medium well drained soil

2 Crops — Mungbean and uridbean

3 Seed rate and time of sowing

(i) Kharif — Onset of monsoon (end of June) seed rate 20-25 kg/ha

(ii) Spring (Summer) — Sowing by end of February to 15th March, seed rate – 25-30 kg/ha

4 Spacing and population

Row to row	— 30 cm
Plant to plant	— 8-10 cm
Depth of sowing	— 5 cm
Plant population	— Kharif 3.5lakh/ha, spring (summer) 4.0lakh/ha

5 Seed treatment

(i) Fungicide	— (a) Thiram 1.5 g + Bavistin 1.5 g or (b) 2.5 g Bavistin, or (c) 3 g Thiram/kg seed
(ii) Culture	— 10 g Rhyzobium/kg seed 10 g PSB/kg seed

6 Fertilizer — N_{20} P_{50} and S_{20} kg/ha through

(i) Superphosphate 300 kg + urea 40 kg/ha **OR**

(ii) DAP 100 kg and Gypsum 100 kg/ha **OR**

(iii) DAP 100 kg and Iron pyrite 100 kg/ha (Gypsum and Pyrite to meet the need of sulphur)

6. PLANT PROTECTION

One or two sprays of insecticide are a must. Thiodan 1200 ml alternated with 700 ml of Monocrotophos/ha in 600 litres of water may be sprayed. Attack of blue beetle within 15 days of sowing may be controlled by Thiodan spray. For the control of Powdery mildew, spray Bavistin 500 g/ha or Kerathian 250 ml/ha in 500 litres of water at initial stage of attack.

7. HARVESTING AND THRESHING

By the time the crop matures, plants and pods dry. The crop is harvested all at a time in black gram but not in mungbean. Crop harvesting (cutting entire plant) at time in black gram compares more favourably than green gram (where pods need to be picked up 2-3 times). Harvesting in black gram is thus a labour saving device than green gram. Point of synchronous maturity in black gram as compared to green gram goes a long way increasing the area under black gram in the country. Threshing of both is traditionally done on the threshing floor either by beating by stick or trampling by bullocks and not by any mechanical thresher.

8. YIELD POTENTIAL

Improved varieties are potentially producing 12-14 quintals/ha.

Table 1. All India area, production and productivity of total pulses (1994-1999)

Year	Area (lac ha)	Production (lac tons)	Productivity (kg/ha)
1994-95			
Total pulses	230.3	140.4	610
Mung	29.0	10.7	369
Urid	28.0	11.4	407
1995-96			
Total pulses	222.8	123.1	552
Mung	27.1	8.2	303
Urid	28.0	13.2	471
1996-97			
Total pulses	224.5	142.4	635
Mung	30.8	13.1	423
Urid	31.5	14.9	473
1997-98			
Total pulses	228.7	131.0	567
Mung	30.2	9.5	316
Urid	30.6	12.9	421
1998-99			
Total pulses	238.2	148.1	622
Mung	30.9	11.2	362
Urid	31.1	14.1	454

Source: Directorate of Economics and Statistics, New Delhi.

Particulars –	Mungbean	Uridbean
Area increase –	29.0 to 30.9	28.0 to 31.1
Production increase –	10.7 to 11.2	11.4 to 14.1
Productivity decrease –	369 to 362	407 to 454
Future prospects –	High potential	Very high potential

Source: Project Coordinators report 1999-2000

9. BOTANICAL DESCRIPTION

I. Plant description

A typical vigna, plant type and floral morphology have been depicted through Fig. 1.

Habit: Urd bean is an erect or sub erect twining herb, annual with plant deeply hairy.

Root: The taproot is well branched and fairly extensive. Smooth and round nodules varying greatly in size develop singly or in group on the root.

Stem: The stem is defuse, furrowed and much branched from the base. It is green or purple and is covered with dense brown hair, hairs pointing downward. The stem is 30-100 cm long occasionally trailing type.

Leaf: The leaves are alternate, trifoliate with pointed leaf lets, subtended by small stipules. There are two narrow stipules at the base of the petiole. The leaves are dark green. The ovate stipules and falcate stiples, leaf

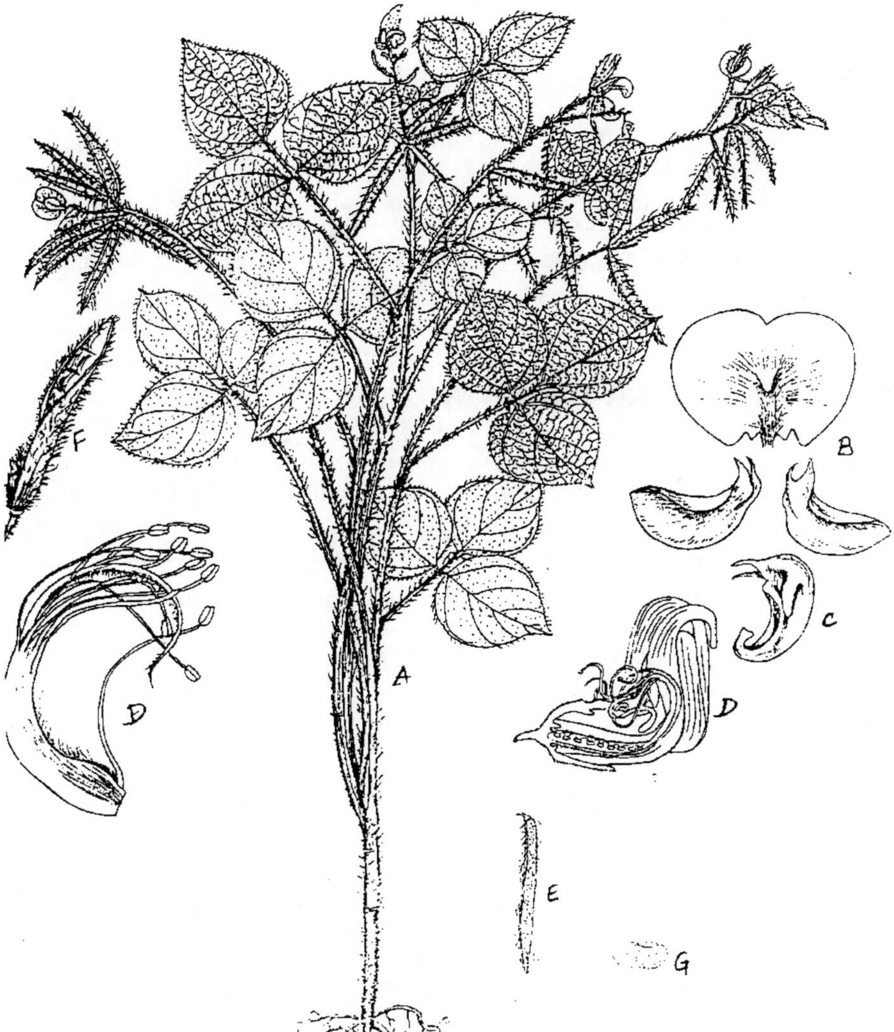

Fig. 1. A typical plant of Vigna spp. and its floral morphology
Vigna Mungo L. Blackgram Urad (Fabaceae)
A : Plant with inflorescence; B. Longitudinal section of flower; C. Corolla; D. Androecium and gynocecium; E. Pod; F. Dehisced pod; G. Seed

peltatively attached, pulvinate, pinnately trifoliate sparsely hairy on both the surfaces, sub coriaceaus, palmately reticulate, lateral leaflet oblique leaf lets are ovate to lanceolate with entire margin.

Inflorescence: The inflorescence is axillary, may have 2-3 branches. There are 5-6 small flowers borne in capitate cluster at the top of hairy penduncle.

II. Floral biology

Flower: The papilionaceous flower is complete pentamerus, small yellow, Bractiate, Bracteolate, Pedicillate, hermaphrodite, hypogynous, zygomorphic cyclic, dichlamydeous, heterochlamydeous perigynous produced in short axillary racemes in clusters of 9 to 15. Bract small deciduous, bracteoles mostly conspeiuous persistent and longer than the calyx.

Calyx: It comprises five sepals (three large and free, two small and fused) gamosepalous, imbricate and calyx lobes are linear. The calyx is campanulate. The lowest tooth being ususlly larger than the rest, the two upper most being subconnate.

Corolla: The corolla consists of five petals papilionaceous (one standard, two wings and two keels united) pale yellow, descendingly imbricate and much exerted. The standard petals are 12 to 16 mm wide and yellow. The keel being prolonged into a very long beak and forming a more or less complete spiral.

Androecium: It is diadelphous (9 + 1). Fillaments are alternately long and short anthers are uniform dithecous, introrse and dorsifixed.

Gynoecium: It is monocarpellary with superior unilocular sessile ovary with a few ovules. The style is terminal, filiform, densely bearded beneath the stigma twisted below the stigmatic surface.

The stigma is oblique and hairy. The placentation is marginal. An annuolar disc is present around the base of the ovary.

Floral formula - | . O K(5) $C_{1+2+(2)}$ $A_{(9)+1}$ G_1

Floral diagram –

SYSTEMATIC BOTANY

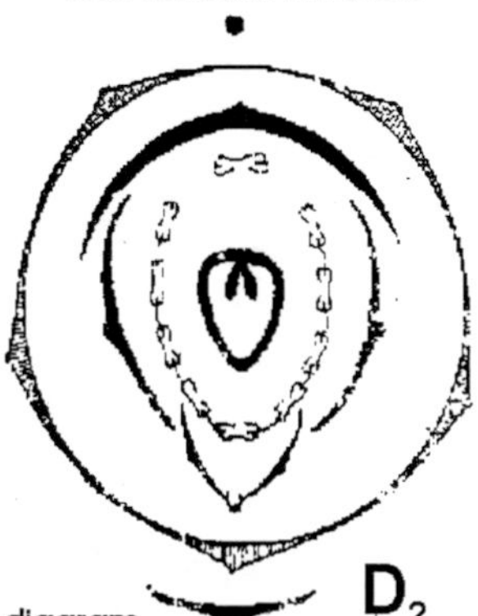

D_2

D_2 - Floral diagram
(Typical Papilionaceae)

Fruit: A legume pod, cylindrical, erect or sub erect, buff to dark brown at maturity. Septate between the seed and covered with long ferruginous hair.

Seed: Seeds are oblong with square ends or ovoid. Mostly black in urid and green in mung. Cotyledons are white.

10. DISTINGUISHING FEATURES OF MUNGBEAN AND URDBEAN

Foregoing description dealt specifically with urd bean and moong bean in general however it is often difficult to distinguish between plants and seeds of both *V. radiata* and *V. mungo.* Table 2 gives distinguishing characters of both the species.

Table 2. Distinguishing characters of moong bean and urd bean

Character	Moong bean	Urd bean
Stem	Green mostly erect or sub erect	Dark green
Epicotyle	Green	Dark green
Height	Tall	Medium
Leaf	Mostly green large	Mostly green medium
Hairiness	Plant sparsely hairy	Plant densely hairy
Hair colour	Slightly brownish	Redish brown
Stipule	Broad	Narrow
Flower colour	Light yellow	Dark yellow
Fruiting habit	Up to seven pods per cluster mostly on upper branches	3-5 pods per cluster fruits on the lower and upper branches covered under leaves
Pod size position and other characters	Large, spreading or reflexed, shatters readily, have short hair	Small, erect or sub erect, do not shatter much, possess long hair
Number of seeds per pod	7 or more	5 or less
Seed colour	Green/dark green/shining green	Black/dark black/shining black
Seed size	Medium	Bold
Seed shape	Globose	Oblong
Seed coat	Many tiny wavy ridges present; some times very faint but never lacking	No ridges
Cotyledons	Yellow, when chewed pasty, sensation absent	Whitish or pale yellow when chewed it gives a pasty taste
Hilum	Flat	Concave

III. Stages of development

There are two phases of development; one is vegetative phase and the other reproductive phase.

Vegetative phase: The germination of seed takes place within 3 to 5 days and full plant growth is completed within 30 to 35 days. Generally, dry season with intermittant slight rainfall is required for good vegetative growth. Some times this is observed that the growth is checked by the attack of blue beetle at initiation of 3rd leaf. As soon as it is observed one spray of Endosulphan is necessary to check the insect attack.

Reproductive phase: The flowering starts after attaining 30-35 days plant growth and reproductive phase continues up to one month till the pods are matured. The maturity of pods comes at 70-75 days and late varieties mature by 85-90 days. The orientation of pods in uridbean is generally covered by the leaves unlike moong where pods appear openly right at the top of plant. Pods at maturity are brown in colour with or without hair. Leaves become dry at the time of harvesting. The harvesting is done by cutting the entire plant a little above the ground level.

11. SELFING AND CROSSING PROCEDURE IN MOONG BEAN AND URD BEAN

(i) Anthesis and pollionation

Flowers start opening early in the morning and are completely open between 7 to 8 AM. The anthers begin to shed pollen in the previous day evening, before the flower opens and anthesis is complete prior to mid night. Black gram and green gram are essentially self-pollinated crops. Since the pollen shedding takes place long before the flower opens, selfing is the rule and cleistogamy prevails up to 40 per cent. Also the stamens and stigma are enclosed by the petals in such a way that the self pollination is ensured.

(ii) Emasculation

For crossing emasculation is a must, which refers to the removal of stamens from the flower. For this, young unopened bud is kept between thumb and forefinger of the left hand. The point of dissecting needle is inserted just under the standard petal in an oblique position along the top of the bud. The left side of the standard and the wing petal is pushed outward and held with left hand. The left side of the keel petal is removed with the forceps. The pistil and stigma are then exposed and the anthers are removed with the forceps (Boling *et al.*, 1961).

Evening emasculation followed by morning pollination gives best results. Pollination is done by gently rubbing anther of male inserting the staminal column and closing it with the standard and wing petal. Since flower shedding is common, putting the butter paper bag is avoided. The emasculated flowers are identified by thread wound around. The crossed pod will be smaller in size with two or three seeds. It has been reported that pollination immediately after emasculation (forenoon or after noon) gives poor pod set as compared to evening emasculation/next morning pollination or morning emasculation/same evening pollination.

12. BREEDING OBJECTIVES

The breeding objectives in moong bean and urd bean may be discussed in order of priority, as mentioned below —

(i) Breeding for high yield by improving yield contributing components like number of pods/plant, pod length, seeds/pod and seed size.

(ii) Breeding for disease resistance particularly for powdery mildew, leaf spot diseases like macrophomina, cercospsora, YMV and others.

(iii) Breeding for insect resistance particularly for white fly, vector for YMV, aphid, pod borer, leaf crinkle, leaf eating caterpillar.

(iv) Breeding for better quality, which includes seed size, seed colour, protein content (25-30%), methionine content (1.17%), cooking quality, percentage of hard seed, dal recovery (70%) and low amount of anti nutritional substances.

(v) Breeding for wider adaptability and stability of yield.

(vi) Breeding for improved plant type with erect growth, upright and determinate habit, vertical leaf with profuse podding and higher protein content, the indeterminate spreading plant types and excessive branching appear to restrict penetration and distribution of solar radiation through the canopy. Erect, narrow leaf cultivar with reduced branching may be expected to improve utilization of solar radiation. Such cultivars can be grown at much higher planting density.

(vii) Breeding for abiotic stresses – Drought, salinity, temperature, water stagnation etc. Nutritional deficiency/toxicity if any, development of herbicidal resistance. Toxicity – Tripsin inhibitor and Hemagglutenins – A thorough evaluation of available germplasm may be rewarding because low or no Tripsin inhibitor/Hemagglutanin genotype may already be found existing, awaiting identification.

(viii) Breeding for shattering resistance – Mutation breeding eliminates shattering and vivipery. Alternatively non-shattering character may be transferred to *V.radiata.*

(ix) Breeding for incorporation of dormancy, seed character, processing character (Dal making), photosynthetic efficiency, photothermo sensitivity, non-podding and lodging problem.

13. BREEDING METHODS FOR CROP IMPROVEMENT (in light of breeding objectives and behavior of crop)

(a) Breeding procedures in black gram

Pulse improvement work before independence was confined to collection, evaluation of land races in different regions. Farmers were growing local cultivars, which lacked genetic purity. Pure line selections from local cultivars were evaluated for yield potential. The promising types were further tested in multilocational trial. On the basis of performance in multilocation trial, the best genotypes were identified and released as varieties. Such improved varieties contributed to productivity of pulses in early phase of pulse improvement.

Breeding procedures, which are commonly used for genetic improvement of grain legume, include Introduction (T-9 from U.P.). Pure line selection CO3 from Alangudi local, CO5 from Musiri local), ADT3 from Thriunesveli local, ADT5 pure line selection from a Kanpur variety.

Hybridization and selection including pedigree method, bulk method and single seed descent and back crossing.

(a) *Intervarietal* — KM2 (derivative from T9 x L64), TMV-1 (derivative from Midhiulundu x KM1), ADT4 (29 x AD2 x 6114) and VBN3 (LBG402 x LBG17).

(b) *Interspecific* — (i) *V. mungo*, var. silvistris at Pantnagar YMV resistant lines obtained. But pods shatter. More number of back crosses suggested (ii) *V. mungo* x *V. radiata* for increasing pod length, digestibility. Sterility is the main problem; a few plants obtained revert back to parental form.

The success of pedigree method or its modification depends to a great extent on the choice of parental cultivars. The important donors for black gram are given in Table 3 (Singh, 1991a).

Table 3. Superior lines/cultivars of black gram to be used in crossing programme (Singh, 1991a)

Character	Line
Erect plant type, early maturing, high yield	Pant U, 19, 26, 30, Type 9, Strain 8
Profuse podding	M 101
Long pod	Strain 4
Bold seed	NP 5, sel.1
High yield	M 101, BR 68, Gwalior, PLU 500A, 501, 576, 639, 727, 730, 767, 769, 792, 801, 818
High protein	PLU 153, 171, 176, 303, 494, 999
High methionine	PLU 579
High tryptophan	IC 11613

Mutation induction has become an established tool in plant breeding to supplement existing germplasm and to improve cultivars in certain specific traits. A few improved varieties have been released through mutation breeding approach in black gram. The effective doses of mutagens for black gram are 0.25 to 0.50% EMS or 20-40 KR of Gama rays. So far as three varieties of black gram have been developed and released through mutation breeding. One such variety is UPU 9-40 released as Manikya in Karnataka in 1988, through treatment (40 KR) and selection from the parental cultivar UPU9 (Singh, 1991a).

Breeding for resistance to diseases

The serious diseases of black gram are mungbean yellow mosaic viruses, leaf crinkle virus, cercospora leaf spot, powdery mildew, root and stem rot and bacterial blight. Singh (1981, 1991) has compiled the sources of resistance to these diseases. A selective list of donors from these sources has been given in Table 4.

Table 4 Sources of resistance to different diseases in black gram

Mung bean yellow mosaic virus (MYMV)	UPU-1, UPU-2, Pant U 19, Pant U 26, Pant U 30, UAH 2, UPAH 2B, PLU 227, PLU 340
Mosaic mottle virus	T 65, T 67
Cercospora leaf spot	3529, 4387
Powdery mildew	UA 2A, UAH 2B, UG 117, HPU 1, HPU 2, HPU 5, P 47, P 163, P 178
Root and stem rot	3115
Bacterial blight	Krishna, 13/10, 13/2, 1369/4, PLU 443

The genetics of MYMV in black gram is fully investigated. Two recessive genes condition the resistance.

The screening against MYMV in black gram is possible in hot spot areas. Susceptible cultivars like No. 55, Khargone-3 or Gwalior 18 may be planted at regular interval to increase inoculum intensity. Artificial infection as a supplement to field screening could be done at pre flowering stage by covering the rows with muslin cloth cages of 60 x 90 x 120 cm dimension and releasing 300 viruliferous white flies in each cage. The scoring for resistance is done on 1-9 scale as 1-completely free, 3-traces of necrotic mottle, 5-moderate necrotic mottle, 7-restrieted yellow mottle and 9-complete yellow mottle (Singh, 1980).

For Cercospora, infected dried leaves soaked and squeezed in water, filtered through muslin cloth and conidia concentration of 500-1000/ml on 45-50 days old plants, may be sprayed to create artificial infection for screening. Infection for powdery mildew is suitably achieved by shaking coindia from heavily infected leaves over young seedlings of the test material.

For screening against root and stem rot, a sick plot is prepared by year after year sowing of susceptible genotypes and repeated incorporation of inoculum.

For developing resistant varieties against the aforementioned diseases, crosses should be made between donors listed in table 4 and those identified as resistant sources. The F_2 size for MYMV should be large enough to isolate resistant plants in good number as only 1/16 F_2 plants may be resistant as two recessive genes are involved in most cases. For other diseases where there are indications of involvement of one dominant gene, relatively smaller F_2 plant population size may be adequate.

Breeding for resistance to insect pests – white fly, aphids, leaf eating larvae and pod borer cause damage to black gram. A few sources of resistance have been reported by Chhabra *et al.* (1984), these are LU 178, LU 190 as resistant to white fly and L 126, LU 178 and LU 196 as resistant to jassids. These should be used in crossing programmes and the material in subsequent generations be exposed and screened against these insects, preferably in field conditions under adequate pressure of insects. In the area of disease resistance in uridbean, biotechnology has been resorted for successful development of plant-lets in inter specific cross (urd bean x moong bean, otherwise almost impossible normally. A lot of meiotic irregularities render seeds/pod and setting percentage highly reduced. It can be said that the application of biotechnology in the present is deeply low. A lot of continuous work in this direction is still awaited.

(b) Breeding procedure in green gram

Breeding methods typical of self-pollinated crops are found suitably applied to green gram also.

So far with pure line selection Co1, paiyur varieties have been developed. Hybridization followed by pedigree, varieties like ADT1, ADT2, Co3 and VBN-1 have been developed. Bulk, back cross and single seed descent methods have also been usefully tried in green gram. The individual plant progenies are planted with spreader rows at regular interval.

The yield component, seeds/pod being negatively corelated with seed weight, a compromise based on market preference of seed size like 3.5 to 4.0 g/100 seeds is mostly preferred

In Indian conditions two generations/year one kharif and other spring/summer are resorted to for faster advancement of generation through single seed descent method (S.S.D.).

Maturity wise duration of 70-80 days in kharif enables the crops to escape being caught in September rain. The crop should favorably mature and harvested in second fortnight of September. Summer crop should have maturity of 60-70 days while in spring 80-90 days duration is considered suitable.

Mutation breeding in mung bean with the use of 40-50 KR of Gama rays or 0.2-0.3% EMS or any other mutagen have been tried. The important donors for desirable agronomic characters are as follows (Singh, 1991)

Early maturity	Baisakhi Type 1, PS 16
More pods	LM 17, LM 64, LM 310, LM 352, LM 294-1
Long pods	LM 14, LM 35, LM 55, LM 89, LM 189, LM 314, EC 16035
More seeds	LM 7, LM 57, LM 95, LM 128, LM 206, LM 617, LM 715, LM 1030
Bold seed	CES-5-1-F 5, Plm 171, IC 8839

Breeding for resistance to diseases

Several diseases reported for green gram are MYMV, cercospora leaf spot, powdery mildew and bacterial leaf spot. Of these powdery mildew yellow mosaic and cercospora leaf spot are three major disease problems of this crop (Nene, 1988)

The methods employed for development of disease resistant lines in green gram are the same as those discussed for black gram susceptible infector or spreader row used in green gram however is variety "Jalgaon 781". The disease scale for rating is 1-9 (after Singh *et al.*, 1988).

Rating scale for MYMV (after Singh *et al.*, 1988)

Scale	Plants/foliage affected	Symptoms	Reaction
1	0.1-5%	Mottling of leaves on less than 1% plants	Resistant
3	5.1-10%	Mottling of leaves on 1.1-10% plants	Moderately resistant
5	10.1-25%	Mottling and yellow discoloration on 10.1-25% plants	Moderately susceptible
7	25.1-50%	Mottling and yellow discoloration of leaves on 25.1-50% plants	Susceptible
9	50.1-100%	Severe yellow mottling on over 50% plants, stunting of plants and failure of flowering and fruit setting	Highly susceptible

Resistance to insect pests

White fly, jassids and mites (Kharif season) and thrips (Spring/summer season) have the potential to cause significant damage in mung bean. Split wooden cages are efficient and reliable for accurate screening and classification of genotypes. Resistant sources as compiled by Singh (1991) and Verma *et al.* (1991) are as follows:

Whitefly	ML 1, 3, 5, 6, 7, 15, 26, 131, 170, 267, 337, 421, 422, 423, 427, 435, 436, 437, 438, P 231, 290 292, 325, 364
Jassid	ML 5, 131, 267, 337, 422, 423, 428
Thrips	G 65, SML 32, 100, 134, 137, 141, 143, 148, 154, 165
Aphids	Angul 3-6, NP 1-8, LM 115, MO 10, MG 50-10 Yellow, MG 500-10A, EG-MG-12, EG-MG-4, CES 55 CES 44, PHLV 18, 13
Pod borer	V 2107, 2109, 2135, 4270
Mites	TR 1-1093
Root knot nematode	V 1412, 1709, 2010, 2773

Biotechnology in green gram

Reports indicate that approach of biotechnology for achieving greater success for recovering viable hybrid embryo in species combinations where prezygotic incompatibility usually prevents zygote formation in *V. mungo* x *V. radiata* crosses (Bhadra *et al.*, 1989). The said combination helps in transfer of non-shattering trait of urd bean in to green gram. Transfer of traits like more seeds/pod and easy digestible proteins in green gram may be introduced in black gram.

With the use of anther culture in green gram, production of haploid at a considerably enhanced frequency in shorter time as compared to the conventional methods of obtaining haploids including delayed pollination, crosses including polyploids, and wide species combination, together with the identification of sponteneous haploids and their induction by physical and chemical treatments (Bhadra *et al.*, 1989). Whilst still in infancy, haploid production in cultivated vigna species offers considerable potential for the improvement of pulse crops.

14. LIST OF VARIETIES UNDER CULTIVATION

(a) Mungbean

JM 721: It is a cross of ML5 x PIMS3 evolved at JNKVV, Campus Indore (M.P.), released in the year 1995, adapted to whole of M.P. for kharif as well as summer. It flowers in 30-35 days, matures in 70-75 days, height is about 55-65 cm. It is moderately resistant to leaf spot diseases and tolerant to powdery mildew. Plant type is erect with good growth. Medium size pods in clusters on top making a straight row. Seeds are medium sized, dumbly shaped, dull green colour, 1000-grain weight is 25-28 g. (Photo Plate 1).

Pusa Baisakhi: It is a selection from Type 44 at IAR1, New Delhi, released in 1975. Adapted to whole of M.P. It flowers in 35 days, matures in 65 days. Height is about 50 cm. It is susceptible to powdery mildew, leaf spot and yellow mosaic. It has mid dwarf, spreading plant type, medium sized pods. Seeds are medium bold dull green. 1000-grain weight is 30 g.

K851: It is a cross of (4453 – 3 x T_1) at Kanpur adapted to whole of the M.P. Preferred in summer for its earliness. It flowers in 30-35 days. It matures in 60-65 days. It is susceptible to powdery mildew and leaf spot. It has dwarf and spreading plant type. Pods are long. Seed is medium bold, shinning green. 1000-grain weight is 30 g.

(b) Black gram

No. 55 Khargone-3, T-9, Pant-U30, Jawahar Urd-2 and JU-3, Gwalior-2, PU-4, IU 8-6, Mash-48.

No. 55: It is a selection from local bulk of Satpuda Plateau. A medium maturing variety (90 days) has black medium size seed. Plant is erect with heavy bearing of pods, a highly adaptable variety with average yield of 1000-1200 kg/ha is recommended for central, southern and eastern M.P.

T-9: It is an introduction from U.P. It is an early maturing variety (70-80 days) with medium size seed; small erect plant has good bearing of pods. It is suitable for double cropping and has potential of 1000 to 1200 kg/ha at higher plant population. Recommended for all over M.P.

Khargone-3: This is a selection from local bulk of Sanawad (Nimar) is a semi spreading medium maturing variety (80-90 days) bold, black seed with good dal recovery with an average yield of 1000-1200 kg/ha. Recommended for Nimar region of M.P.

Gwalior-2: It is a selection from local material. It is late and takes 105 days to mature. Seeds are medium size and black. Its yield on an average is 1000-1200 kg/ha. Recommended for Northern M.P.

JU-2: It is selected from a cross T-9 x Reva 1-1. This is short statured (27 cm). Plant with main stem bearing, seed medium bold in size with black shining colour. It matures in 65-80 days. This variety is moderately resistant to YMV. Suitable for rainfed areas, black cotton soil and timely sown conditions as well as for closure spacing, yields 1300 kg/ha.

JU-3: It is selection from a cross (mash 48 x T-9) x (T-9 x UPU-2). This is medium statured (45 cm) plant with four branches on main stem, seed bold in size and black in colour. It matures in 70 days. It is tolerant to macrofomina, cercospora leaf spot and YMV, suitable for whole of M.P. double-cropped areas and also as a cash crop in summer, yields 1300 kg/ha.

15. REQUIREMENT OF SEED PRODUCTION

Self-pollination is a rule in black gram and green gram, hence variety maintenance is not much of a problem. An isolation of 20 metres from other fields of the species is sufficient for seed production. An integration of seed production chain is however required for production of breeder seed and its efficient conversion into certified seeds, suitable involvement of private seed companies may be explored for the purpose. Newly evolved high yielding varieties must be popularized and other varieties to be phased out at a rapid targeted pace. The policy on seed subsidy and seed pricing of pulse sector is gradually phased out.

16. MAJOR ACHIEVEMENTS IN THE CROP THROUGH BREEDING

Under International programme moong bean is an important crop at the Asian Vegetable Research and Development Centre (AVRDC) with a mandate for vegetable crop research and development in humid and sub humid tropics. Established in 1971, its activities include germplasm collection, storage and enhancement, varietal improvement, production technology improvement, environmental and nutritional studies, Technology transfer and training. AVRDC lines have been officially released as new cultivars 25 times in 15 countries including Pusa 105 (VC-1137-2-B) in India.

Under national programme the All India Co-ordinated Pulse Improvement Project commissioned in 1965-66 and subsequently upgraded to "Project Director" in 1977 and to Directorate of Pulses research in 1984. And after wards known as (I.I.P.R.) Indian Institute of Pulses Research with headquarter at Kanpur (U.P.) is coordinating the National Research Programme on pulses including moong bean and urd bean. In 1995 AICPIP was trifurcated and MULLRP project thus came in to being having the mandate of research and development of six pulse crops viz., moong bean, urd bean, lentil, lathyrus, rajmash and pea.

Under this project, there are about 15 main centres and 10 sub centres for regional testing. Besides these there are several voluntary centres, which cooperate in National testing programme. Most of the centres are located at the state Agricultural Universities and ICAR institutes. Area under moong bean and urd bean during kharif, summer and spring increased because the development of synchronous, early maturing, YMV and powdery mildew resistant and high yielding cultivars took place.

Varieties of moongbean, urdbean developed at IIPR, Kanpur have been given in Table 5.

Table 5. Varieties developed at the Indian Institute of Pulses Research, Kanpur

Crop	Variety	Year of identification	Area of adoption	Special features
Mungbean	PDM 11	1985	CZ	Recommended for spring season, YMV resistant
		1987	NEPZ	Recommended for spring season, YMV resistant
	PDM 54	1987	NEPZ	Recommended for summer season, YMV resistant
		1988	SZ	Recommended for rainy (Kharif) season
		1989	NEHZ	Recommended for kharif season
	PDM 84-139	1992	NEPZ	Recommended for summer season, YMV resistant
	PDM 84-143	1990	EZ, CZ, SEZ	Recommended for kharif and spring seasons, YMV resistant
Urd bean	PDU 1	1986	NWPZ, NEPZ, CZ	Recommended for spring season, YMV tolerant
	DPU 88-31	1993	NEPZ	Recommended for kharif season, YMV resistant

NWPZ - North-western plains zone; NEPZ - North-eastern plains zone;
CZ - Central zone; SZ - Southern zone;
EZ - Eastern zone and SEZ - South-eastern zone

REFERENCES

Boling, M., Sander, D.A., Mattock, R.S., 1961. Mungbean hybridization technique. *Agron. J.* 53: 54-55.

Bhadra, S.K., HAMMATT, N., Davey, M.R., 1989. Prospects of the use of in vitro techniques in improvement of Vigna pulses. *SABRAO Journal*, 21: 75-91.

Chandel, K.P.S., 1984. Role of wild vigna species in the evolution and improvement of mung (*V. radiata* L.) wilezek and uridbean (*V. mungo* (L.) Hepper) annals agric. *Res.* 5: 98-11.

Karmakar, P.G., Dana, S., 1987. Cytogenetic identification of a vigna sublobata collection. *The nucleus*, 30: 47-53.

Nene, Y.L., 1988. Multiple disease resistance in grain legumes *Ann. Rev. Phytopathol.* 26: 203-217.

Singh, D.P., 1980. Inheritance of resistance to yellow mosaic virus in black gram (*Phaseolus mungo* (L.) Hepper) Theor. *Appl. Genet.* 57: 233-235.

Singh, D.P., 1981. Breeding for resistance to diseases in green gram and black gram. *Theor. Appl. Genet.* 59: 1-10.

Singh, Gurdip, Kapoor, S., Singh, Kuldip, 1988. Multiple disease resistance in mungbean with special emphasis on mungbean yellow mosaic virus. PP. 290-296. In S. Shanmungasundaram (ed.) *Proc. Second Intern. mungbean, sympo*, 16-20 Nov., 1987, Bangkok.

Singh, D.P., 1991a. Genetics and breeding of pulse crops. PP. 307-344, Kalyani Publ., New Delhi, Ludhiana.

Verdcourt, B., 1971. Tribe Phaseolae, Leguminosae. Part 4, in E. Milne-Redhead & R.M. Polhill (eds.), Flora of Tropical East Africa, Crown Agents, London.

Verma, M.M., Sandhu, S.S., Brar, J.S., Gurdip Singh, 1991. Mungbean breeding for tolerance to biotic stress. Golden Jubilee celebrations-symp. Grain legumes. Feb., 9-11, IARI, New Delhi. Proceedings. PP. 127-147. *Indian Soc. Genet. Plant Breed.*, IARI, New Delhi.

Zucovaskiji, P.M., 1962. Cultivated plants and their wild relatives. *Commonwealth Agric. Bureau*, London.

CHAPTER - 9

SOYBEAN

A.N. Shrivastava, C.B. Singh and M.K. Shrivastava

1. INTRODUCTION

Soybean is an efficient producer of the two most scarce items in the World food economy that is high quality protein (40%) and oil (20%). The Versatile plants fix 50kg nitrogen per hectare. For centuries soybean has served as an indispensable component in East Asian nutrition and cuisine as *miso* (soy paste), *Shoyu* (soy sauce), *tofu* (soy curd), soymilk (fluid extracted from soybeans) and *tem peh* (fermented cake like product).

At present, soybean ranks first among the major oilseed crops in the World. In India, it ranks third after groundnut and rapeseed and mustard. It has also revolutionized the Indian economy by raising standard of farmers and earning valuable foreign exchange through exports of de-oiled cake.

The vernacular name of soybean includes *Bhat, Bhatmas, Ramkulthi, Ramkurthi* etc. in Hindi, *Gargkilay* in Bangla, *Bhut* in Punjabi, *Petanijokara* in Assami and *Uraimbai-kutang* in Khasi (Anon. 1956).

2. TAXONOMIC CLASSIFICATION, ORIGIN AND CHROMOSOME NUMBER

Taxonomically, the genus *Glycine* Willd. is a member of the family Leguminosae syn. Fabaceae, subfamily Papilionoideae, tribe phaseoleae, and subtribe Glycininae. As currently delimited, the genus *Glycine* Willd. is divided into two subgenera, *Glycine* and *Soja* (Moench) F.J. Herm. The subgenus *Glycine* is currently composed of 16 wild perennial inbreeding species indigenous to Australia (Hermann 1962; Newell and Hymowitz 1980; Tindale and Craven 1993). The subgenus *Soja* includes two annual and diploid (2n=40) species viz., *Glycine max* (cultivated) and *G. soja* (wild) which can readily hybridize. *G. gracilis* a weedy intermediate form between *G. max* and *G. soja* only occurs in Northeast China (Skvortzow, 1927). Later it was removed from the species rank and incorporated into *G. max* (Hermann, 1962; Wang, 1976). Details of different species of both subgenera are mentioned in Table 1.

Morphology, cytogenetics, seed protein, phytoalexins, restriction endonuclease fragment analysis (REFA) of mitrochondrial DNA, ribosomal RNA, chloroplast DNA and sequence from the internal transcriber spacer of nuclear ribosomal DNA evidences show that *G. soja* is the wild ancestor of the *G. max* (Doyle and Beachy 1985; Shoemaker *et al.* 1986; Hymowitz and Singh 1987; Keen *et al.* 1986, Doyle 1988; Kollipara *et al.* 1997). The cultivated soybean *G. max* was differentiated by a reciprocal translocation from *G. soja* (also known as *G. ussuriensis* and *G. formosana). G. max* and *G. soja* carry similar genomes and were assigned genomic symbol GG.

The origin of soybean was described in eastern Asia and later, Hymowitz (1970) and Cuzin (1976) suggested its origin in north and central China, where its domestication was started in the Chou Dynasty (Hymowitz, 1970) and completed during the Shang Dynasty (1700-1000 BC). In the ancient dictionary of China "*Kouangia*" soybean has been described as *Ta-teon* and Ta-tou. In Chinese script, it is referred as 'Shu' or 'Sou' and it is probable that the word soy, soja or soya might have been derived from these. From China, it spread to Southern China as well as Japan, India, Nepal, Russia, around the first century (Gai, 1996).

Table 1. Details of the species of the genus *Glycine* Willd.

Sr	Species	Chromosome no. (2n)	Genome	Code	Pl	Distribution
A.	SUBGENUS *GLYCINE*					
1	*G. albicans* Tind. and Craven	40	II	ALB	–	Australia
2	*G. arenaria* Tind.	40	HH	ARE	505204	Australia
3	*G. argyrea* Tind.	40	A_2A_2	ARG	505151	Australia
4	*G. canescens* F.J.Herm.	40	AA	CAN	440932	Australia
5	*G. clandestena* Wendl.	40	A_1A_1	CLA	440958	Australia
6	*G. curvata* Tind.	40	C_1C_1	CUR	505166	Australia
7	*G. cyrtoloba* Tind	40	CC	CYR	440962	Australia
8	*G. falcata* Benth	40	FF	FAL	505179	Australia
9	*G. hirticaulis* Tind and Craven	40 80	H_1H_1 --	HIR HIR	-- --	Australia Australia
10	*G. lactovirens* Tind. and Craven	40	I_1I_1	LAC	--	Australia
11	*G. latifolia* (Benth.) Newell & Hymowitz	40	B_1B_1	LAT	378709	Australia
12	*G. latrobeana* (Meissn)Benth.	40	A_3A_3	LTR	483196	Australia
13	*G. microphylla* (Benth.) Tind.	40	BB	MIC	440956	Australia
14	*G. pindanica* Tind. and Carven	40	H_2H_2	PIN	--	Australia
15	*G. tabacina* (Labill.) Benth.	40 80	B_2B_2 Complex	TAB	373990 --	Australia Australia, West Central and South Pacific Islands

16	*G. tomentella* Hayata.	38	EE	TOM	440998	Australia
		40	DD		505222	Australia, Papua New Guinea
		78	Complex		--	Australia, Papua New Guinea
		80	Complex		--	Australia, Papua New Guinea, Indonesia Philippines, Taiwan
B.	SUBGENUS *SOJA* (MOENCH) F.J. HERM.					
1	*G. soja* Sieb.and Zucc.	40	GG	SOJ	81762	China, Russia, Japan, Taiwan, Korea
2	*G. max* (L.) Merr.	40	GG	MAX		Cultigen

G. tabacina tetraploid (complex)= Allopolyploids (A and B genome) and segmental Polyploids (B genome).
G. tomentella tetraploid (complex)= Allopolyploids (2n=78) (D and E, A and E, or any other unknown combinations).
G. tomentella tetraploid (complex)= Allopolyploids (2n=80) (A and D genome or any other unknown combinations).

From the first century AD to the age of Discovery (15th –16th century) soybeans were introduced and land races were established in Japan, Manchuria, Mangolia, (south east Asia and south central Asia). These regions comprise the secondary gene centre for soybean.

The soybean migrated very late to the West. In Netherlands before 1737 as Linnaeus described in the Hortus cliffortianus. The soybean seeds sent by Missionaries in 1739 were planted for the first time in 1740 in Jardin des plantes, Paris, France and in 1790 in Royal Botanical Garden, Kew, England. Soybean was introduced to North America in 1765 (Hymowitz and Harlan, 1983). The first mention of soybean in the United States literature is in 1804 by Dr. Jammes Mease stating that it bears the climate of Pennsylvania very well.

In ancient literatures of India, soybean finds no mention. Possibly, it is being cultivated since long in Northern India, i.e., hilly area of Assam, Bengal, Manipur, hills of Khasi and Naga and at 6000-ft height in Kumayun region (Anon. 1956). Black seeded soybean was grown successfully at Nagpur Farm erstwhile central provinces. Soybean has been described in the book "Flora of British India Vol. II) written by Sir, J.D. Hooker in 1879. Soybean appears to have been introduced first in Northern India either from China through Himalayan mountain ranges or from Java through Burma.

3. CLIMATIC REQUIREMENTS, AREA AND DISTRIBUTION

The optimum temperature for germination is 30°C (20°C soil temperature) with 40-60 percent soil moisture whereas temperatures below

5°Cand above 40°C results no germination. Warm and moist climatic conditions are most suitable for soybean crop. The temperature range of 25-28°C is found suitable for soybean growth. The threshold temperature is 29°C and increase of 2°C from the threshold temperature causes reduction in yield (Burke and Evett, 1999). Low relative turgidity (below 90%) of the top leaves reduces photosynthesis. Flowering is delayed below 24°C while no flowering occurs below 10°C. It is a photosensitive short day plant. Flowering initiation and duration, podding period and height are affected by day length. For flowering 13 hours critical day length is required and most of the varieties flower, during less than this photoperiod after determined number of days (7-26 days).

A well-drained soil with pH between 6 to 7.5 has been found suitable for cultivation of soybean. Very light, sandy, sodic and alkaline soils are not suitable for this crop.

It is the most important legume in the World in terms of total production and international trade. India ranks fifth in the World in area and production after USA, Brazil, China and Argentina. In India, area under soybean cultivation is around 6 million hectares (1999-2000) in comparison to 32000 hectares in 1970-71. It experienced one of the highest phenomenal growth rates in the recent past i.e., 15-20% per annum. The significant contribution of M.P. (75-80%) in area and production of country's total rendered its recognition as **'Soya State'**. The other predominant states are Maharastra and Rajasthan. Soybean is also getting place in Karnataka, Tamil Nadu, Andhra Pradesh, Uttar Pradesh, Uttaranchal, Punjab, Chhattisgarh and Himachal Pradesh.

4. SALIENT FEATURES OF THE CROP CULTIVATION

Soybean is grown as rainy season crop (*Kharif*) under rainfed situations. The predominant area is vertisols and associated soils. It occupied rainy season fallow land and replaced sorghum, minor millets, arhar and further upto some extent paddy too. The favourable time of sowing for soybean is between 25th June to 7th July. Seed rate ranges from 80-100 kg/ha depending upon seed size with 70% germination. The nutrient requirement is 20 kg nitrogen; 60-80 kg phosphorous and 20 kg potash per hectare in addition to 10 tonnes FYM /ha. Application of 20kg sulphur/ha and 2.5 to 5 kg Zinc/ha is recommended in deficient soils.

Seed treatment with fungicides and biofertilizers protects seeding diseases and help in fixation of atmospheric nitrogen, respectively. Efficient drainage, weed management, and plant protection measures are required for good crop production. Soybean-wheat and soybean-chickpea are major crop rotations while other *rabi* crops are also being taken after soybean as per situations. Intercropping of soybean with maize, pigeonpea, sorghum, finger millet and sugarcane is recommended.

5. BOTANICAL DESCRIPTION

Plant Description Soybean is an erect, bushy and annual plant with leafy growth. The plant varies from 0.5 to 2 meter in height. The branches may be spreading or ascending depending on growing conditions. The soybean cultivars are grouped into two types based on growth habit i.e., *Indeterminate* (flowering starts before stem elongation ceases and flowers are borne on axillary racemes and flowering period is longer and extensive) and *Determinate* (Flowers are born on both axillary and terminal racemes, stem elongation ceases with differentiation of terminal bud and the flowering period is shorter and more intense). Different parts of plant are depicted in Fig.1.

Seed

Its shape varies as spherical, spherical-flattened, elongated and elongated-flattened. Seed has small scar of linear to oval shape known as hilum, which marks attachment of the seed to the pod. The area surrounding the hilum may be of different shades of black, brown, gray or yellow. A tiny hole formed by integument is present near the one end of the hilum known as micropyle. The hypocotyl-radicle axis is present just below the micropyle. On the other end of hilum, raphe is present as a small groove extended upto the chalaza. The mature seed consists of an embryo and a seed coat (a leathery textured structure). The embryo consists of two large fleshy cotyledons, a plumule with two well-developed primary leaves and a hypocotyl-radicle axis. The seed is devoid of endosperm. The seed coat may be of yellow, yellow green, green, brown, black or variegated colour. The colour of cotyledon may be yellow (common), greenish yellow or green.

Root

The primary root grows as main root axis and penetrates soil more or less straight downward called tap root. The secondary, lateral rootlets are branched from the taproot, whereas multi-branched adventitious roots also emerge. The bacterial root nodules appear after 8-10 days of planting which begin nitrogen fixation about two weeks later.

Stem

It is usually erect in cultivated types while prostrate/ trailing/twining in wild types. The structure of the stem is solid and cylindrical with wider base to tapering towards tip. The part of stem between primary root to cotyledonary node known as hypocotyl, elongates during germination and push the cotyledons above the soil. Hypocotyl may be of voilet (Anthocyanin) or green in colour. The stem may be glabrous or pubescent with gray or tawn hairs.

Leaves

Following kinds of light green, medium green or dark green coloured leaves with pubescent or glabrous surface are found in soybean:

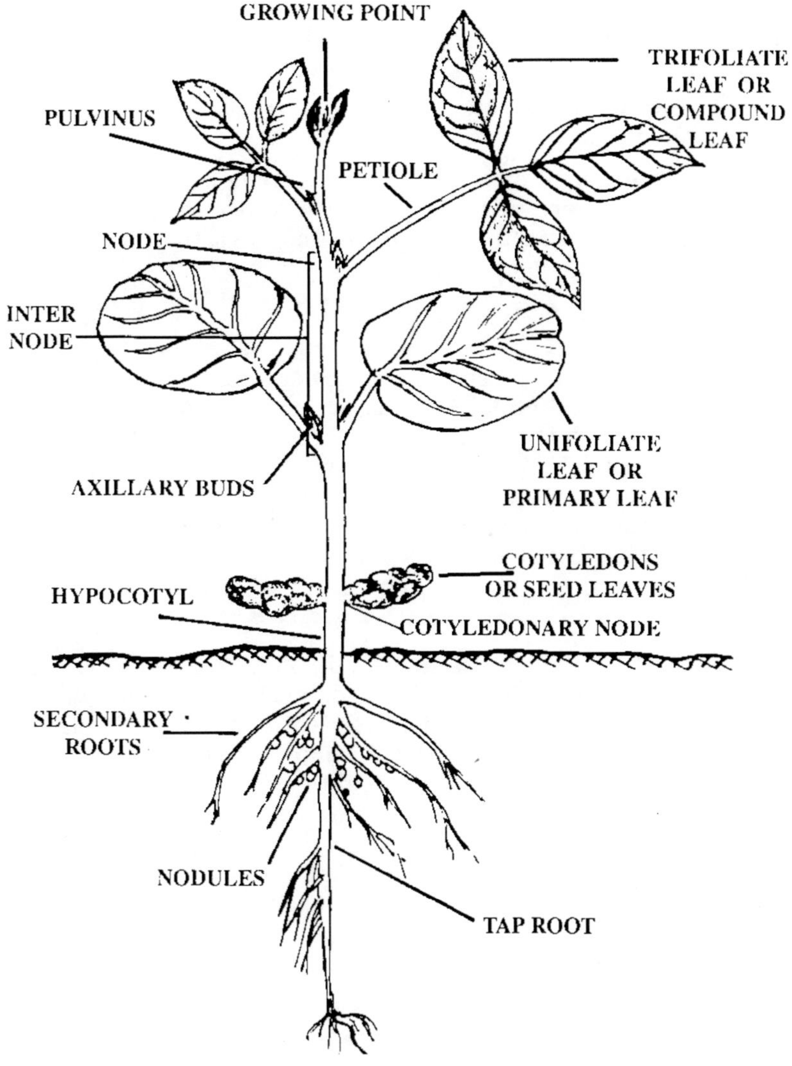

Fig. 1. Different Parts of Soybean Plant.

1. *Cotyledons or seed leaves*: With the growth of hypocotyl the two cotyledons emerged above the soil which separate, become green and take part in photosynthetic activity in addition to supply nutrition to the plant.

2. *Simple primary leaves or unifoliate leaves*: The first true leaves are unifoliate (single blade per petiole) appear on first node and both are formed at opposite sides.

3. *Trifoliate leaves or compound leaves*: All the leaves above first node produced on main stem or branch are trifoliate (three blades per petiole) and arranged on alternate nodes one by one in opposite sides.

In some varieties trifoliate to pentafoliate leaves of trangular, pointed ovate, rounded ovate or lanceolate shapes are found.

4. *Prophylls:* At the base of each lateral branch and below the inflorescence a tiny pair of simple leaves lack both petiole and pulvini are called prophylls.

At the base of the petiole, two distinctive structures (i) *stipule* (a pair of spear-or lance-shaped modified leaves formed on the petiole stem junction) and (ii) *pulvinus* (an integral part of the junction made up of a large cell group at the very base of the petiole) are present.

Flowers

The inflorescence is raceme with flowers mainly borne on axillary racemes (few flowers are formed on terminal raceme) of indeterminate type and both on axillary and terminal racemes of determinate types. Flowers are arranged in cluster of 3-15 per node on peduncle, a stem like branch. Axillary buds develop into flowers expand first towards the base of the main stem and then progress towards the tip. The flower initiation generally starts from the fourth node. Many flowers shed without forming pods.

Floral biology

Flowers are 6-7 mm long, bisexual, self-pollinated, pedicellate, zygomorphic, pentamerous, bracteate and typical papilionaceous with a tubular calyx of five unequal lobes (Fig. 2). Calyx is persistent and attached with pods. The corolla of violet (purple) or white colour comprises of a posterior banner or standard petal, two lateral wings petals and two anterior keel petals. The keel petals are not fused. The 10 stamens (androecium) in diadelphous pattern form a tube around the pistil by the fusion of nine stamens. The tenth stamen remains free. Anthers are bilobed, dorsifixed and introse. The single pistil (gynoecium) is unicarpellary, superior and unilocular with style curved towards the free posterior stamens. The pistil has one to four campylotropous (curved) ovules with marginal placentation. Hairs are present on outer surface of ovary or calyx tube.

Blooming starts in early morning but in prolonged cooler temperatures flowers do not open and behave as cleistogamous. In normal conditions anthesis occur prior to opening of flower. The pollens are heavy and seldom wind-borne. Natural crossing may occur to the extent of 0.5 to 1%.

Pods: Flattened or cylindrical shaped and pubescent (grey or tawn hairs) or glabrous pods are borne on main stem or a branch in bunches. At maturity, it turns into brown, tawn or black colour. It may bear 1-5 seeds per pod however, 2-3 seeds per pod are common.

Stages of development

A standard terminology has been described by Fehr and Caviness (1977) for various stages of soybean development. The vegetative and reproductive stages are determined by node identification.

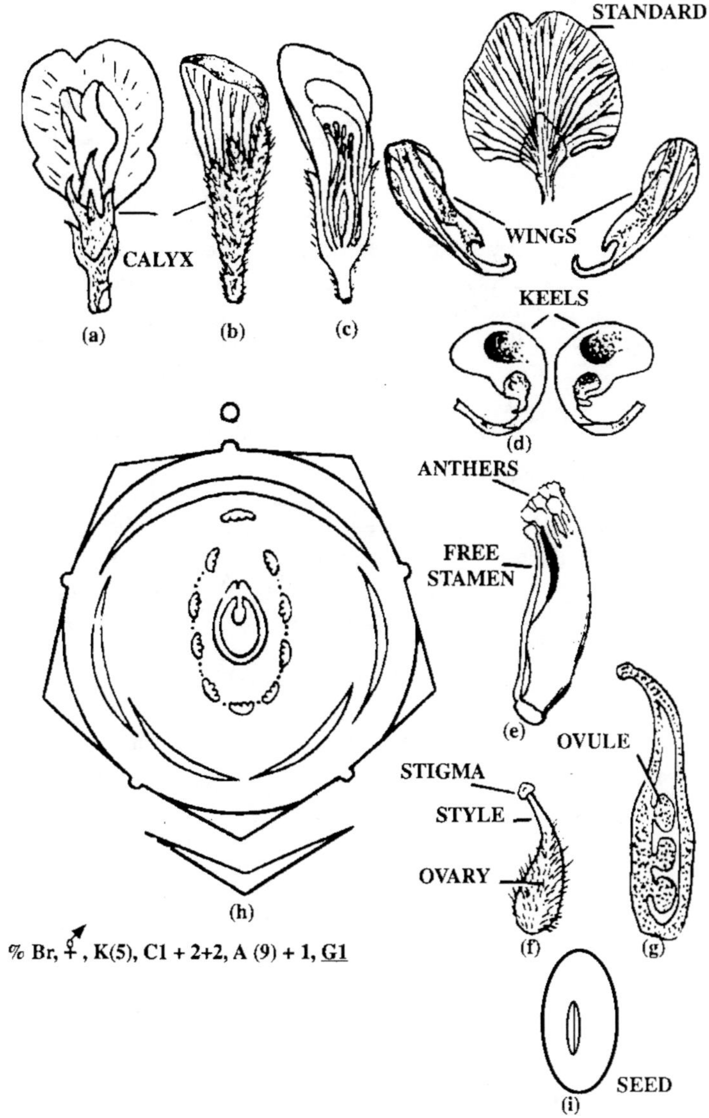

Fig. 2. Soybean Flower Structure and Seed
(a-b) External Structure, (c) Internal Structure, (d) Corolla, (e) Androecium, (f) Gynoecium (External parts), (g) Gynoecium (Internal parts), (h) Floral diagram & Floral formula, (i) Soybean Seed.

1. *Vegetative stages:* Vegetative stages denoted by letter 'V' (Table 2) are counted from the time of plant emergence from soil.
2. *Reproductive stages*: Reproductive stages denoted by letter 'R' starts from flowering onwards till maturity (Table 3).

Table 2. Vegetative stages of soybean

Stage No.	Stage Title	Description
VE	Emergence	Cotyledons emergence above the soil surface
VC	Cotyledon	Unifoliate leaves unrolled sufficiently so the leaf edges are not touching.
V1	First node	Fully developed two simple or unifoliate leaves at the nodes (two nodes closely located directly opposite to each other, immediately above the cotyledonary nodes).
V2	Second node	Fully developed trifoliate leaf at node above the unifoliate nodes.
V3	Third node	Three nodes on the main stem with fully developed leaves beginning with the unifoliate nodes.
Vn	n^{th} node	n number of nodes on main stem with fully developed leaves beginning with the unifoliate nodes.

Table 3 Reproductive stages of soybean

Stage No.	Stage Title	Description
R1	Beginning bloom	One open flower at any node on the main stem.
R2	Full bloom	One open flower at one of the two upper most nodes on the main stem with a fully developed leaf.
R3	Beginning pod	At one of the four uppermost nodes, on the main stem with a fully developed leaf, bearing 5mm long pod.
R4	Full pod	At one of the four uppermost nodes, on the main stem with a fully developed leaf, bearing 2cm long pod.
R5	Beginning seed	At one of the four uppermost nodes, on the main stem with a fully developed leaf, bearing 3mm long seed in a pod.
R6	Full seed	At one of the four uppermost nodes, on the main stem with a fully developed leaf, pod containing a green seed that fills the pod cavity.
R7	Beginning maturity	One normal pod on the main stem that has reached its mature pod colour (yellow or brown).
R8	Full maturity	Ninety five per cent of the pods reached mature colour (turned yellow or brown).

6. SELFING AND CROSSING TECHNIQUE

Special care should be taken for successful hybridization programme due to small floral structure. Parental cultivars are grown in pots or in fields (pair rows) at different intervals to achieve sufficient number of crossed pods and crossing between early and late types of cultures. Swollen floral bud with corolla visible through the calyx of female parent is ready for emasculation. The other buds are removed gently with forceps without any injury to the selected bud. Sepals are removed with the help of forceps by pulling down the lobes one by one. Then the exposed corolla and 10 stamens are pulled upward with the help of forceps. Stamens left are also removed with the help of forceps. Any injury to the pistil should be avoided.

Pollination is done just after emasculation between 8-10 am. Fresh open flowers are collected from the desirable male parent and corolla are removed with the help of forceps. The pollens are transferred to the stigma of the emasculated bud by brushing anthers holding the male flower at the base by forceps. Pod is visible in about seven days in case of successful pollination and fertilization. The bud is marked by tagging with the name of parents and date of pollination. Other pods from the same axis are removed. The scar made by removal of the calyx helps in identifying the crossed pods. An efficient person may have 50% success. However, normally success rate varies from 8 to 20 percent depends upon climatic conditions.

7. BREEDING OBJECTIVES

The important objectives in soybean are high yield, maturity period suitable for different agroclimatic zone, resistance to loding and shattering, high germinability and longevity, tolerance to abiotic stresses, resistance/ tolerance to diseases and insect-pests and seed quality. Some special purposes like use as vegetables or for culinary purpose in some regions have also been taken into account.

High Yield: High productivity is the most important objective of any crop. In soybean at present varieties have potential of 25-30 q/ha which should be raised to 40-45 q/ha. Number of pods per plant, number of seeds per pod and seed size are the major components of yield (Prabhakar and Tiwari, 1993). Incorporation of four-seededness has shown a greater promise to increase the number of seeds per pod. The actual yield increase in soybean is realized slower in comparison to cereals due to difference in seed composition i.e., protein (40%) and oil (20%) content. It is mainly due to the complex pathway of protein and oil because it requires double amount of photosynthates to form the same amount of carbohydrate or starch in cereals.

Maturity for the area of adaptation: Varietal development programme involves suitability of maturity period for a given environment or zone. In Central Zone, and Southern zone varieties with 85 to 105 days of maturity period are suitable whereas in Northern Plains and Hills 110-120 days of maturity period is suitable. The late maturity is dominant and governed by E_1, E_2, and E_3 genes whereas the fourth gene E_4 is sensitive to long photo period.

Resistance to Lodging: The lodging is causing significant losses in yield. No absolute limits for lodging have been established. The 1-5 scale is described as 1 for erect, 3 for leaning 45□ angle and 5 for prostrate plants. The dwarf, determinate and strong stem characters are required for lodging resistance.

Resistance to Shattering: It is required to retain seeds in pod after maturity and less losses of seeds in the process of harvesting and lifting. Susceptibility to pod shattering is dominant/partial dominant with additive gene action (Tiwari and Bhatnagar, 1991).

High germinability and longevity: It is among the least storable crop with inherent problems of viability, germination and emergence. Smaller seed size and longer hypocotyl length could be reliable parameters for these traits. Speed of germination has also been found to be a criterion of seed vigour to predict field emergence.

Resistance to abiotic stresses: Abiotic stresses viz., water stress, temperature and pH and salinity of soil have major influence on morphological, physiological and biochemical processes of the plant. Higher root density, turgidity of the leaves, frequent stomata closure, stability in nitrate reductase activity, protein synthesis, proline accumulation and lesser increase in abscisic acid. Genetic variability for various characters contributing towards resistance for abiotic stresses has been observed and exploited.

Resistance to diseases: With the increase of area, diseases are becoming serious problem. About 40 diseases are reported to occur in different regions, however diseases of economic importance are presented in Table 4.

Table 4. Major diseases of soybean

S.No.	Disease	Causal organism
A	Seeding and stem rot	
(i)	Charcoal rot or ashy or stem blight or dry root rot	Conidial stage: *Macrophomina phaseolina* (Tassi) Goid. Sclerotial stage: *Rhizoctonia bataticola* (Taub.) Butler.
(ii)	Collar rot or sclerotial blight	*Sclerotium rolfsii* Sacc.
B	Foliage/Pod diseases	
(i)	Rust	*Phakopsora pachyrhizi* Syd.
(ii)	Rhizoctonia root and stem rot & Aerial blight	Telelial stage: *Corticium solani* (Prill and Delchr) Bourd and Galz. Sclerotial stage: *Rhizoctonia solani* Keuhn
(iii)	Myrothecium leaf spot	*Myrothecium roridum* Tode ex. Fries
(iv)	Frog eye leaf spot	*Cercospora sojine* Hara
(v)	Pod blight or Anthracnose	*Colletotrichum dematium* (Fr.) Grove. Syn. *Colletotrichum truncatum* (schw) Andrus & Moore.
(vi)	Yellow mosaic	Bean yellow mosaic virus (BYMV)
(vii)	Soybean mosaic or common mosaic	Soybean mosaic virus (SMV)
(viii)	Bacterial pustules	*Xanthomonas axonopodis* p.v. glycines (Smith) Vauterin

Resistance to insect-pests: In seventies only a dozen of insect-pests were problem, whereas at present about 270 insect-pests have been found damaging to soybean crop. The major insect-pests of economic importance have been mentioned in Table-5 for which resistant/tolerant genotypes are required.

Seed Quality: It contains 40% protein and 20% oil in the seed. The protein is used for poultry live stock and industrial products and the oil for cooking purpose. Therefore, breeding to develop cultivars with high protein and oil with good quality and marketability is an important objective.

Table 5. Major insect-pests of soybean in India.

S.N	Common name	Scientific name
A.	Seed & Seedling feeders	
(i)	Field cricket	*Gryllus* sp.
(ii)	Seed maggot	*Delia platura* (Meigen)
(iii)	Cut worms	*Agrotis ipsilon* (Huf.)
(iv)	Blue beetle	*Cneorane* sp.
B.	Stem borers	
(i)	Stemfly	*Melanagromyza sojae* Zehnt.
(ii)	Girdle beetle	*Obereopsis brevis (*Swed.)
C.	**Foliage feeder**	
(i)	Green semilooper	*Diachrysia orichalcea (*Fabr) *Chrysodeixis acuta* (Walker)
(ii)	Brown striped semilooper	*Mocis undata* (Fabr.)
(iii)	Leaf minor	*Aproaerema modicella* Deventer
(iv)	Tobacco caterpillar	*Spodoptera litura* (Fabr.)
(v)	Linseed caterpillar	*Spodoptera exigua* (Hb.)
(vi)	Bihar hairy caterpillar	*Spilosoma obliqua* (Walker)
(vii)	Red hairy caterpillar	*Amsacta moorei* (Butler)
(viii)	Gram pod borer	*Helicoverpa armigera* (Hb)
D	Sap Suckers	
(i)	White fly	*Bemisia tabaci* (Genn)
(ii)	Green jassids	*Empoasca terminalis* (Distant*)*
(iii)	Thrips	*Caliothrips indicus* (Bag)
(iv)	Green stink bug	*Nezara viridula* (L.)

8. BREEDING METHODS

All the breeding methods implied for self-pollinated crops viz., Introduction, Mass Selection, Pure Line Selection, Bulk Method, Pedigree Method, Back Cross Method and other special procedures are used for soybean improvement.

Introduction: In India, large number of germplasm from USA were introduced in 1963-64 and onward in which about 3500 lines were obtained from USDA and evaluated at JNKVV, Jabalpur and GBPUAT, Pantnagar. Some of the introductions like Bragg, Clark 63, Hardee, Lee and Monetta were released as variety.

Mass selection: Improved segregates can be achieved for some characters like seed size practicing mass selection in association with the bulk method. Indirect mass selection for specific gravity has been used to improve oil and protein content (Smith and Weber, 1968).

Pure line selection: This method has been utilized by selecting individual plants from introduced germplasm lines/locally available land races. In 1943, number of selections was made from lines received from USA and other sources, of which T49 (yellow) and T1 (black) at Kanpur, Pb1 (selected from Nanking variety of China) in Punjab were released. Several varieties like JS71-05, JS2, MAUS 2, NRC7, Shilajeet, VLS2, VLS21, CO1, ADT1 have been released from pure line selections.

Pedigree method: This is the most popular breeding method used in the improvement of soybean. The selection for desirable aims as yield, maturity, number of pods and seeds, disease resistance, seed quality etc. is subjected from F_2 onwards. The selection procedure is the same as in case of other self-pollinated crops. Most of the varieties popular at present are developed through Pedigree method.

Modified pedigree method was given by Brim (1966) have been suggested in soybean as Single Seed Descent Method (SSD), Multiple Seed Descent Method (MSD) and Single Hill Procedure.

Bulk method: Desirable plants from segregating generations are selected and harvested together and sample of seed is used to plant the next generation. In F_5 and F_6 individual plant progeny rows are grown. Superior rows are harvested as bulk for yield evaluation. Natural selection can cause change in the characteristics of bulk population during inbreeding.

Back cross method: As in other crop breeding, this procedure is used to transfer desirable trait from donor parent to well adapted cultivar. Use of limited back crossing has been found quite successful followed by pedigree method.

Special breeding methods: Some special breeding methods has been suggested along with conventional methods viz., early generation testing, population improvement, and Index selections.

Early Generation Testing: The aim of early generation testing is to identify desirable heterogeneous population or lines from which superior homozygous lines can be selected. Selection can be practiced among populations or within populations (Fehr, 1987).

Population Improvement: This can be practiced in soybean due to discovery of genetic male sterility. Brim and Stuber (1973) have described a recurrent selections scheme utilizing genetic male sterility.

Index Selection: Keeping multiple breeding goals, there is a need to make selection based on several traits simultaneously to improve final product i.e. yield. Independent culling, Tandem selection, and Index selection are the methods under the procedure of Index selection.

Biotechnological approach: Molecular biology offers new procedures of transferring desirable DNA from any source. A new class of genetic markers for DNA finger printing of soybean germplasms and commercial cultivars is provided by restriction fragment length polymorphism (RFLP) and random amplified polymorphism DNA (RAPD). The molecular genetic marker's have potential uses in plant breeding revealing genetic relationship among lines. Soybean biotechnology in India is at its nascent stage needs refinement.

Developing efficient regeneration protocols would be tile basic step in kick starting the Indian soybean biotechnology programme. This will enable insertion of various gene constructs available for efficient exploitation of important traits like resistance to biotic and abiotic stresses and quality improvement. Genome mapping would facilitate tagging, isolation and transfer of desired QTLs to decrease yield gap. Identification of tightly linked marker(s) of desirable traits will be helpful for molecular breeding in gene tailoring and gene pyramiding.

9. MAJOR ACHIEVEMENTS

China, the origin, of soybean has been involved since ancient times in maintenance, development and spread of soybean throughout the World. Prior to 1913, only large number of germplasm maintenance and cultivation of primitive (land races) varieties were in practice. After that, modern breeding started and between 1923-1995, 651 soybean varieties were released. The intensive breeding had major concern of yield potentials, resistance to diseases and insect pests and seed quality. The verietal replacement rate is raised to 7 times in some of the zones of china (Lijuan *et al,* 1999). Evaluation and characterization of 23578 germplasm lines for, morphological characters, seed characters, seed quality, resistance to biotic and abiotic stresses and molecular markers (Zhanyou, *et al.* 1999) have been performed. The breeding systems in the improvement of resistance to soybean mosaic virus, soybean cyst nematode, rust, downy mildew, frog eye leaf spot, phytophthora, stem borers, and tolerance to drought and salinity have been well established.

Soybean introduced in 1765 in USA and grown as minor crop mostly as forage crop till 1930. Regarding genetic, cytogenetic and plant breeding research, minor attention was paid before 1936 in comparison to wheat, rice and maize. Afterward release of shattering and lodging resistant varieties, area increased very fast. Major breeding advances for soybean have been made in USA conducting genetic research and development of potential cultivars. The new releases offer greater protection from Phytophthora root rot, cyst nematode, charcoal rot and sclerotium rot. The semi-dwarf soybeans are being bread and released for cultivation. The first genetic map of soybean based upon DNA markers was published in 1990. It contains 150 RFLP (restriction fragment length polymorphism) markers with a total length of about 1200 recombinant units (Keim *et al.* 1990). The recent version of the soybean linkage map is based upon a frame work of 600+SSR (simple sequence repeat) markers and contains 20 consensus linkage groups with a total of about 1400 DNA markers (RFLP, SSR, RAPD and AFLP) along with number of isozyme and classical loci (Cregon, 1999). Soybean contains about 80,000 – 1,00,000 genes. To know the function of several genes at a time a functional genomics programme of soybean has been planned at Washington and Illinois (Vodkin *et al.,* 1999). The major and quantitative genes QTL governing agronomic, physiological, pest resistance and seed composition traits have been mapped using molecular markers (Boerma and Mian, 1999). Similarly, large-scale screening, breeding and QTL mapping for abiotic stresses have been established. Round up ready soybeans resistant to herbicides is popular in USA to over come the problem of weeds (Reese *et al.*, 1999).

The major problems of diseases in Brazil and Argentina have been controlled through genetic resistance (Almeida, 1999). Tolerance to drought and aluminum and insensitivity to photoperiod also get proper attention in Brazil (Carter *et al.,* 1999).

In India, soybean research and cultivation had minor contribution before 1965, though the soybean is known since ages. Earlier farmers grew it in water logged area for green manuring and seed as animal feed. From the year 1963-64 encouraging results of exotic lines (USA) like Bragg, Clark 63, Lee etc. (yellow seeded) demonstrated the feasibility of soybean cultivation. The inherent defect of poor germinability restricted the cultivation. However, with no choice, farmers shifted to local cultures viz., Kalituar and T49, which were also associated with indeterminate habit, lodging and shattering like deleterious traits. Further intensive research conducted for improvement in germinability, lodging and shattering resistance with high yield and adaptability, favoured considerable increase in acreage (Singh *et at.*, 1999). Improving genetic architecture of soybean varieties congrous to different agro climatic zones, led to unprecedented growth rate. The contribution of research conducted at JNKVV, Jabalpur has been always significant in this phenomenal expansion. Till date 65 varieties of soybean have been released for cultivation. Source for high yield, wide adaptability, good germinability and resistance to major diseases (YMV, root rot, rust and other foliar diseases), insects-pests and abiotic stresses have been identified and being utilized to evolve future cultivars of soybean.

10. REQUIREMENT OF SEED PRODUCTION

The availability of high quality seed of improved varieties is a pre-requisite for exploitation of production potential of soybean like other crops. Poor germination and seed viability being problems need greater attention in production and processing. Seed production programme should be taken in the area of varietal adaptation with full packages for sound seed production. The selected variety should be sown in a field in which no other varieties of the same crop or lower grade of the same variety was grown in previous year/season. To avoid genetic contamination 3-4 meter isolation distance should be maintained. Off types and diseased plants should be rogued out at flowering and maturity. Admixtures and injury to seed during sowing, harvesting, threshing and storage should be avoided. Soybean has no designated objectionable seed borne diseases and weeds at field and seed levels. Seed should be stored in a cool and moisture proof store to maintain viability, germination and vigour.

11. ORGANIZATIONS INVOLVED IN THE CROP IMPROVEMENT

Linkage, coordination and execution arrangement for the research and development of soybean from the Asian Vegetable Research and Development Centre (AVRDC), Taiwan; International Soybean Programme (INTSOY), University of Illinois, 61821,USA; Brazilian Agriculture Research Enterprises; National Soybean Research Centre, EMBRAPA; Chinese Academy of Agriculture Sciences, Beijing, China; Soybean Institute, Jilin, Academy of Agriculture Science, Jilin, Province, China; National Agriculture Research Centre, Konnondai, 3-1-1, Tsukuba, 305, Japan; Soybean Production Research, USDA, ARS, Stoneville, Mississipi, 38776,USA and IOWA State University, USA have important role. At national level research is organized by All India Coordinated Research Projects on Soybean through different centres. JNKVV, Jabalpur has played a major role in soybean improvement in India. National Research Centre for Soybean (NRCS), Indore is established to coordinate and provide strength for soybean research activities at various centres in India.

12. VARIETIES

In India, several varieties have been identified for different agro-climatic zones (Table 6). The brief descriptions of the important varieties are mention in Table 7.

Table 6. Soybean varieties suitable for different agro-climatic zones of India

Sr.	Zone	Area covered	Suitable varieties
1	Northern Hill Zone	Himachal Pradesh & Hills of U.P.	Bragg, PK 262, PK 308, PK 327, PK 416, Pusa-16, Pusa 20, Pusa 24, Shilajeet, Shivalik, VL soya1, VL soya 2, VL soya 21, VL soya 47, NRC 2.
2	Northern Plain Zone	Punjab, Haryana, Delhi, and Plains of UP and Western Bihar	Alankar, Ankur, Bragg, PK 262, PK 308, PK 327, PK 416, PK 564, PK 1024, Pusa 16, Pusa 24, Shilajeet, SL 4, SL 96.
3	Central Zone	M.P., Bundelkhand Region of U.P., Rajasthan, Gujarat, Northern Orissa, Northern Part of Maharastra.	Bragg, JS 72-44(Gaurav) JS 72-280 (Durga), Gujrat Soybean 1, Gujrat Soybean 2, Indira Soy 9, MAUS 47, JS 75-46, JS 71-05, JS 2, JS 80-21, JS 335, JS 90-41, MACS 13, MACS 57, MACS 58, Monetta, PK 472, Pusa 16, Pusa 24, Punjab 1, NRC 2 (Ahilya 1), NRC 12 (Ahilya 2), NRC 7 (Ahilya 3), NRC 37 (Ahilya 4)
4	Southern Zone	Karnataka, Tamil Nadu, Andhra Pradesh, Kerla, Maharastra.	CO1, Hardee, KHSb 2, MACS 124, Monetta, PK 471, PK 472, PK1 029, Pusa 37, Pusa 40, MACS 450.
5	North Eastern Zone	Assam, West Bengal, Bihar, Meghalaya, Sikkim, Arunanchal Pradesh, Nagaland, Tripura.	JS80-21, Birsa soybean 1, Bragg, PK 472, Pusa 16, Pusa 22, Pusa 24, MACS 124.

Table 7. Brief description of the important released varieties of soybean.

Sr. No.	Variety	Pedigree/ Source	Year of release	Notificati on Status	Duration (Days)	Yield (q/ha)	Important features
1	JS72-44 (Gaurav)	D60-9647 X EC7034	1980	S.O.19(E) dt.14.1.82	104-106	20-25	Yellow testa, black to brown hilum, voilet flower, tawny pubescence, indeterminate, tall plant, brachytic stem
2	JS72-280 (Durga)	EC14437 X Bragg	1981	S.O.19(E) dt.14.1.82	102-105	20-22	Yellow testa, dark black hilum, White flower, tawny pubescence, semi determinate, tall plant
3	JS2	Selection from Tehri Garhwal material	1982	S.O.19(E) dt.14.1.82	90-95	18-20	Yellow testa, light brown hilum, voilet flower, tawny pubescence, pods with dense brown pubescence, determinate, highly shattering.
4	JS75-46	Improved Pelican X Semmes	1984	S.O.165(E) dt.6.3.87	100-106	25-30	Dark voilet flower, grey pubescence, yellow testa, pinkish brown hilum, semi determinate,. tall plant
5	JS76-205	Bragg X Kalitur	1984	S.O.386(E) dt.15.5.90	105-110	18-20	Voilet flowers, brown pubescence, black testa, buff coloured hilum, medium tall plants, indeterminate.
6	JS80-21	JS75-1 X PK 73-94	1990	S.O.527E dt.16.8.91	105-110	25-30	Voilet flowers, tawny pubescence, yellow testa, brown/ black hilum, high seed germinability, semi determinate.
7	JS71-05	Selection from Lee type exotic material	1990	S.O.527E dt.16.8.91	90-95	20-25	Voilet flowers, very small and sparse pubescence on stem and leaves, glabrous pods, yellow testa, light black hilum, dwarf, determinate.
8	JS335	JS78-77 X JS 71-05	1992	S.O.636F dt.2.9.94	95-100	25-30	Voilet flowers, very small and sparse pubescence on stem and leaves, glabrous pods, yellow testa, blackish hilum, semi determinate.

Sr. No.	Variety	Pedigree/ Source	Year of release	Notificati on Status	Duration (Days)	Yield (q/ha)	Important features
9	JS90-41	PS 73-7 X Hark	1999	S.O.425E dt.8.6.99	87-98	25-30	Voilet flowers, tawny pubescence, greenish yellow testa, lanceolate leaves, four seeded pods, blackish hilum, semi determinate.
10	NRC 2 (Ahilya 1)	Mutant of Bragg	1996	S.O.360E dt.1.5.97	103-106	25-30	White Flower, tawny pubescence, yellow testa, gray to black hilum, determinate
11	NRC 12 (Ahilya 2)	Mutant of Bragg	1996	S.O.360E dt.1.5.97	96-99	25-30	Voilet flower, tawny pubescence, yellow testa, light to dark brown hilum, determinate.
12	NRC 7 (Ahilya 3)	Selection from S 69-96	1996	S.O.360E dt. 1.5.97	90-99	25-35	Voilet flower, grey pubescence, yellow testa, brown hilum, determinate.
13	PK 472	Hardee X Pb 1	1986	S.O.258E dt.14.5.86	100-105	30-35	White Flower, grey pubescence, yellow testa, brown hilum, determinate.
14	Pb 1	Selection from Nanking variety	1975	S.O.440E dt.21.8.75 S.O.13(E) dt19.12.78	95-100	20-22	Voilet Flower, tawny pubescence, yellow testa, light brown hilum, semi determinate.
15	MACS 58	JS 2 X Improved Pelican	1989	S.O.599E dt.31.7.89	90-100	25-35	Voilet Flower, tawny pubescence, yellow testa, light brown hilum, semi determinate
16	MACS 124	JS 2 X Improved Pelican	1992	S.O.860E dt25.11.92	95-105	25-30	Voilet Flower, tawny pubescence, yellow testa, dark brown hilum, semi- determinate, resistant to lodging.
17	PK1024	PK308 X PK317	1994		115-118	25-30	White flower, tawny pubescence, narrow leaves, determinate, brown hilum, resistant to pod shattering.
18	PK1042	Bragg X PK416	1996	S.O.360E dt.1.5.97	110-119	30-35	Tawny pubescence, white flower, yellow testa, brown hilum, tolerant to pod shattering, determinate.

Sr. No.	Variety	Pedigree/ Source	Year of release	Notification Status	Duration (Days)	Yield (q/ha)	Important features
19	PK1029	PK262 X PK316	1995	S.O.360E dt.1.5.97	90-95	25-30	White Flower, tawny pubescence, yellow testa, black hilum, determinate, resistant to pod shattering.
20	MAUS-2 (Pooja)	Selection from SH84-14	1995 State 1996 Centre	S.O.360E dt.1.5.97	105-110	25-35	White flower, semi determinate, grey pubescence, yellow testa, light brown hilum, resistant to pod shattering.
21	KB 79 (Sneha)	Hardee X Monetta	1995	S.O.360(E) dt.1.5.97	85-93	17-18	Voilet flower, grey pubescence, determinate, yellow testa, brown hilum.
22	PK 416	UPSM 534 X S38	1985	S.O.258(E)	115-120	30-35	White Flower, tawny pubescence, yellow testa, brown hilum, semi-determinate.
23	Pusa 20 (DS 74-20-2)	Bragg X Lee	1988	S.O.10(E) dt.1.1.88	110-115	25-30	White Flower, tawny pubescence, yellow testa, black hilum, compact plant.
24	Pusa 16	CNS X Lee	1987	S.O.834(E) dt.18.9.87	110-120 (Hills) 100-105 (Pains)	25-30	Voilet Flower, tawny pubescence, yellow testa, grey hilum, compact plant, good germinability
25	PK 564	(UPSM 534 X S 38) X Bragg	1991	S.O.793(E) dt22.11.91	105-115	30-35	White Flower, tawny pubescence, yellow testa, dark brown hi lum determinate and sturdy plant type
26	Monetta	An exotic variety EC2587	1985	S.O.540(E) dt.24.7.85	80-85	20-23	Voilet Flower, tawny pubescence, brown pod, yellow testa, light brown hilum, determinate, early, suitable for intercropping, resistant to pod shattering.
27	MACS 13	Hampton X EC7034	1984	S.O.295(E) dt.9.4.85	90-100	25-30	Voilet flower, tawny pubescence, yellow testa, brown pod, black hilum, indeterminate.
28	VL Soya 2	Selection from VHC 856007	1989	S.O.915(E) dt.6.11.89	104-116	25-30	Voilet flower, sparse grey pubescence, black hilum, light coloured pods, yellow seed, determinate.

Sr. No.	Variety	Pedigree/ Source	Year of release	Notificat-ion Status	Duration (Days)	Yield (q/ha)	Important features
29	MACS 450	Bragg X MACS111	1999	S.O.425(E) dt. 8.6.99	90-95	25-40	Voilet flower, tawny pubescence, black hilum, yellow testa, determinate.
30	VL Soya 21	Selection from VHC3055	1995	S.O.1(E) dt.1.1.96	120-122	20-25	White flower, tawny pubescence, determinate yellow testa, brown hilum, good germinability.
31	Indira Soya 9	Selection from JS 80-21	2001	S.O.92(E) dt.2.2.2001	106-110	20-25	Voilet flower, grey pubescence, semi-determinate, yellow testa, brown hilum
32	MAUS 47	PS73-7 X Hark	2000	S.O.821(E) 13.9.2000	85-90	20-25	Voilet flower, tawny pubescence, determinate, yellow testa, brown hilum
33	NRC 37	Gaurva X Punjab 1	2001	S.O.92(E) dt.2.2.2001	100-105	25-30	White flower, tawny pubescence, semi-determinate, yellow testa, dark brown hilum

REFERENCES

Almeida, A.M.R. (1999) Pathogen variability and soybean resistance in Brazil. Proceedings World Soybean Research Conference VI Aug. 4-7. Chicago, Illinois, USA pp 88 - 91.

Anonymous (1956) The Wealth of India vol. IV F-G. Raw materials CSIR, New Delhi pp 142-150.

Boerma, H.R. and Mian, M.A.R. (1999) Soybean quantitative trait loci and marker assisted breeding. Proceedings World Soybean Research Conference VI Aug. 4-7. Chicago, Illinois, USA pp 68-83.

Brim, C.A. and Stuber, C.W. (1973). Application of genetic male sterility to recurrent selection schemes in soybean. Crop Sci 13:538-530.

Brim, C.A. (1966) A modified Pedigree Method of selections of soybeans. Crop Sci 6:220

Burke, J.J. and Evett, S.R. (1999). Identification of the temperature threshold for soybean irrigation. Proceedings World Soybean Research Conference VI Aug. 4-7. Chicago, Illinois, USA pp-603

Carter, T.E., Desouza, P.I. (Jr.) and Purcell, L.C. (1999) Recent advances in breeding for drought and aluminum resistance in soybean. Proceedings World Soybean Research Conference VI Aug. 4-7. Chicago, Illinois, USA pp 106-125.

Cregon, P.B. (1999) DNA markers, maps and techniques. Proceedings World Soybean Research Conference VI Aug. 4-7. Chicago, Illinois, USA pp 46-55.

Cuzin, V.F. (1976) Soviet Great Encyclopedia. (third ed.) Moscow Publishing Company Soviet Encyclopedia. 24:281-282.

Doyle, J.J. (1988) 5S ribosomal gene variation in the soybean and it's progenitor.

Theor. Appl. Genet 75:621-624

Doyle, J.J. and Beachy, R.N. (1985) Ribosomal gene variation in soybean (glycine) and it's wild relatives. Theor. Appl. Genet. 70:369-376.

Fehr, W.R. (1987) Breeding methods for cultivar development. In W.R. Wilcox (ed). Soybeans: Improvement, Production, and Uses. Amer. Soc. Agron. Madison, WI, USA pp 249-293

Fehr, W.R. and Caviness, C.E. (1977) Stages of soybean development. Special reports 80. Cooperative Extension Service. IOWA State University Aimes, IOWA .

Gai. J. (1996) Soybean breeding. In. J. Gai (ed.) Plant Breeding. Crop Sciences, China Agriculture Press Beijing.

Hermann, F.J. (1962) A revision of the genus Glycine and its immediate allies US Dep Agr. Tech. Bul.1268

Hymowitz, T. and Singh, R.J. (1987) Taxonomy and speciation. In J.R. Wilcox (ed) Soybeans: Improvement, Production and Uses. Second edition Am. Soc. Agron. Monograph 16, Madison, WI pp 23-28.

Hymowitz, T. (1970) On the domestication of the soybean Econ. Bot. 24 : 408-421

Hymowitz, T. and Harlan, J.R. (1983) Introduction of soybean to North America by Samuel Bowen in 1765. Econ Bot. 37: 371-379.

Keen, N.L., Lyne, R.L. and Hymowitz. T (1986) Phytoalexin production as a chemosystematic parameter with *Glycine.* Biochem. System. Ecol. 14:481-486.

Keim, P., Diers, B. W., Olson, T. C., Shoemaker, R.C. (1990) RFLP mapping in soybean : association between marker loci and variation in quantitative traits. Genetics 126:735-742.

Kollipara, K.P., Singh, R.J. and Hymowitz, T. (1997) phylogenetic and genomic relationship in the genius *Glycine* Willd. based on sequence from the ITS region of nuclear rDNA. Genome 40:57-68.

Lijuan, Q., Ruzhen, C., Jiangying S., Xianghua, L. Zhanglin, C. and Zenglu, L. (1999) The history and use of primitive varieties in Chinese soybean breeding. Proceedings World Soybean Research Conference VI Aug. 4-7. Chicago, Illinois, USA pp 165-172

Newell, C.A. and Hymowitz, T. (1980) A taxonomic revision in the genus *Glycine* subgenus *Glycine* (Leguminosae) Brittonia 32:63-69

Prabhakar and Tiwari S.P. (1993) Multiple regression analysis for seed yield in soybean. Soybean Genet Newsl. 20: 70-72.

Reese, P., F. Flint, J.C. (Jr) and Horak, M.J. (1999) The roundup ready soybean system, Weed control timing and efficacy, crop safety and yield. Proceedings World Soybean Research Conference VI Aug. 4-7. Chicago, Illinois, USA pp 65

Shoemaker, R., C. Hatfiled, P.M., Palmer, R.G. and Atherly, A.A. (1986) Cholroplast DNA, variation and evolution in the genus *Glycine* subgenus *Soja*. J. Hered. 77:26-30

Singh, C.B., Shrivastav, A.N. and Gour, V.K. (1999) Breeding soybean for adaptation in India. Proceedings World Soybean Research Conference VI Aug. 4-7. Chicago, Illinois, USA pp 452.

Skvortzow B.V. (1927). The soybean wild and cultivated in Eastern Asia. Proc. Manchurian Res. Soc. Pub. Ser. A. Nat. History Sec 22:1 - 8.

Smith R.R. and Weber C.R. (1968). Mass selection by specific gravity for protein and oil in soybean populations, Crop Sci. 8:373 - 377.

Tindale, M.D. and Craven, L.A (1993), *Glycine pindanica* (Fabaceae Phaseolae), a new species from West Kimberley, Western Australia. Austral. Syst. Bot. 6:371 - 376.

Tiwari, S.P. and Bhatnagar P.S. (1991) Genetics of pod shattering in soybean. Soybean Genet Newsl. 13:150-153.

Vodkin, L., Khanna, A., Slough S., Erplending, J., Coryell, V., Polacco, J., Young, N., Retzel, E. and Keim, P., Shoemaker, R. (1999) Functional genomics and expression. Proceedings World Soybean Research Conference VI Aug. 4-7. Chicago, Illinois, USA pp 61-62.

Wang, C.L. (1976) Review on the classification of soybeans(in chinese) Acta Phytotaxon Sin. 14:22-30.

Zhanyou, X., Ruzhen, C., Lijuan, Q, Ying, S.J and Xianghua, L. (1999) Evaluation of soybean germplasm in china. Proceedings World Soybean Research Conference VI Aug. 4-7. Chicago, Illinois, USA pp 156-165.

CHAPTER - 10

GROUNDNUT

Arun R. Khire

Groundnut *[Arachis hypogaea]* an oil seed crop is popularly known as *Mungfali* and *Chiniya Badam*.

1. TAXONOMICAL CLASSIFICATION, CHROMOSOME NUMBER AND ORIGIN

The genus *Arachis* belongs to the family "Fabaceae", Sub-family "Papilionaceae" has more than 70 wild species, of which only *A. hypogaea* is commonly cultivated (Rudd, 1981). It is a segmental amphidiloid [2n=4x = 40] with a basic chromosome number (x) of 10, but it behaves cytologically like a diploid. *Arachis hypogaea* is classified as Subspecies *hypogaea* (Krapovickas and Rigoni, 1960); Variety *hypogaea,* the Virginia group (Gregory *et al.,* 1951); Variety *hirsuta (*Kohler, 1898); Subspecies *fastigiata* (Waldron, 1919); Variety *vulgaris* the Valencia group (Gregory *et al.*, 1951) and Variety *Vulgaris* the Spanish group (Gregory et *al.*, 1951]. The peculiarity of the genus lies in its peg formation and geocarpy.

Bolivian region (South Balivia-north western Argentina) as the primary centre of origin of cultivated groundnuts is suggested by Krapovickas (1968). The eastern foothills of Andes have a great range of ecologically distinct environments suitable for natural hybridization and establishment of recombinant type. The diversity of groundnut and archaeological remains found in northwestern Argentina provide evidence of antiquity of groundnut cultivation.

Close resemblance between the Variety "Indigenous" grown in Mysore to varieties grown in South America, Philippines, China and Indonesia was reported by Badami (1936) and he proposed the introduction of groundnut to India from one of these sources. Krapovickas (1968) concluded that the

Peruvian type (Var. *hirsuta*) was first transported along the American West coast to Mexico and then across the pacific to Philippines from where it spread to China and India in addition to Malaysia and Indonesia. However, Merrill (1954) was of the opinion that groundnut was first introduced form Brazil to Goa and then to the rest of the Asia by the Portuguese in the 16th century.

2. CLIMATIC REQUIREMENT, AREA AND DISTRIBUTION

Semi arid and arid regions are suitable for groundnut cultivation as compared to humid tropics. It is commercially cultivated between the latitude 40□ North and 40□ South up to the altitude of 1065 metres. Soil temperature is a critical factor for germination, seedling emergence and early plant growth. Soil temperature below 18□C leads to poor germination while above 54□C destroys the embryo (Anon, 1972).

At global scale, it is grown over an area of 23.71 million hectares with the production and productivity of 30.92 million tonnes and 1304 kg/hectare respectively. In India, it is cultivated over an area of 8 million hectares with the production of 8 million tonnes. Among the groundnut producing countries, India sows the highest area with the second largest production next to China. Around 6.8 million hectares area of the crop in India is rainfed where productivity fluctuates between 500 to 1500 kg/hectare. It is grown in the *rabi* and summer season, with an area of 1.5 million hectares and the productivity of 1500 kg/ha. The groundnut (about 89% area) is mainly grown in the states of Andhra Pradesh, Gujarat, Karnataka, Maharashtra and Tamil Nadu. The rest of the area is scattered mainly in the states of Rajasthan, Uttar Pradesh, Madhya Pradesh, Punjab and Orisa.

3. SALIENT FEATURE OF CROP CULTIVATION

Being a C3 plant it is sensitive to photoperiod with main influence on flowering (Hudgens and Mc Cloud, 1974). It grows well in the areas lying in the rainfall ranges between 50-125 cm. It is cultivated in rainfed areas with well-distributed rainfall. Groundnut cannot withstand frost or water logging. Fertile and well drained, light, loose, friable, sandy loam soils well supplied with calcium and adequate amount of organic matter are ideal for its cultivation.

Rabi groundnut is grown in the areas where the winter is not severe and night temperature do not fall below 15□C with sowing from September onwards. Summer groundnut i.e., with assured irrigation facilities occupies an area of 20 lakh hectares in India.

4. BOTANICAL DESCRIPTION

Sub sp. *hypogaea: Habit* procumbent, decumbent or erect; branching alternate, inflorescence simple and never borne directly on the main axis, first branch on the cotyledonary lateral always vegetative; 2 or 4 seeds per pod; pod beak pronounced, moderate or absent; pod constriction prominent, moderate or absent; pod very large (>20 mm) or small (<10 mm); testa colour commonly tan but red, white, purple and variegated forms exist; seed dormancy usually present; foliage dark green.

> Variety *hypogaea* (the Virginia group): *Habit* procumbent, decumbent or erect; main axis is short and procumbent forms (not exceeding 40-50 cm); stem is not very hairy; medium-late maturing. This group consists of plants that have spreading (runner), spreading bunch or upright (erect bunch) growth habits and an alternate branching pattern; they lack flowers on main stem leaf axil, pods are 2 seeded, possess fresh seed dormancy, flower longer, and mature later than those of the Valencia or Spanish groups.

> Variety *hirsuta*: *Habit* procumbent; main axis may exceed 100 cm; stem hairy; pods strongly beaked with 2-4 seeds; very late maturing.

Sub sp. *fastigiata*: *Habit* erect to decumbent; branching sequential; inflorescence simple or compound, always present on main axis; first branch on cotyledonary lateral is reproductive; seed dormancy usually absent; foliage usually lighter in colour than sub sp. *hypogaea*.

> Variety *fastigiata* (the Valencia group): Vegetative branches on primaries are absent or regularly placed at the distal nodes; inflorescence usually simple; medium to small sized pods with no beak have 2-4 (rarely 5) seed; colour of the testa is tan, red, white, yellow, purple or variegated. Plant is mostly upright in growth habit.

> Variety *vulgaris* (the Spanish group): Vegetative branches occasional and irregularly placed; inflorescence compound; medium to small sized pods without prominent beak have 2 seed pods; testa is of tan, red, white or purple coloured. Plants are mostly upright in growth habit.

With extensive inter and intra-subspecific hybridization the intermediate forms with true breeding have been established through selection, thus blurring the differences between groups.

5. PLANT DESCRIPTION AND STAGES OF DEVELOPMENT

The seed of the groundnut is of different size [length 7 to 21 mm, diameter 5-13 mm and weight 0.17 to 1.24 g/seed] and shape. The testa is thin and papery. Seed coat colour is an important distinguishing character because of various expression viz., off white, yellow, pale tan, light tan, dark tan, rose, gray, orange, red, light purple, dark purple and variegated (more than one colour). Cultivars belonging to variety *hypogaea* have larger and heavier seeds than variety *fastigiata* and *vulgaris*. Each seed consists of two cotyledons, and a prominent embryo. It germinates as epigeal.

In moist soil (Temperature 27°C) primary root emerges within 24-36 hrs. Lateral root develops during 2-5 days. Development of nodules starts after 15th day of germination. It harbors nitrogen-fixing bacteria of the genus *Rhizobium*. Nodules appear more in the cultivars of variety *hypogaea* than the variety *fastigiata*.

The stem is angular, pubescent and solid with large pith. In later stages, it becomes hollow, cylindrical, branched and puberulent. The carriage of laterals is an important character, which determines the growth habit. Two distinct forms of growth habit, namely spreading (runner trailing, procumbent and prostrate) in *Virginia* runner type and erect (upright, erect and bunch) in Virginia bunch, Valencia or Spanish bunch type have been reported.

Peripinnate, with two pairs of opposite, subsessile, obvate (variable), shortly mucronate leaflets with entire ciliate margin leaves are borne spirally in a 2/5 phyllotax. Stipules are prominent, linear and adnate to some length and become free of the petiole at the pulvinus. Foliage of subspecies *hypogaea* is darker green than subspecies *fastgiata.*

Inflorescence: It is reduced monopodium type, simple or compound. It consists of three, more flowers occur in the axils of cataphylls, or foliage leaves. The reproductive axis does not occur at the same nodes as the vegetative axis. In variety *hypogaea,* it is simple and expands to little extent during maturity, while in variety *vulgaris,* it is compound and expands moderately. In variety *fastigiata,* the internodes of simple inflorescence may elongate to form a conspicuous branch, which may have two leaves at the end, such branch bears inflorescence, with numerous scale leaves at the basal nodes.

Flowers: Papilionoid corolla is inserted on the top of the hypanthium, covering the staminal column. Flower is protected by two bracts. One of them is simple subtending a short peduncle and other is bifid, subtending the pedicel. Flower is sessile with elongated tubular hypanthium. One of the five lobes of calyx is separated whereas the others are fused. Calyx is yellow, orange, creamy or rarely white in colour with a basal central area the crescent. The wings are usually yellow marked at the tip with diffuse band of deep yellow, orange or brick red colour. The keel is hyaline, colourless to light yellow and covers the staminal tube.

Monoadelphous stamens (ten) form a staminal column, which surrounds the ovary. Two stamens are usually sterile, four are globose and uniloculate alternating with four oblong anthers of which three are biloculate and the one opposite the standard is uniloculate. The filaments are fused upto 2/3 of their length and form acute angels with their fused bases.

Single ovaried pistil is surrounded by the base of hypanthium. The ovary is about 1.5 mm long and 0.5 mm wide with 2-4 ovules. The long filiform style has two sharp bends, which is wrapped with upward slanting hairs near the summit on the surface facing the standard. The stigma is club shaped or clavate generally at anther level or protruding slightly above.

Flowering starts after 20-30 days of sowing. Usually one flower of an inflorescence reaches to anthesis per day. However, in variety *vulgaris* more than two flowers may reach to anthesis per day. In hot scorching sun, the flower withers within 5-6 hours. Stigma becomes receptive a day before anthesis.

Strict self-pollination due to cleistogamy is a rule in groundnut. The fertilization is completed in about six hours after pollination and then floral parts wither within 5-6 hours. After fertilization, an intercalary meristem below the ovary is activated. The green ovary turns purplish from the tip to downwards. The rapid cell division of the meristimetic tissue leads to the formation of stalk like structure (peg) that bears fertilized ovules at the tip. All These developments are completed within seven days after fertilization. The movement of the peg is positively geotropic.

6. SELFING AND CROSSING

Selfing is not a problem for groundnut due to clesitogamous nature. For crossing seeds of female parents are sown on raised beds (30 - 45cm). Male parents are also sown nearby to get adequate pollens. Green house facilities can also be used.

Emasculation procedure is carried out after 4 p.m. The flower bud on the maternal parent is emasculated by removing the lower lip of calyx, then wing and lastly keel. Then the exposed anthers are removed with the help of a forceps. After emasculation stigma is covered and a coloured nylon thread is tied on the calyx tube. Pollination is done in the morning between 7am to 10 am. In the next day morning standard is expanded and stigma becomes visible.

The healthy flower from the male parent is plucked for pollination. The pollens are squeezed from the anthers with the help of forceps and transferred over stigma. A wet towel-paper of 12cm x 12cm is placed over flower carefully without touching the stigma. The elongated peg usually becomes visible in 7-10 days after pollination. The developing peg carries a withered flower with a thread still attached.

In an alternative method, a cut is made on the concave side, below the tip of the flower and the calyx and corolla are pulled out softly by holding the tip of flower bud. In this way, the only Keel remains on the bud. The bundle of stamens is removed by forceps. A small piece of straw tube (4-5 cm length) with its banded upper end is kept on the emasculated bud. Next day morning pollens are dusted on the stigma and again covered by a straw tube (Prabhakar Reddy *et al.*, 1970). Patil (1971) suggested a circular cut with a fine razor blade on the bud, about 2mm from its base to remove the anthers. The emasculated buds could easily be recognized by the excised corolla and no lebelling after emasculation is needed.

7. BREEDING OBJECTIVES

High yield coupled with high shelling percentage and high oil content is the major consideration.

Nearly 80% of the groundnut is cultivated in rainfed condition therefore, early maturity and drought tolerance are to be taken into the consideration while developing a variety.

Fresh seed dormancy in *Spanish-Valencia* group.

Resistance to insect pest viz., aphids, jassids, thrips, leaf miner, *spotoptera*, leaf folders white grubs etc.

Resistance to disease viz., early and late leaf spots, rust, collar rot, stem rots, bud-necrosis, etc.

Management responsiveness.

Resistance to abiotic stresses viz., drought, water logging and cold tolerance.

So far, 100 varieties have been released in India for general cultivation, still the average productivity is low (700-900 Kg/ ha) in comparison to world (1100 Kg/ha). It is mainly due to lack of substantial genetic diversity for wide range of agronomical and physiological attributes and adherence to several unstable-characters (viz., pod number and weight per plant) as sole selection criterion of productivity. Further segmental alloploid nature exhibiting complex tetrasomic inheritance for several agronomic characters, adversely affect the breeding programme (Prasad, 1993).

8. BREEDING METHODS

The methods used for genetic improvement are of traditional and modern concepts.

Traditional methods

Introduction : Day length insensitiveness of the crop has made it suitable for the introduction to wide range of geographical regions. The successful introduction is Ah 32 and Ah 334, MK 374, Kadari 2, Big Japan and early runner. At present National Bureau of Plant Genetic Resources, New Delhi looks after the exchange of genetic sources with other countries.

Mass selection and pure line selection: Many of the present day varieties are the outcome of reselection from established varieties or germplasm lines brought from different parts of the world. Mass selection is good for simply heritable traits, while pure line selection works well for the complex and less heritable attributes like yield and quality. However, use of these methods has built a narrow genetic base and could not create variability.

Hybridization: It is one of the most effective tool to create variability. Parents from the diverse geographic origin are crossed with each other to bring desirable genes together in F_1 and to get desirable genotypes in segregating population. Many varieties have been developed through hybridization followed by pedigree method of selection viz., J11, GAUG1. GAUG10, GG2, GGII, TG17, TMV9, Dh3-30 etc.

Mutation breeding: The advantage of mutagenesis to increase variability for characters of economic importance in groundnut have been demonstrated by Gregory (1955). Among the physical mutagens X-rays, fast neutrons and ó rays were extensively used in groundnut. Certain alkylating agents such as diethyl sulphonate (DES) and ethylmethane sulphonate (EMS) have also been used. A combined approach of mutation cum hybridization is also being employed where crossing with irradiated pollen is done or F_1 plants are subjected to irradiation to get more recombination. Varieties bred through mutation are TAG 24, TG 26, TG1, BG1, BG2, MI3 etc.

Interspecific hybridization: The wild species of groundnut is a huge source of valuable genes. *A. villosullicarpa* and *A. chacoence* have found resistant against leaf miner and bud necrosis, respectively. The cross-compatibility and sterility are the major problems in inter specific hybridization. At ICRISAT 3 wild species, namely *A. chacoense, A. cordenassi,* immune to *Cercosporidium personatum* and *A* sp. HLK 410 resistant to both fungi were crossed with a range of *A. hypogaea* parents, The triploid ploidy level of the F_1's (sterile) were doubled and it was back crossed with *A. hypogaea,* to obtain desired resistant tetraploids (Gibbons *et al.,* 1981). The incompatibility barriers can be breaked by producing hybrid calli and plant lets.

Modern concepts

Synthetic varieties: Synthetic varieties are the mixture of two or more pure-lines. It is more stable in performance over different environments, in comparison to pure line due to broad genetic base. A mixture of 2 Virginia type varieties, [Va 56R and Fla 393 (as booster)] at 2/3 and 1/3 ratio yielded more in comparison to its component varieties over several locations for several years (Emery *et al.,* 1970). However, it may pose problems in processing due to differences in seed size and maturity.

Multi-line varieties: These are early generation composite of four to ten sister lines selected in the F_4 to F_8 generations. All these individual lines are maintained separately. Multi line varieties have wide adaptation and stable yield over environments but their uniformity is lesser then the pure-lines. Six popular varieties of U.S.A namely Dixie runner, Early runner, Florispan. Florigint, Florunner and Early bunch, are the outcomes of this method only.

Single seed descent (Accelerated generation increase): Late generations of F_2 plant are advanced by a single seed and then selection is practiced among essentially homozygous lines (Wynne and Isleib, 1980). This method also takes care of attributes with low heritability. With the use of greenhouse, phytotron and off-season nursery facility, one can obtain the F_5 generation within 2 years. The popular variety NC 6 of groundnut in U.S.A. has been evolved through this method.

Early generation testing: It provides early generation yield trials to test large number of F_2 lines from a single cross (Coffelt and Hammons, 1974). This method is useful in retaining acceptable breeding lines. The NC2 variety of U.S.A. is an outcome of this method.

Convergent (or cyclic) hybridization: The traditional approaches like pure line leads to homozygocity rapidly, but it looses genetic variability. Multiple crossing system such as convergent cross involves a systematic series of crossing cycles. After the first parental single cross combination each F_1, producing the preceding generation is continuously crossed in a pyramidal fashion. This method breaks the linkage group, increases genetic potential of source population, and creates new recombination in relatively less time (biblio). However, time and labour involved are directly proportional to number of parents and crossing cycles. Involvement of more number of parents reduces the chances of obtaining desirable genes. No genetic control over cross population and formation of undesirable recombinants are the some disadvantages.

Selective diallel mating system: It allows the breeder to use a selected group of parents and provide opportunity for multi-parental generation combination (Wynne and Isleib, 1983). As many parents are involved in this method, the crossing programme becomes tedious. The efficiency of convergent crossing programme is more in producing number of high yielding families, in comparison to this method.

Modified composite cross: This method exploits the variation released by combining complexly characters (Hammons, 1976). In F_2 -F_5 progenies of the heterogenous bulk population, mechanical sizing of the pods is done whereas for agronomic suitability line, selection is adapted in later generations. This method yielded several good populations with high agricultural value.

9. ORGANIZATIONS INVOLVED IN CROP IMPROVEMENT

Groundnut is one of the mandatory crops for International Crop Research Institute for Semi Arid and Tropics, Hyderabad. In south Asian countries, the responsibility of groundnut research lies over National Research Centre for Groundnut. In India, the task of initiation and coordination of research lies with the All India Coordinated Research Project on Groundnut and National Research Centre for Groundnut, Junagarh.

10. RECOMMENDED VARIETIES

List of the groundnut varieties suitable for cultivation in different agro-climatic conditions and their important features are presented in Table 1. Details of the important varieties are —

JYOTI : It is a Spanish bunch type variety developed through selection from exotic 7. Pods are medium in size, usually two seeds with rosy testa. The average pod yield is 1600 kg/ha with high shelling percentage [77] and oil content (53%). It matures in 105-107 days and suitable for summer.

SBXI or J II : Spanish bunch type, developed from a cross Ah 4218 x Ah 4354. Its average yield is 1500Kg/ha with high shelling percentage (75) and oil percentage (48.5%). It matures in 106-109 days, which can be grown in summer. Medium size pod is two seeded with light brown testa.

Ak 12-24: A Spanish bunch type developed through selection from local material suited for medium to heavy soil and summer season. It yielded 1300 Kg/ha pod of high shelling percentage [75] and oil content [48.5%] that matures in 105-107 days. A moderate beak and a reticulation may be observed on the pods. Pods are medium in size with two rose colour seed.

Table 1 List of the important varieties and important features

	Features	Varieties
1	Suitable for high yield (2000 kg for k and 2500 kg for r/s)	Gangapuri, Spanish Improved, T-64, M-145, M-13, POLLARCHI-2, TG-1, Dh-3.30, NH-1, CHANDRA, KADIRI-3, NH-2, BG-2, TG-17, GG-2, CO-2, Dh-8, GG-11, SG-84, BG-3, GIRNAR-1, ICGS-44, RG-141, RG-138, ICGS-1, ICG (FDRS)-10, ICGS-37, RSHY-1, TIRUPATI-2, CSMG-84.1, ICGS-5, BAU-13, TKG-19A, B-95, ICGV-86325, DRG-17, DRG-12, TG-26
2	Suitable for high shelling per cent (75 per cent)	AK-12-24, TMV-2, TMV-4, J-11, SB-XI, TMV-7, PLLACHI - 1, M-145, TMV-9, TMV-10, JYOTI, GAUG-1, POLLACHI-2, DH-3.30, KADIRI-2, KADIRI-3, JL-24, CO-2, VRI-3, TIRUPATI-2.
3	High oil content	TMV-10 (54.4%), GG-20(50.67%), ICGV-90208 (55.6%), ICGV-90173 (54.0%), ICGV-90201 (54.6%), TG-30(53.4%)
4	Suitable for rabi/summer cultivation	J-11, JL-24, AK-12-24, TG-17, TAG-24, GG-4, GG-6, TG-3, ICGS - 37, ICGS - 44, K 134, DRG -12, VRI - 2, VRI - 3, SG - 84.
5	Suitable for local table consumption and hps exports	JL - 24, GG - 20, TAG - 19A, J - 11, GG - 2, GG - 2, GG - 5, TG - 17, TG -1 SOMNATH, GG - 11, M - 13, BAU - 13, B-95, ICGS - 86564, ICGS - 49M, ICGV - 90375, ICGV - 90089, ICGV -90212, ICGV - 90173, ICGV - 90208.
6	High protein (22%) and sugar (10%)	ICGV - 90312, ICGV - 90308, TG - 31.
7	Resistance to foliar diseases	ICGV - 88256, ICGV - 89402.
8	Resistance to rust and lls	ALG - 63, DH - 74, TG - 29, PHRG - 95, R-9248
9	Resistance to stem rot	JL - 287, JL - 24, SOMNATH
10	Tolerance to pbn	ICGS - 76, TG - 31, TG - 33, TMV - 256, SB - XI
11	Insect pest resistance	ICGV - 91167, ICGV - 90201.
12	Resistance to thrips and spodoptera	DH - 7, R - 9727
13	Suitable for river bed rice fellow area	JL - 24, AK - 12-24, TAG - 24, TG - 26, RSHY - 1
14	Suitable for acidic soil	JL124, GIRNAR - 1, TS - 34, TG - 3, TAG - 24, ICGS - 44, ICGV - 94357, ICGV - 92242.
15	Suitable for salt affected soil	GIRNAR - 1, ICGS - 44
16	Suitable for cultivation in residual moisture	RSHY - 1, VRL - 3, Dh - 8, TG - 3, JL - 24, ICGS - 44, TAG - 24.

JGN 3: A Spanish bunch type (Jyoti x JL24) drought tolerant variety with an average yield of 1700 Kg/ha. It matures in 104-107 days with high shelling percentage [69%] and oil content [49%]. Pods are medium in size, moderate reticulate with two roundish oval rose colour seeds.

Jl 24 or PHULE PRAGATI : A Valencia group with bunch habit is a selection from E.C. 94943. It is a well-adapted variety throughout the country. Its average yield is 1800Kg/ha, with shelling and oil percentages of 75.0 and 50.7 respectively. It matures in 95-100 days. Plant has dark green foliage. Pods are medium in size two seeded with moderate reticulation.

GANGAPURI: A Valencia group with purple coloured big and bold, three seeds per pod. It yields up to 1800 Kg/ha. Shelling percent is around 65-68 and oil is 49.5 percent.

11. SEED PRODUCTION

The loose friable loam soils with plenty of organic matter and calcium is suitable for seed production. Adequate irrigation during dry spells is essential. Proper drainage of the soil for protecting the crop from water logging is a pre requisite. Fertilization @20 kg Nitrogen, 40-80 kg, Phosphorus and 20 kg Potash per hectare with application of Gypsum @500 Kg/ha, at the time of pick flowering is beneficial. An isolation distance of two metres should be maintained during seed production programme.

12. MAJOR ACHIEVEMENT

As many as 100 varieties have been released so far. A large collection (4385) of germplasm accessions from 68 countries is maintained at National Research Centre for Groundnut, Junagarh. A multiple disease resistant variety Girnar 1 has been released for general cultivation. Partially fertile interspecific hybrid between *A. hypogae* x A. chacoense. Autotetraploids of *A. duranensis* and A. *otavioi* have been developed.

At global level, China ranks first for groundnut production. Its productivity was just 1200Kg/ha in 1970, which touched the level of 2200Kg in 1990 (Wenguang *et al.,* 1995) by the release of several good varieties viz., virginal bunch (Luhua 4, 9, 10, Hua 17, Xuzhou 68-4, Haihua 1, Shuangji 1, 8130 and 79266) and Spanish bunch [Baisha 1016, Lunua 8 and 13]. United State, a leading country in mechanized cultivation of groundnut has touched the new heights of productivity i.e., 3000kg/ha (Isleib and Wynne, 1991) by release of several good varieties in all the three groups.

REFERENCES

Badami, V.K. 1936. *Arachis hypogaea* Linn. Groundnut or Peanut- Original habitat and its distribution in the world. *Journal of the Mysore Agricultural and Experimental Union.* 15(4): 141-54.

Coffelt, T.A. and Hammons, R.O. 1974. Early generation yield trials of Peanuts. *Journal of Heredity* 65:94-6.

Emery, D.A. Benson, J.A. and Wynne, J.C. 1970. Four empirical experiments with Virginia peanut seed mixtures. *Oleagineux* 25: 275-8.

Gibbons, R.W., Nigam, S.N. Moss. J.P., Nevil, D.J. and Dwivedi, S.L. 1981. Disease resistance breeding at ICRISAT. Proceedings of Peanut Breeding symposium. American Peanut Research and Education Society. Richmond, Virginia 16 July 1981 PP. 39-43.

Gregory, W.C. 1955. X-ray breeding of peanuts *(Arachis hypogaea L.) Agronomy Journal. 47*:396-9.

Gregory, W.C. Smith, B.W. and Yarbrough, J.A. 1951. Morphology, genetics and breeding. *(In) The peanut-The Unpredictable Legume*, PP. 28-85. National Fertilizer Association, Washington, DC, VSA.

Hammons, R.O. 1976. Peanut breeding strategy: Modified composite cross.

Journal of American Peanut Research and Education Association 8:48-53.

Hudgens, R.E. and Mc Cloud, D.E. 1974. The effect of low light intensity on flowering, yield and kernel size of florunner peanuts. *Soil Crop Science Society. Florida Proceedings.* 34: 176-8.

Isleib, and Wynne, J.C. 1991. Groundnut production and research in North America. Proceedings of an International Workshop at ICRISAT. Patancheru, Andhra Pradesh. India. 25-29 Nov. 1991. PP. 57-76.

Kohler, E. 1898. *Medizinal - Pflanzen;* Blearbeit. M. Vogtherr and M. Murke, Band III Gera.

Krapovickas, A. 1968. Origin, Variability and spread of the groundnut *(Arachis hypogaea). (In) The Domestication and Exploitation of plants and Animals.* P.P. 427-41. Ucko, P.J. and Falk, L.S. (Eds). General Duckworth Co. Ltd, London, 1969.

Krapovickas, A. and Rigoni, V.A. 1960. La nomenclatura delas subspecies variedades de *Arachishypogaea L. Revista De Investigaciones agricolas* 14: 197-228.

Merrill, E.D. 1954. The botany of cook's voyages. *Chronica botanica* 14: 161-354.

Patil, S.H. 1971. A modified crossing technique for groundnut. *Indian Journal of Genetics and Plant Breeding* 31: 126-7.

Prabhkara Reddy, G., Reddy, P.S. and Murthy, A.N. 1970. An important crossing technique in groundnut *(Arachis hypogaea L.) Andhra Agricultural Journal* 17(4): 124-7.

Prasad, M.V.R. 1993. Genetic enhancement of groundnut for higher productivity based on canopy development. Proceedings of the National seminar on Oil Seed Research and Development in India. Status and strategies. 2-5 August 1993. Hyderabad, (A.P.) India PP. 2-5.

Rudd, V.E. 1981. Aescynomeneae. *(In) Advances in Legume Systematic,* Part I, PP. 347-54 Polhill, R.M. Raven, P.H. (Eds.) Royal Botanical Gardens, Kew, England.

Waldron, R.A. 1919. The peanut *(Arachis hypogaea).* Its history, histology, physiology, and utility. *Pennsylvania University Botany Laboratory Communication* 4: 301-38.

Wenguang, Hu, Shufen, Duan. and Quingwei, sui. 1995. High Yield Technology for Groundnut. *Inter national Arachis News letter.* 15: 2-4.

Wynne, J.C. and Isleib, T.G. 1980. Use of single seed descent and population improvement methods. Proceedings of the *American Peanut Research and*

Education Association 12: 48 (abstract).

Wynne, J.C. and Isleiib, T.G. 1983. Comparison of convergent cross and selective diallel mating system for utilization of exotic germplasm of peanuts. *Agronomic abstracts, American Society of Agronomy,* PP. 86.

CHAPTER - 11

LINSEED

S.R. Shrivas

1. INTRODUCTION

Linseed or flax [*Linum usitatissimum* L.] is cultivated in a number of countries for centuries. The genus *Linum*, belongs, to the family Linaceae which is comprised of fourteen genera and about 150 species. The vernacular name of linseed (English) include in Alsi, Tisi, Chikna, Bijri, (Hindi), Alshi (Gujarati), Javasa (Marathi), Mosina (Bangali), Pesu (Oriya).

In India, linseed is primarily used as an oilseed crop. Every part of the linseed plant is utilized commercially. About 20% of the total oil produced is used for cooking while 80% is used for manufacture of paints, varnish, oil cloth, lenoleum, pad-ink, printing ink etc. The oil cake is a good feed for milch cattle. The stem yields fibre of high strength and durability that is blended with wool, silk, cotton, etc. Strong twines, canvas, suiting, shirting and various indispensable products for defense purpose are manufactured from it.

2. TAXONOMICAL CLASSIFICATION, ORIGIN AND CHROMOSOME NUMBER

Taxonomy: Linseed along-with three other genera belong to the tribe Eulineae of the family Linaceae. The cultivated species, *Linum usitatissimum* has 30 somatic chromosomes (Tammes, 1928) whereas, the somatic chromosome number of the other species varies from 16 to 86 (Gill, 1966 and Harris, 1968). Majority of the species studied have either 18 or 30 chromosomes. The genus Linum consists of 100 species is karyotypically heterogeneous. The species differ in the number and morphology of chromosomes.

Origin: Linseed was introduced in India from Egypt. *Linum angustifolium* Huda. was cultivated in some parts of Europe from prehistoric times till the Aryan era. Vavilov on the basis of polyphyletic origin reported three centres of origin in two geographical areas viz., Central Asia for seed type, Mediterranean centre for fibre types and Near eastern centre for dual purpose type (Richharia, 1962). As per the second theory of origin, flax is originated from a single species *L. angustifolium* (Tammes, 1928) in Mediterranean region, Iran and Canary Island. In these places, it is retaining the identity of distinctive traits.

3. CLIMATIC REQUIREMENT, AREA AND DISTRIBUTION

Linseed is grown under varied soil and climatic conditions. Its production depends principally on rainfall and moderately cool climate. It grows in Madhya Pradesh and Maharashtra in heavy, deep, moisture-retaining black-cotton soils, whereas in Uttar Pradesh and Bihar in the lighter Gangetic alluvium and in Himanchal Pradesh (Kangra) and Punjab (Gurdaspur) in paddy lands.

The crop makes its best growth on well-drained, fertile medium heavy soils, especially silty loam, clay loam and silty clays. Light sandy soils are unsuited for the seed crop particularly in the regions of deficient rainfall. The crop is also grown in the paddy areas as *utera* cultivation. It thrives well on soils with pH ranging between 5.0 and 7.0 but it is sensitive to saline and alkaline soils.

On the global scenario, India ranks first in terms of area under linseed cultivation followed by the USSR, Argentina and Canada. However, productivity of India (344 kg/ha) is very low in comparison to world average (720 kg/ha). With the release of number of double purpose varieties, linseed becomes economically competent with other crops.

India, USSR, USA, Canada and Argentina constitute major linseed production regions in the world. The major linseed growing states are Madhya Pradesh, Maharashtra, Uttar Pradesh and Bihar. In addition to the above it is also grown in Rajasthan, Orissa, Karnataka, West Bengal, Assam, Andhra Pradesh, Himanchal Pradesh, Jammu and Kashmir, Punjab and Nagaland.

4. SALIENT FEATURES OF THE CROP CULTIVATION

Plough the land 2-3 times and bring the soil to a fine tilth with 1-2 harrowings. For cultivation in rainfed condition, plough the land in the summer at the time of pre monsoon to conserve moisture. Farmyard manure at the rate of five tonnes per hectare is applied during land preparation whereas fertilizer is applied at the time of sowing. In irrigated condition, half the dose of nitrogen and full amount of phosphorous is applied as basal dose while the remaining nitrogen is applied with the first irrigation. The sowing time varies (1st week October to 15th November) in different states, depending upon the agro-climatic conditions of the regions. Early sowing escapes the crop from attack of rust, powdery mildew and bud fly (*Dasyneura lini*). 20-25 Kg seed/ha is sown at the row to row spacing of 25-30 cm. Shallow sowing i.e., 2-3 cm is always advantageous. The seed is treated with Carbendazim + Thiram [1 gm + 2g/ Kg seed] or Thiram [3g/Kg seed] to protect the crop from seed and Soil borne diseases. Two irrigation i.e., first at about 35-40 days after sowing and the second at about 65-70 days after sowing is recommended to increase the double yield. Cultivation of disease and pest resistant varieties recommended for the zone is advocated (Table 1). The crop is harvest when the leaves are dry, capsules have turned brown and the seeds have become shiny. A relative humidity of 70% with 8% moisture content in the seed are the best conditions for the storage.

Table 1. Agronomical practices adapted in different crop growing situations

Varieties	Sowing time	Spacing (cm)	Seed Rate* (Kg/ha)	Fertilizers* NPK Kg/ha	State
Sweta, Subhra, LCK 9313, Garima	Mid of October	25	30 R 25 I	40-20R 80-30 I	Assam, UP, WB, Bihar
Himalini, Surbhi, Janki, LC-54	Mid of October	25	30 R 25 I	40-20 R 60-80-30 I	Haryana H.P.
R. 552, Kiran, Triveni, J-23	1st week of October	30 30 25	35-40 U 30 R 25 I	10-20 U 40-20 R 60-80-30 I	M.P. M.S.
Kiran, J-23, LMH-2 Himalini	Mid of October	25	25-30 R	40-20 R 60-80-30 I	Orrisa, Punjab Rajasthan

*R = Rainfed; U = *Utera*; I= Irrigated condition
Note :

LCK 9216,LCK 9313,LCK 9211 LMH-62, RL 993 Shikha and Gaurav are double purpose varieties.
2. R-552 recommended for *Utera* cultivation
3. Laxmi 27, J-23 Garima, shubhra, Himalini and Triveni are recommended for irrigated conditions.

Cropping systems: Linseed can be grown as a sole crop under irrigated conditions with adequate fertilizer application. Double purpose linseed competes well with other crops in terms of net monetary return.

On rainfed lands, inter-cropping of linseed particularly with chickpea, lentil, durum wheat, barley and safflower and under irrigated condition with wheat, potato and rape seed and mustard is popularized. When linseed is inter-cropped with chickpea, the incidence of wilt and pod-borer damage in chickpea is reduced and the inherent high risk of chickpea cultivation is minimized. The quantity of the fertilizer to fulfil the requirement of chickpea and lentil is reduced when inter-cropped with linseed. Some of the efficient inter-crop combinations for various states are presented in Table 2.

Table 2. Inter-crop combinations of various crops with linseed

State	Situation	Intercropping system	Row ratio
Uttar Pradesh (Excluding Bundelkhand)	Rainfed	Linseed + Chickpea/lentil Linseed + Wheat	3:1 or 1:3 4:2
	Irrigated	Linseed + Wheat Linseed + Mustard Linseed + Potato	4:2 5:1 3:3
Bundelkhand of U.P.	Rainfed	Linseed + Chickpea/lentil Linseed + Wheat	3:1 or 1:3 1:3
Madhya Pradesh	Rainfed	Linseed + Chickpea	3:1
Bihar	Rainfed	Linseed + Chickpea Linseed + Mustard	3:1 5:1
West Bengal	Irrigated	Linseed + Mustard Linseed + Potato	5:1 3:3
Maharashtra and Karnataka	Rainfed	Linseed + Chickpea Linseed + Safflower	3:1 or 1:3 Different row ratios
Punjab and H.P.	Rainfed	Linseed + Chickpea	3:1
	Irrigated	Linseed + Mustard Linseed + Wheat	5:1 4:2

Paira or *Utera* cropping Linseed is broadcasted in the standing field when the rice crop is at seed dough stages for utilizing the residual moisture when tillage is difficult. About 25% of the linseed area (0.25 million hectares) is under *utera* cropping.

5. BOTANICAL DESCRIPTION

5.1 Plant description

Linseed (*Linum usitatissimum* L.) is an annual herbcaceous plant grown as a winter crop in warm climates from October-November to March-April. Two distinct morphological types, namely, flax and linseed, are recognized in this cultivated species. The flax types [tall with straight culms and less number of secondary branches] are usually grown for extraction of fibre. The linseed types, which are predominantly grown in India, are meant for extraction of the oil. The plants of linseed types are relatively short and possess more of secondary branches. The linseed types grown in Northern and Central India are tall and small seeded with low oil content. The Peninsular group has shorter plants with bold seeds and high oil content.

5.2 Floral biology

The linseed flower has five petals, which may be of white, blue, violet, purple, or pink colour. The flower has five anthers and a pistil with five slender styles. The linseed capsule is five chambered with two seeds in each chamber (Fig. 1).

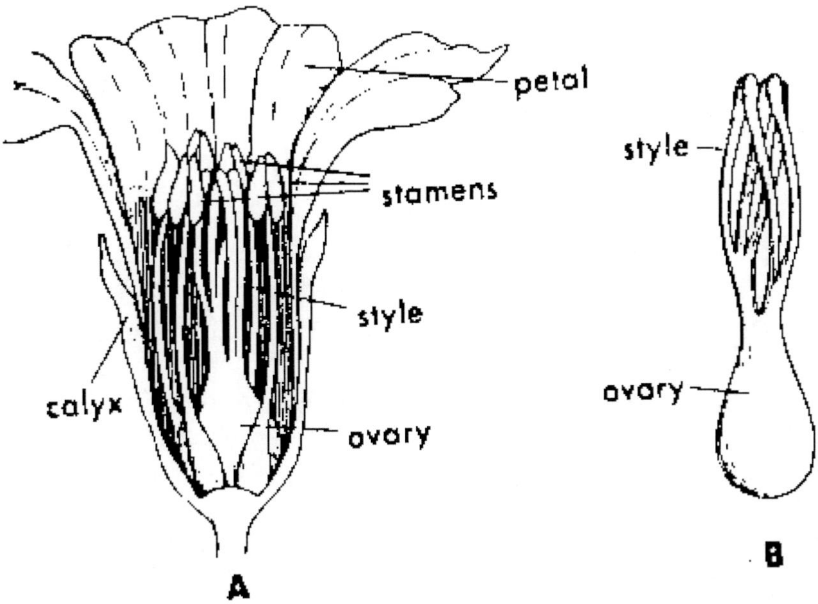

Fig. 1. Flower of Linseed A: Longitudinal section B. Pistil with five styles

6. SELFING AND CROSSING

It is a self-pollinated crop although 0.3 to 2 per cent natural crossing may occur. The flower begins to open and the anther shed pollen shortly after sunrise. The flower is fully open by 7 am and the petals fall before noon.

Emasculation: The linseed flower forms a coloured cone on the preceding day of the flower opening and it is emasculated in that afternoon. The cone of petals is removed by pulling it gently with the thumb and index finger. Then one or two sepals are rolled back and held down while the five anthers are removed with a fine-pointed tweezers. Generally, emasculated flowers are not covered, as insects are not attracted due to absence of petals. The emasculated flower is marked by a coloured string tied around the petiole.

Pollination: It is made in the next morning after emasculation before 8 am. Delay in pollinating may result in a poor seed set. Petals are removed from the flowers selected to supply the pollen, and their anthers are brushed lightly over the stigma of the emasculated flower. One male flower may pollinate two or three emasculated flowers. The linseed plant usually blooms over a long period. Therefore, emasculation and pollination may continue for a month period time.

7. BREEDING OBJECTIVES

1. *Wider adaptability*: To develop high yielding varieties for different agro-climates.

2. Improvements in seed yield.
3. *Suitability in cropping system*:

   To develop high yielding variety for late and irrigated conditions.

   For rainfed, normal and sole cropping system.

   For inter-cropping systems.

   For *utera* cultivation.

   For double purpose under irrigated conditions.
4. *Early maturity*: It is required for double cropping and cropping sequence.
5. *Fertilizer responsive*: Suitable agronomic types, responding to management practices may bring a break through in yield of linseed.
6. *Resistance to abiotic stresses*: Rise in temperature during the growth period results into considerable shedding of the flowers and capsule of late maturing cultivars.
7. Improvement in oil quantity and quality.
8. To incorporate high seed yield oil content and fibre quality in one genotype.
9. *Resistance to diseases and pests*: Genetic resistance to major disease such as rust, wilt and powdery mildew should be incorporated into the high yielding cultivars (Joshi and Jeswani, 1964).
10. Resistance against gall midge (*Dasyneura lini)* is necessary to improve the economic yield particularly to western part (hot climatic area) of India viz., Chhatishgarh, Orissa and West Bengal.

8. BREEDING METHODS

Linseed improvement work in early days between the year 1923 and 1937 was started at Pusa (Bihar), Kanpur (U.P.) and Nagpur (M.P. now Maharashtra). A large amount of local types was collected at these centres and the superior types were recommended for extensive distribution for cultivation. Genetic improvement of linseed has been carried out through conventional breeding methods, i.e. introduction, selection, and hybridization.

Introduction: The introduction of high-yielding new types from similar agro-climatic areas constitutes the easiest and most rapid method of crop improvement. The exotic collections were extensively used in crossing programmes with well-adapted indigenous cultivars to develop the breeding material with high yields of seed and oil as well as with resistance to various diseases.

Pure-line selection: The method was used in the earlier stages of linseed improvement. Almost all the cultivars developed up to 1940 were pure-line selections. eg., NP-12, NO-12, 1 NP-124 and T-1 in Uttar Pradesh, B-67 and B-96 in West Bengal and S-48 in Karnataka.

Hybridization: The breeding materials developed through hybridization are mostly handled through the pedigree method. Modified pedigree method suggested by Brim (1966) is most useful in crosses where both parents are well adapted and the number of inferior offspring is small. Back-cross method is used to incorporate genes for rust resistance into susceptible varieties.

Disease resistance: Work on rust and wilt resistance was started as early as 1939 by crossing NP types with resistant varieties obtained from Australia. Because of this, some of the promising rust resistant types like NP (RR) 9, NP (RR) 10 and NP (RR) 68 were selected. Rust and wilt resistant variety 'K-2' with high yield and bold seed was developed through hybridization in the year 1945.

9. ORGANIZATIONS INVOLVED IN CROP IMPROVEMENT

Under the umbrella of AICRI (ICAR) the linseed research is evolving high yielding, multiple resistance cultivar. In this series, variety Himalini, Janki, Surbhi (H.P.), Garima, Sweta, Shubhra, laxmi-27 (U.P.), Triveni, Sheetal (Rajathan), R-552, Kiran, Jawahar-1 and Jawahar-23 (MP) and Gaurav, Jeevan, Nagarkot, Shikha, RL-993 (Northern India) were released for different states (Table 3 and 4). At present research is towards the development of high yielding genotypes under *utera* cultivation; Incorporation of high quantity of quality oil and fibre quality in one genotype (Dual purpose); and incorporation of resistance to gall midge and *Alternaria* with main emphasis to exploit heterosis through cytoplasmic male sterility system with open corolla.

Table 3. Description of linseed varieties with their salient features

Variety	Pedigree	Year of release	Maturity (days)	Yield q/ha	Salient features
T-397	T-491/T 1193-1	1984	122	1 I	Small seeded; spreading type; tolerant to rust, wilt, drought; oil 44%
Neelum	T-1/NP (RR)-9	1978	143	1.5 I	Medium tall, erect; bold seed; tolerant to wilt; rust; oil 43%.
Himalini	K2/Kangra Local	1985	160	1.3 I	Medium seed; resistant to wilt, rust and powdery mildew; oil 24%
Chambal	Local/RR-45	1978	120	0.9 I	Bold seed; tolerant to rust, wilt and powdery mildew, oil 44%
LC-54	K2/Kangra local	1982	160	1.3 I	Medium seed, resistant to wilt, rust and powdery mildew, oil 42%
R-552	No.-55/B-67	1984	117	0.9 R	Medium bold seed, resistant to wilt, rust and powdery mildew, oil 44%
Pusa-2	Selection from BS-12	1985	120-155	1.4 I 0.7 R	Medium seed, tolerant to rust and drought, responsive to fertilizer, oil 43%
Pusa-3	K2/T-603	1985	150	1.3 I	Medium bold seed, oil 42% responsive to management

Jawahar-23	EC-9832/Hira	1985	120	1.0 I	Medium bold seed, resistant to rust, wilt and powdery mildew, oil 43%
Jawahar-1	Selection from K.P.29	1982	117	0.9	Medium seed, resistant to rust wilt and powdery mildew, it escape to gall fly, 44%
Garima	T-126/Neelum	1985	127	1.5 I	Medium seed, resistant to rust, powdery mildew, Altern- aria blight and wilt, oil 42%
Sweta	Mukta/T-1206	1985	133	0.8 R	Uneven light, medium seed, resistant to blight, oil 44%
Shubhra	Mukta/K-2	1985	133	1.4 I 0.8 R	Medium seed, resistant to rust, and Alternaria blight, oil 45%
Kiran	Afg-8/R-1//Afg-8	1988	122	0.7 R	Rlight seed, resistant to rust, wilt, powdery mildew, oil 43%
Janki	New River / LC 216	1988	167	1.2	Erect, violet flower, medium seed, resistant to rust, wilt and powdery mildew, oil 42%
Surabhi	LC-216/LC-185	1995	169	1.0 U	Dwarf, erect, creamy white flower, yellow small seed, resistant to lodging, rust, wilt and powdery mildew, oil 44%
Triveni	Chambal/Neelam	1993	123	1.4 I 1.1 R	Fairly resistant to wilt and powdery mildew, oil 44%
Sheetal	Bengal-46/R-5567/ /Afg-14	1994	114	0.8 R	Erect, white flower, medium capsule, tolerant to wilt, rust and powdery mildew, oil 43%
LMH-62	EC-41628/EC-7759 /DPL-20/ Neelam	1997	123	1.0 R	Blue flower, moderately resistant to rust, wilt and powdery mildew oil 43%
LCK-9211	Gaurav/Janki	1998	155	1.5 R	Erect, deep blue flower, shining seed, resistant to rust, wilt, oil 41%
LCK-9313	Kaxnu0-27/EC-13 87	1998	137	1.5 I 920 R	Violet blue flower, shining seed, resistant to powdery mildew, rust, wilt, Alternaria blight and bud fly, oil 43%

10. SEED CERTIFICATION STANDARDS

A crop of linseed shall not be eligible for certification if planted on land on which the same kind was grown in the previous season unless the crop in the previous season was grown from certified seed of same variety. A minimum of two inspections shall be made the first during flowering and the second between capsules and maturity stages. A minimum isolation distance of 50 meters for foundation seed class and 25 meters for certified seed class shall be provided. Off-types upto 0.10 percentage for foundation and 0.2 percentage for certified seed is permitted. Seed is certified as per the standards given in Table 5.

Table 4. Description of double purpose type (Seed and fibre yielding) linseed varieties

Variety	Pedigree	Year of release	Maturity (days)	Yield q/ha	Salient features
Gaurav	Selection-3/ EC- 1552	1986-1987	137	1.0 S 1.0 F	Erect, compact and tall, blue flower, medium seed, resistant to rust and wilt, Alternaria blight; oil 43%.
Jeevan	Summit/LC-216	1987 1988	177	1.0 S 1.0 F	Medium seed, resistant to rust, wilt, powdery mildew, frost and lodging, oil 45%.
Nagarkot (KL-31)	New river/LC-216	1988 1995	173	1.2 S 1.2 F	Erect, compact and medium tall, blue flower, medium seed, resistant to wilt, powdery mildew, oil 43%.
Shikha (LCK-8528)	Hira/CRISTA	1989 1997	140	1.2 S 1.0 F	Erect, compact and tall, blue flower, medium seed, resistant to rust, wilt Alternaria blight, oil 42%.
LCK-9216	Gaurav/Janki	1997	140	1.0 S 0.7 F	Very light blue flower, tall medium seed, resistant to powdery mildew, rust, wilt, Alternaria blight and bud fly, oil 41%.
RL-993	Laxmi-27/EC-1387	1998	138	1.0 S 1.0 F	Voilet blue flower, resistant to rust, wilt, powdery mildew, Alternaria blight, bud fly, oil 42%.

Table 5 Seed standards for certification of linseed seed

Factor	Maximum permitted(%)*	
	Foundation	Certified
Pure Seed (Minimum)	98.0	98.0
Inter matter (Maximum)	2.00	2.00
Other crop seeds (Maximum)	0.10	0.20
Weed seeds (Maximum)	0.10	0.20
Germination (Minimum)	80.0	80.0
Moisture (Maximum)	7.00	7.00

11. FLAX (FIBRE)

The varieties of linseed (*Linum usitatissimum,* Linn.) when grown for their fibre are known as fibre flax and when cultivated for their seed are called oil flax or seed flax. Flax fibre ranked to the third place after cotton and jute assigning a reason that fibre is not classified in the fibre trade separately. The fibre flax varieties are characterized by tall, smooth, compact, few branching habit with low seed yielding ability otherwise of seed flax or linseed types. Breeding efforts for combining the double purpose traits in inherently superior genotypes are continue and number of high yielding double purpose varieties viz., Gaurav, Nagarkot, Shikha and LCK 9216 have been released which can compete with yield of other crops.

Fibre properties: The features of flax fibres are their great strength, fineness and durability. They are lustrous and resistant to environmental fluctuations than cotton and jute. The various chemical constitutions and quality specifications for flax fiber standards in Indian industries are presented in Table 6.

Table 6. Chemical constitutions and quality specifications for flax fibers

S.N.	Chemical constituents	Percentage	Quality particular	Specification
1	Cellulose	76.0	Length	50 cm
2	Lignin and pectin	10.5	Strength	28-34
3	Ash	1.0	Denier	4-12
4	Moisture	9.0	Moisture regain	12-14%
5	Extractives	3.5	Elongation	1.5-2.5%

Fibre scutching machine: In European countries, scutching is industrialized. In India, also the power-operated machines have been developed by which a large quantity of stalk can be scutched within a short time.

Hilli regions, Indo-Gangatic alluvium and central part of the country are predominantly under double-purpose linseed cultivation. Gradually the area under double-purpose linseed cultivation is increasing in the country due to availability of high yielding double purpose varieties. The additional profit through fibre will be a great incentive to the growers. This can also catch up the substantial world market.

Manpower and power fibre scutching machines developed need to be popularized in the linseed growing areas offering scope for self-employment and for producing articles of domestic consumption. The government should also consider for developing a system to collect and supply fibre to industries for establishing a flax fibre trade.

REFERENCES

Brim, G.A. (1966) A modified pedigree method of selection in soybeans. *Crop Sci. 6:220.*

Gill, K.S. (1966) Evolutionary relationship among Linum species pp 243 Ph.D. *Thesis University of California River side.*

Harris, B.D. (1968) Chromosome number and evolution in North *American species of Linum Am. J. Bot. 53-1197-204.*

Joshi, A.B. and M. Jeswani (1964) Disease resistant varieties of linseed ICAR. Publ. pp 17-18.

Richharia, R.H. (1962) Linseed monograph. pp 17-18.

Tammes, T. (1922) Genetic analysis, scheme of co-operation and multiple allelographs of *Linum usitatissimum* L. *Genetics, 12:219-26.*

CHAPTER - 12

RAPESEED - MUSTARD

S.S. Bhadouria

1. NAME OF THE CROP

English – Rapeseed, Mustard; Hindi Rai – *Sarson*; Local Rai *Sarson.*

2. TAXONOMICAL CLASSIFICATION, CHROMOSOME NUMBER AND ORIGIN

In the family cruciferae, the genus *Brassica* is the largest important group, which includes number of economic plants, grown for oilseeds, edible roots, stems, leaves, buds or seeds etc. The genera *Raphanus* and *Sinapis* follow in importance, the former is grown for its edible roots and latter as a source of condiments and industrial oil. *Eruca* is also included in this group. The tribe *Brassiceae* is important as many of its wild relatives have a potential value as source of oils, condiments and other products. There is also the possibility to utilize these wild relatives in breeding programme of crop species, either as donors of useful nuclear genes or to supply cytoplasmic and sterility. The scientific names, common names and the chromosome numbers of important species are mentioned in Table 1.

Morinaga (1934) put forward the hypothesis that the high chromosome *Brassia* spp. *B. napus* L (n=19, AC) *B. juncea* Coss (n=18, AB) and *B. carinata* Braun (n=17, BC) originated as amphidiploid hybrids from combinations by pairs of the low chromosome spp. *B. nigra* Koch (n=8,B), *B. oleracea* L (n=9,C) and *B. campestris* L (n=10, A). This hypothesis was verified by U (1935). The findings of Morinaga and U were verified by Frandsen (1943). The diagramatic presentation of derivation is given in Fig. 1.

Table 1. Scientific names, common names and the chromosome numbers of important species

Scientific name	English name	Hindi name	Chromosome number (n)	Genome
Brassica campestris:				
Var. *Sarson*	Yellow *Sarson*	*Sarson*	10	A
Var. *Toria*	Indian rape	*Toria*	10	A
Var. *dichotoma*	Brown *Sarson*	Kali *Sarson*	10	A
Brassica nigra	Black mustard	Banarsi Rai	8	B
Brassica oleracea	Cauliflower	Gobhi or Phool gobhi	9	C
Brassica carinata	Ethopian mustard	Karan Rai	17	AC
Brassica juncea	Indian mustard	Rai/Laha	18	AB
Brassica napus	—	Gobhi Sarson	19	BC
Eruca sativa	Rocket	Taramira or Sohan	11	–

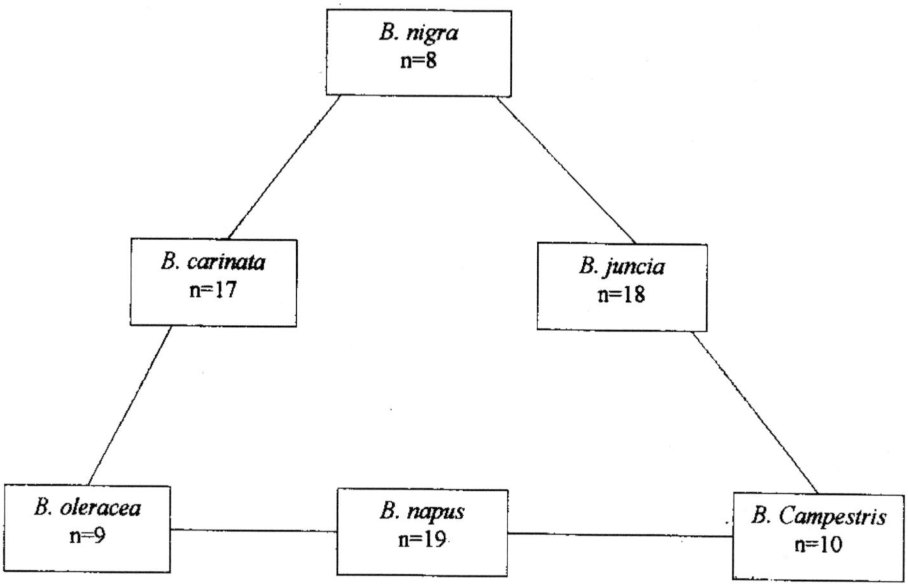

Fig 1 The derivation of *Brassica* spp.

Origin

There are several views about the origin of different species of *Brassicas*. Among the three ecotypes of *B. campestris*, yellow *Sarson* appears to be the oldest. The earliest reference to it is in ancient Sanskrit literature, the Upanishads and Brahamanas (ca. 1500 BC) where it is mentioned as 'Siddhartha' (Prakash, 1961). Wild *B. campestris* is regarded as the progenitor of the different forms as reported by Sinskaia, (1928); Vavilov, (1949). Zeven and Zhukovsky, (1975). However, its evolution and domestication is still not clear. Burkill (1930) was of the opinion that *B. campestris* originated somewhere in Europe and it was originally biennial but through selection, annual forms evolved. Sinskaia (1928) and Vavilov (1949) regarded Central Asia, Afghanistan and adjoining part of India as centre of origin of oliferous *B. campestris*. It is believed that brown *Sarson* is the oldest type and *Toria* is developed from this, perhaps in the foothills of Himalayas, as ecological requirements for short duration type. Yellow *Sarson* probably emerged in the Indo-Gangetic plains as a mutant of superior quality, In the opinion of Alam (1945) *B. campestris* forms are nearest to Indian brown *Sarson* and *Toria* and these were introduced in India through the northwest and extended to east.

Brassica juncea is also mentioned in oldest references in Sanskrit literature as 'Rajika' (Prakash, 1961). Allchin (1969) reported that carbonized seeds of this species have been excavated from the ancient sites of chan-hu-daro of the well-known Harappan civilization (ca. 2300-1750 BC). Vavilov (1949) opined that Afghanistan and adjoining areas are the primary centre of origin. Central and western China, Eastern India as well as Asia-minor through Iran formed the secondary centers. While Sun (1970) ruled out the possibility of China as the centre of origin on the basis that ancestral spp. *B. campestris* and *B. nigra* have never been reported there. Wild forms of B. juncea are found in Middle East, on the plateaus of Asia Minor and Southern Iran (Tsunoda and Nishi, 1968). Vaughan, Hemingway and Schofeld (1963) indicated that there might be two races of *B. juncea* i.e. Indian, with entire leaves and Chinease with foliage. Hence, it may be assumed that Indian mustard (*B. juncea*) forms originated in India itself and there is possibility of its migration from Middle East with human populations. Further differentiation in to forms might be occurred due to hybridization and cryptic structural changes.

3. CLIMATIC REQUIREMENT, AREA AND DISTRIBUTION

Rapeseed-Mustard crops require tropical, subtropical as well as temperate climate with relatively cool temperatures for normal growth and yield. This is grown in Rabi season i.e. September–October to February-March. It grows well in the areas of scanty (250 mm) to medium (750 mm) rainfall. *Sarson* and *taramira* preferred in low rainfall areas, where *raya* and *Toria* are grown in medium and high rainfall areas. The crop can be grown well in light to heavy loam soils. Indian mustard (*raya*) can be cultivated in all types of soils, but *Toria* does better in loam to heavy loam soils where as *taramira* is mostly grown in poor and light soils.

4. SALIENT FEATURES OF THE CROP CULTIVATION

Generally, Indians cultivate it in two different cropping patterns i.e., under stress conditions as a mixed crop, and under optimum conditions with advanced technology as a pure crop.

Land preparation: A fine tilth is required to ensure good germination and growth. At the time of soil preparation, planking is to be given after each ploughing for preservation of moisture.

Seed rate and seed treatment: For pure crop, seed is sown at the rate of 5Kg /ha after treatment with Apron SD35 @ 6 g/Kg seed or Thirum @ 3 g/Kg seed.

Sowing period and method: Indian mustard is sown in October (2nd to 3rd week), *Toria in* September and Taramira in October to November. Seed is drilled in rows at 30cm apart at depth of 2.5 to 3.0 cm with tractor/ bullock drawn seed drill.

Fertilizers: For *toria* (Irrigated), 60 N + 30P + 20K, Mustard (Rainfed), 40 N + 20P + 10K Mustard (Irrigated) - 80 N + 40P + 20k + 25 S and for Taramira (rainfed) 30 N Kg/ha is recommended.

Irrigation: One pre-sowing irrigation (if required) and one irrigation at the time of flower initiation for *Toria* i.e. 30-35 days after sowing; One pre sowing irrigation (if needed), second at the time of flower initiation i.e. at 40-45 days after sowing and third at the time of siliqua formation i.e. 75 to 80 days after sowing for Mustard is recommended.

Inter-culture operations: Weeding and Thining at 15-20 days after sowing is done to maintain plant distance of 10-12 cm. Spray of Fluchloaralin @ 1 Kg a.i./ha at the time of seed-bed preparation to control the weeds.

Plant protection: Alternaria blight, white rust, downey mildew and stem rot are major diseases, which can be control in economical and effective way by adapting seed treatment, spray of Diathane M-45 @ 0.25% at 50 DAS and its repetition after 15 days interval for two times and removal of stalks after harvesting and long rotations to be followed.

Painted bug and sawfly are predominant pest in *Toria* whereas mustard aphid in mustard. Painted bug and sawfly (T*oria*) and aphid (mustard) can be managed by sowing of seed at proper stage, spray of Thiodone 35EC @ 1200 ml in 600 litre of water or dusting with 2% folidol powder @ 20-25 Kg/ha. Spray of Metasystox @ 900 ml in 600 litre of water is effective to control painted bug and aphid.

Harvesting and threshing storage : Crop is harvested at physiological maturity, when 75% siliquae are matured and leaves dropped down. Harvesting of over matured crop causes loss in yield due to shattering etc. Store the crop after drying to keep about 5.8% moisture in seed.

5. BOTANICAL DESCRIPTION

This group belongs to annual herbs with 0.50 to 2.00m plant height. Roots are long and tapering. Roots of *Toria* are short and are considered as surface feeder, whereas brown *Sarson* has long roots with a limited lateral spread. Yellow *Sarson* has both extension and lateral spread. The branches arise at an angle of 30° to 40° in *Toria* and brown *Sarson*. In yellow *Sarson* the branches arise laterally at an angle of about 10° to 20° and give a plant a narrow and pyramidal shape. The inflorescence is a corymbose raceme. In yellow *Sarson*, the four petals are spread apart, while *Toria* and brown *Sarson* have overlap or part placed petals depending upon the flowers have a hypogynous syncarpous ovary. In *Toria* and Brown-*Sarson* the ovary is bicarpellary. In Yellow-*Sarson* the ovary may be tri or tetra carpellary. The fruit is a siliqua with two, three or four- carpelary. The flowers begin to open from 8:00 a.m. and continue upto midnight.

6. SELFING AND CROSSING

Anthesis starts 3 days before the flower opening and thus bud pollination is recommended. Pollination is carried out in nature by bees, which visit the flowers for nectar. In self-compatible group cross-pollination to an extent of 15 to 30 percent is reported. Steps for controlled pollination in *Brassica* group are

Selection of buds and their emasculation: First or second order lateral shoot of full bloomed plant is to be selected, which should not have any fruits set. Preferably three days old flowers are to be removed with the help of forceps and by snapping at the base with out any damage to shoot. Small buds which are expected, not to open with in two days, should also be removed by the help of bent scissors or pair of forceps. Six to eight buds expected to open on the following day or a day after that to be chosen for emasculation.

Two fine slits are made in two directions of the bud with the help of forceps, the sepals and petals are separated and the six stamens plucked with their filaments. After removal of six stamens the opened buds are gently closed with the help of forceps along the sepals in an upward direction. Then buds are labeled by marking identity of female parent, the sign 'X' and a space for male parent to be filled after pollination with date of emasculation/muslin pollination. The buds are then covered by thin muslin bag. This bag is to be supported internally by means of a stick.

Pollination: On the following morning, fresh flowers are collected in the Petridishes and pollinate the opened emasculate flowers of previous day by taking one anther held by forceps by means of filaments. The well dehisced one anther is enough to pollinate about eight flowers. After pollination the label is filled in with the identity of male parent. The bag is laid down and tied at the bottom. Complete record of pollinated buds etc is to be maintained in a field book or register to avoid errors. A second dusting of pollens may be done if there is any doubt about first pollination due to a cloudy day or cold night. Fertilization is effected in about 20 to 24 hours after pollination.

Self-pollination: It is also carried out in the similar manner as above at the bud stage except that emasculation is not done. Buds that are likely to open 2-3 days after are best for this purpose. Actually, this stage is so effective that bud pollination is successful even in self- incompatible group. As soon as there is reasonable development in fruits the muslin bags are removed to allow fruits development under natural conditions.

7. BREEDING OBJECTIVES

Collection, evaluation, maintenance and characterization of germplasm for utilization in breeding programme.

To evolve/ develop high yielding varieties with early duration, bold seed size and high oil content.

To develop varieties for biotic and abiotic stress resistance/ tolerance.

To develop varieties for quality requirements as per international standards i.e. low erucic acid in oil and low glucosinolates in meal.

To develop varieties suitable for double/ multiple cropping patterns as well as responsive to fertilizers and irrigation.

8. BREEDING METHODS FOR CROP IMPROVEMENT

Several breeding methods of varying efficiency are identified as per the need of location, the availability of facilities and the type of material etc. Important methods are —

Mass selection: In this method, plants are chosen based on the phenotypes and the harvested seeds are bulked with out progeny testing. It is therefore, a method where the selection is based entirely on phenotype. It does not take into consideration the ancestral history of the plants selected, nor the nature, phenotypic and genotypic makeup of the plants. Hence the performance of the selection remains unpredictable. So this method can be adopted where facilities for employing more involved methods are not available or the breeder's material does not have selection history.

Single plant or pure line selection: A variety developed by pure line selection may be uniform than developed by this method. In this method single-plant, selections are made from self-pollination progeny or segregating population. In either case, progeny testing is essential part of it and serves to evaluate the breeding behaviour of the selected plant. Pure lines become impure because of mechanical mixtures, natural crossing and mutations. Eve with limitations it is a useful for self-compatible group.

Diagrammatic presentation of pedigree method of selection is presented below:

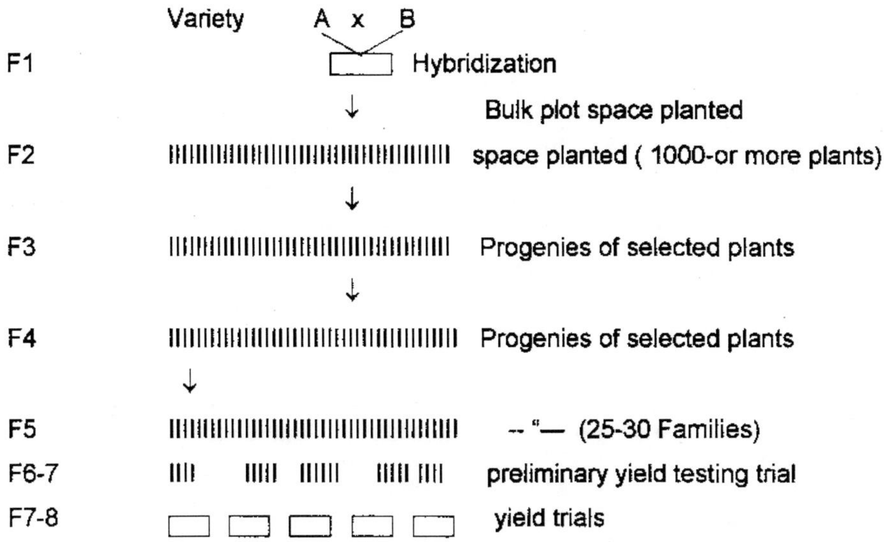

Back cross method: It is a recurrent hybridization which adds a desirable gene into an otherwise desirable variety. To transfer the desirable gene, two varieties are selected i.e., one a adapted productive variety (recurrent parent) which lacks a gene for a superior characteristic that is present in the second variety (donor). Beginning in the F1, the hybrid material is successively back-crossed to the recurrent parent. The donor is involved in the initial cross only. This method is suitable for simply inherited, dominant characters like disease resistance etc.

Recurrent selection method: The frequency of genes of particular quantitatively inherited characters is increased by repeated cycles of selection. It requires identification of superior genotypes for the specific quantitative characters being improved and the subsequent inter mating of the superior genotypes to get new gene combinations.

Polyploidy breeding: Brassica spp. is a natural polyploid. Artificial synthesis of alloploids in *Brassicas* to create positive variability in the breeding material for crop improvement have been reported by Olson (1960), Shiga (1970), Macky (1977) and Prakash and Hinata (1980).

Mutation breeding: A remarkable feature of oil *Brassicas* with reference to their response to ionizing radiations is their high radio resistance. The irradiated material is grown and selected upto M6 generations, then compared with standard variety for genetic variability and performance for other parameters, and then registered.

Hybrid varieties: CMS- male sterility is found effective through various systems. Search to establish restorer gene is under investigation and confirmation.

9. REQUIREMENT OF SEED PRODUCTION

Isolation distance (m)		Off type plants		Inseparable other crop plants		Objectionable weeds [*Argemone maxicana]*	
FS	CS	FS	CS	FS	CS	FS	CS
50*	25**	0.10	0.50	–	–	0.05	0.10

* - Self-compatible same spp./ variety not confirming varietal purity standard.
** - Self incompatible

10. LIST AND DESCRIPTION OF EXISTING VARIETIES

Toria

S.No.	Name of variety	Duration (Days)	Yield potential Kg/ha	Remarks
1	T-9	95-100	1400-1500	
2	Bhawani	85-90	1000-1200	
3	Jawahar Toria-1	85-90	1600-1800	
	Indian Mustard			
1	Kranti	135-138	1500-1800	
2	Pusabold	130-135	1500-1800	
3	Varuna	130-132	1500-1750	
4	Rohini	135-140	1600-1800	
5	Jawahar Mustard -1	125-127	1800-2000	White rust resistant
	Gobhi Sarson			
1	GSL-1	150-155	1800-1900	Sweet oil & hybrid
2	Hyola 401	145-150	1800-2000	OO quality
	Taramira			
1	T-27	130-135	500-600	Suitable for rainfed situation

REFERENCES

Alam, Z. 1945 Ind. J. Agric. Sci., 15, 173-181.

Allchin. F.R. 1969. In the Domestication and Exploitation of plants and Animals eds; P.J. Ucko and G.W. Dimbledy, London. 323-329.

Burkill, I.H; 1930 Gad's Bull, 5, 99-117.

Mackay, G.R. 1977. Euphytica 26,511-519.

Olsson,G. 1960. Hereditas 46,351-396.

Frendson, K.J.; 1943. Dansk Botanisk Arkiv, 11, 1-17.

Prakash, O. 1961. Food and drinks in ancient India. Munshi Ram Manoharlal, New Delhi 265-266.

Prakash, S. and Hinata, K. 1980. Opera Botanica.

Shiga,T. 1970. J.A.R.Q.5,5-10.

Sinskaia, E.N.; 1928. Bull Appl. Bot. Genet. Pl. Breed; 19, 1-148.

Sun, V.G.; 1970. J. Agric. Assoc. China. 71; 41-52.

Tsunoda, S. and Nishi, S. 1968. Proc. XII, Int. Congr. Genet, 2, 77-88

U. N. 1935. Jap. Bot., 389-452.

Vavilov, N.I.; 1949. Chron. Bot . 13, 1-364.

Voughan, J.G., Hemingway, J.S. and Schofeld, H.J. 1963. Bot. J. Linn. Soc. 58. 435-447.

Zeven, A.C.; and Zaukovsky, P.M.; 1975. Dictionary of cultivated plants and their centers of Diversity, Pudoc. Wageningen, 29-30.

CHAPTER - 13

NIGER

S.S. Duhoon

1. INTRODUCTION

The vernacular names of niger (English) include guizotia (French), gingellikraut (German) and sorguja (Assamese), sarguze (Bengali), alashi (Oriya), vevinuvvulu (Telgu), payellu (Tamil), mechellu (Kannada), karala or kharasani (Marathi), ramtil or Kharsani (Gujrati), ramtil or Jagni (Hindi), ramtil (Punjabi) in Indian languages and nog, nuk, nook, noog (the plant) and nehigue (the oil) (Amharic); neuk, nuhk, nahug, nahuk, (Tigrionya and Sahinya); nuga, nughi (Orimigna and Galignya); nughio (kaffinya); gizkoa (Gumuzinya) in Ethiopian language. Ramtil, ramtilla, Jagni, parbati or parbatia are the vernacular names of niger in Madhya Pradesh and Chhattisgarh.

Niger [*Guizotia abyssinica* (L.f.) Cass.] is an important crop in terms of its 35 to 45% content of quality oil and 25 to 35% protein. The crop is capable of giving good yield even under poor crop management, soil fertility and moisture stress. Niger has good degree of tolerance to insect pests, diseases, attack of animals and birds with added advantage of yielding oil and seeds completely free from any toxin. Niger has great potential for soil conservation, land rehabilitation and as a bio-fertilizer. These attributes favour its cultivation in hilly, sloppy areas and marginal and sub-marginal lands in and around forestlands where other crops hardly grow. Niger in India is primarily grown on the denuded soils in hilly and tribal pockets on an area of about 5.42 lakh hectares with annual production of about 1.62 lakh tones and productivity of approximately 299 kg/ha. It has yield potential of 500-600 kg/ha under well managed growing conditions. So much so, the yields up to 1000-1300 kg/ha are recorded with very low inputs required to provide optimum growing conditions for niger.

2. ORIGIN, TAXONOMIC CLASSIFICATION AND CHROMOSOME NUMBER

The genus *Guizotia* is native to tropical Africa. The evidence suggests that niger originated in the highlands of Ethiopia, north of 10 N latitude, particularly west of the Rift Valley (Baagoe, 1974). Harlan (1975) considers niger to be among the earliest of the domesticated crops in Ethiopia and migrated on to the East African highlands. It is believed that during the second millenium BC, it may have moved across to India with other crops as part of what Harlan describes as the "Savannah complex". According to a legend, an Ethiopian Queen, in remote past, occupied a vast territory of Gujarat in India and made Ethiopians emigrate to India who might have taken niger to India (Getinet and Sharma, 1996).

The genus *Guizotia* belongs to the family Asteraceae or Compositae, tribe Heliantheae, subtribe Coreopsidinae. Baagoe (1974) described six species within the genus *Guizotia i.e. G. abyssinica* (L.f.) Cass; *G. scabra* (Vis.) Chiov.; *G. arborescens* I. Friis; *G. reptans* Hutch.; *G. villosa* (Sch. Bip.) and *G. zavattarii* Lanza. *Guizotia scabra* contains two subspecies, *scabra* and *schimperi*. *Guizotia scabra* subsp. *schimperi* is a common annual weed in Ethiopia. The taxonomic status of *G. abyssinica* and *G. scabra* subsp. *schimperi* is controversial (Murthy *et al.,* 1995). Both the species are annual and morphologically similar with 2n=30 chromosomes having similar karyotype. *G. scabra* subsp. *schimperi* is closer to *G. abyssinica* than to its perennial sister species *G. scabra* subsp *scabra* and forms 15 bivalent in the pollen mother cells of hybrids.

Guizotia abyssinica and other species of this genus has a complement of 2n = 30 chromosomes (Richharia and Kalamkar, 1968; Murthy, 1990; Hiramath and Murty, 1992 and Dagne,1994a). During meiosis, 15 bivalent were generally observed with an occasional trivalent and tetravalent association (Chavan, 1961). Hiramath (1985) has found that niger shows considerable pollen sterility associated with univalent formation and subsequent meiotic abnormalities. Chromosome length varied from 26.66 to 63.05/um with considerable variation in arm length ratios ranging from 0.48 to 1.00 (Patel *et al.* 1983). Cultivated niger (*G. abyssinica*) has been reported to hybridize easily with *G. scabra* spp. *schimperi; G. scabra* subsp. *scabra* and *G. villosa* and sets viable seeds in both the directions (Dagne, 1944a).

3. CLIMATIC REQUIREMENT, AREA AND DISTRIBUTION

The crop is normally planted in July, about one month after the onset of monsoon, but may be sown any time between May and September. Flowering begins when the rain ceases in early October. In central India, the crop is normally sown on light, gravelly and shallow soils common to the hills of this region. The crop is successfully raised on heavy clay soils and grows equally well on either well-drained soils or waterlogged clay soils (Prinz, 1976). Soils with pH between 5.2 to 7.3 are suitable. The crop matures in 95 to 150 days, depending on the elevation and time of sowing. In the Eastern Ghat region of Maharashtra state, a single crop of niger is grown in the short, intense rainy season from June to September when up to 3000 mm of rain is expected. In the hilly areas (400 to 1500 m elevation), niger may be planted as a sequence crop following the June to September harvest of *Panicum miliare* and *P. miliaceum*. It is mainly a *kharif* crop but can be grown in *rabi* season also.

Niger, in general, is a crop of cooler parts in the tropics. In India, a rainfall between 1000 to 1300mm is optimum, even a well-distributed rainfall of 800mm is reasonably good (Sharma, 1990b). The plant can withstand high rainfall during vegetative phase but a rainfall of over 2000mm may depress the growth. Rainfall during seed setting and maturity leads to shattering and hence low yield. It can tolerate salinity but flowering is delayed with increasing salinity (Abebe, *et al.,* 1978).

It is an important oil crop in Ethiopia and parts of India, where it is grown mostly by poor farmers, in traditional agriculture systems. In Ethiopia, niger provides about 50 to 60% of that country's edible oil, while in India, niger accounts for almost 2% of total oilseeds production of the country. Niger is grown as a minor oilseed crop in the Sudan, Uganda, Tanzania, and Malavi (Baagoe, 1974; Purseglove, 1979). In Zimbabwe, the crop is grown for silage and green manure (Vaughan, 1970). The major niger growing states in India are Madhya Pradesh (38.8%), Orissa (31.2%), Maharashtra (12.7%) and Karnataka (7%). However, the important states contributing to the total production of the country are Orissa (46.3%), Madhya Pradesh (21.1%), Maharashtra (9.6%) and Bihar (9.6%) (Table 1). Besides, the crop is also cultivated to some extent in hilly areas of Rajasthan (Southern Rajasthan mainly in Udaipur and Banswara district), Uttar Pradesh, Gujarat, Tamil Nadu, Arunachal Pradesh and Assam in North Eastern Hill region. The crop can be grown successfully on all types of soils preferably well drained, loamy soils with good tilth and texture. Madhya Pradesh is the most important state of India for niger cultivation where its cultivation is concentrated in the tribal pockets.

Table 1. Area, Production and Productivity of niger in major growing states of India (1997-98)

S. No.	States	Area '000 ha'	Production '000 tones'	Productivity (kg/ha)	Percent share to All India	
1	Madhya Pradesh	210.3	34.2	163	38.8	21.1
2	Orissa	169.0	75.0	444	31.2	46.3
3	Maharashtra	69.1	15.5	224	12.7	9.6
4	Karnataka	37.5	7.2	192	7.0	4.4
5	Bihar	24.4	15.4	631	4.5	9.6
6	Andhra Pradesh	15.5	6.2	400	2.8	3.8
7	West Bengal	6.5	3.5	538	1.2	2.2
8	Others	9.7	5.0	515	1.8	–
	All India	542.0	162.0	299	100.0	100.0

4. BOTANICAL DESCRIPTION

Niger is an annual dicotyledonous herb with a well-developed root system. Germination is epigeal and seedlings have pale green to brownish hypocotyl and cotyledons (Seegeler, 1983). Leaves are light to dark green in colour, opposite on the lower and alternate on the top of the stem, smooth, 10-20 cm long and 3-5 cm wide with pointed to smooth margin.

The stem smooth to slightly rough, hollow and weak, dark purple to green in colour, about 1 to 2 cm in diameter at the base, moderately to well branched; branches 5-12, plant height vary considerably from 35 to 200cm. Flower is yellow and rarely, slightly green. The heads are 15-50 mm in diameter with 5-20 mm long ray florets. Two to three capitulae (heads) grow together each having ray and disk florets. The receptacle has a semi spherical shape and is 1-2 cm in diameter and 0.5-0.8 cm high. The receptacle is surrounded by two rows of involucral bracts. The capitulam consists of six to eight fertile female ray florets with narrowly elliptic ovules. The stigma has two curled branches about 2 mm long. The hermaphrodite disk florets, usually 40-60 per capitulum, are arranged in three whorls. The disk florets are yellow to orange with yellow anthers, and a densely hairy stigma. The achene is club shaped obovoid and narrowly long, black with white to yellow scars on the top and base with a hard testa. The head produces about 40 fruits.

Floral biology : Two to three buds are formed close together at the tips of the branches, followed by the formation of buds in the axis of the leaves. A single plant often bears 20 to 40 heads. Flowering begins 50 to 90 days after sowing and can continue for 4 to 6 weeks in a field. The head or inflorescence consists of a receptacle, 1 to 2 cm in diameter, surrounded by two rows of involucral bracts. Generally, eight ray florets form a showy outer whorl, light yellow to orange in color. The ray florets of niger are female and normally produce seeds. The disc florets, up to 60 per head are usually arranged in three whorls and are hermaphrodite. Each floret consists of a tube of five lobes (fused petals) with a distinct upper section surrounding the style and stamens, and a lower section surrounding the ovary. The style bears a stigma consisting of two hairy arms.

The florets open about two hours after sunrise. The anthers liberate their sticky pollen and the styles elongate bearing the sigma arms well above the anthers, so that there is no contact between pollen and stigma of the same floret. The outer whorl of ray florets flowers first, with flowering proceeding concentrically over a period of about eight days. Niger is largely or entirely a cross-pollinated crop and highly attractive to bees and other insects at the time of flowering. The florets start drying three days after flowering. The seed (achene) is usually shiny black or dark brown, 3 to 5 mm long and 1.5 to 2 m wide. Seed weight ranges from 3 to 4.8 g/1000 seeds (Chavan, 1961 and Seegeler, 1983).

The physical structure of the niger flower forces cross-pollination but this appears to be reinforced by a sporophytic self-incompatibility system. Little or no seed set has been obtained under bagging with muslin bags, even when the bagged heads of the same plant are rubbed together. When pollen from the same plant is applied to stigmas of one to three-day old buds, seed set is increased over that found in bagged heads, but if the buds are over four days old, no seed is set on selfing (Naik and Panda, 1968; Ramachandran and Menon, 1979). However, some self-compatible plants have been identified in Ethiopia. The progenies on these plants show little or no inbreeding depression.

5. SELFING AND CROSSING

Self-incompatible nature and exclusively emtomophyllus mode of pollination make selfing a difficult process in niger. Very high inbreeding depression upto 91% (Ramachandran and Menon, 1979) associated with very poor seed setting is observed with the advancement of generation from S_3 onwards. With the initiation of flowering, selfing or sib-mating is done by covering two or more representative adjacent plants of a line by a bag made of fine muslin cloth or mosquito net to exclude insects. Shaking of bag/plants is done on alternate days to ensure pollination. At maturity, the seeds within the bag will be S_0 to give rise S_1 generation.

For crossing, Shrivastava and Sharmwanshi (1974) suggested to select well-developed capitulum of desirable female parent at flower initiation stage. Remove all the disk florets, which are hermaphrodite, from the inner whorls of the selected capitulum. Remove all other heads from the plant for proper setting of cross seeds. Dust the female ray florets with the pollen from selected male parent after 1-2 days of opening when the stigma becomes receptive. Bag and tag the heads with proper label after dusting.

6. BREEDING OBJECTIVES

Improvement in yield

Low average yields of 200 to 400 kg/ha and maximum yields of 800 to 1300 kg/ha obtained in niger, indicates its relatively low yield potential and therefore improvement in seed and oil yields will continue to be the primary aim in niger breeding to make it competitive with other oil seeds/ crops.

Improvement in oil content

Oil content in niger seed is low ranging from 30 to 45%. Improvement in oil content has good scope in niger and will continue as one of the major objective in breeding.

Reduction in hull thickness

A sharp increase in oil content could be achieved through selection for thin hull types, as successfully done in sunflower and safflower. Reduction in hull thickness will reduce fiber levels in the meal to make it more desirable as a livestock feed.

Determinate habit

Being indeterminate, niger is prone to seed shattering. The selection for determinate flowering habit will indirectly select for non-shattering types for safe harvesting in this crop.

Resistance to lodging

The soft - weak stem of niger is prone to lodging. Shorter plants with strong, solid stems will induce resistance to lodging and improve fertilizer responsiveness.

High harvest index

The harvest index in niger is very poor and needs major improvement.

The other important researchable issues are the development of bold seeded/capitulum types, early maturing types to fit into various cropping sequences, fertilizer responsiveness and enhanced allelopathic ability. Niger crop helps to reduce weed infestation in subsequent crops. Enhanced allelopathic ability in niger to suppress weeds will be of great utility to the farmers to solve the weed problem and to reduce the cost of cultivation. Fortunately, insects and diseases are not usually serious pests in either major producing country. Thus, presently these appear to be little need to breed for insect or disease resistance.

7. BREEDING METHODS

Genetic improvement of niger has been limited, mainly due to its self-incompatible nature, meager efforts, limited research support and being a minor oilseed for poor people. So far, selection of plants from local collections has been the usual method of developing improved varieties. Following methods are used for niger improvement.

Mass selection

In mass selection, individual plants with desirable traits are chosen and bulked together to grow the next generation without progeny test. It is based on phenotype of maternal plant and there is no control over pollination. The purpose is to increase frequency of superior genotypes in a random mating population. The success depends upon the accuracy with which the phenotype reflects the genotype. In niger, this technique has been successfully used for developing early and short plant type varieties.

Progeny selection

In this method, progenies are grown in individual plots to determine breeding behaviour of selected plants to distinguish genetic variability from environmental variability. Phenotypically similar and superior progenies are bulked to get next generation.

Population improvement

In cross-pollinated crops like niger, the population is heterogeneous and the plants are heterozygous. The aim of the breeder is to increase the frequency of favourable alleles in a population. For a breeder, the genotype of the gametes that the plants contribute to the next generation is more important than the genotype of the individual plants. Various methods of population improvement like recurrent, reciprocal recurrent selection etc. can be successfully applied in niger.

Half-sib recurrent selection

This procedure, is very similar to Pustovoit's "Method of Reserves" used in sunflower which involves selection of several hundred single plants from a random mating isolation plot (Fick, 1978). Part of the seed from each selected plant is grown in a two-row plot and part is kept in reserve. Progenies are evaluated for yield, oil content, maturity and seed shattering relative to the check. The best 25 to 30 lines are identified, and equal amount of the reserve seed of these selections are recombined into another random mating bulk, which is grown, next year in isolation. Progenies with improved performance are retained for evaluation as potential varieties. This system requires no selfing and the general combining ability of the selections is assessed in the progeny row trials. Random mating of the best selections is assured by the self-incompatible nature of the crop.

Disruptive selection

The divergence between Indian and Ethiopian material can be usefully exploited through disruptive selection procedure where improved lines or populations from Ethiopia and India are inter-mated. Rapid advance from selection could be expected in such intercrossed progenies.

Development of synthetic/ composite varieties

The identification of self-compatible types and development of inbred lines in Ethiopia, raises the possibility of producing synthetic varieties from inbreds with good combining ability. Advance generations of multiple crosses made among the lines with high combining ability are designated as Synthetic varieties. Composites are made by mixing seed in equal quantities of lines having uniformity in morphological and other important observable traits especially maturity, size, shape and colour of seed and oil content etc.

Cultivar development

In Ethiopia, niger improvement started only around 1980 and one variety "*Sendafa*" was recommended for the central highland areas. In India under All India Coordinated Research Project on Sesame and Niger of the ICAR, development of improved varieties has been underway for many years and a number of varieties have been recommended for cultivation in different niger growing areas of the country. The salient features of recommended varieties are given in Table 2.

Table 2. State-wise recommended varieties and their characteristics

State	Recommended Varieties	Yield (kg/ha)	Days to maturity	Oil (%)	Salient features
Madhya Pradesh	Ootacamund	500	110	43	Thermo and photo-periods sensitive.
	N-5	450	95-100	40	Thermo and photo-periods sensitive, small, black and sickle shaped seeds.
	IGP-76	400	120-125	39	Tall, black seeded, early maturing.
	JNC-6	550	95	40	Bold seeded, large capitula profusely branched.
Maharashtra	IGP-76*	475	105	40	Sensitive to thermo and photo periods, early maturing.
Bihar	Birsa Niger-1*	600-700	95-100	41	Light pinkish stem at maturity.
Orissa	GA-5 (Bhawani)	400	120-125	40	See under M.P.
	GA-10 (Shiva)	450	115-120	42	Tall, deep black coloured seeded.
Karnataka	Shrilekha	500	87	43	High yield, black bold seed and high oil.
	RCR-317	500	90	40	Black bold seeded, early maturing, high yielding.
Andhra Pradesh	No.71	475	95	42	See under Karnataka.
Rajasthan	IGP-76	475	105	40	See under Maharashtra.
	RCR-317	500	90	40	Bold seeded, early maturing.
West Bengal	GA-5 (Bhawani)	400	120-125	40	Tall, black seeded. See under Orissa.
	GA-10 (Shiva)	450	115-120	42	Tall deep black colored seeded. See under Orissa.
Tamil Nadu	Paiyur-1	500	90	40	Black seeded, early maturing.
North Eastern Hill Region	IGP-76	475	105	40	See under Maharashtra.
Gujarat	G N-1	580	120	41	Tall, profuse branching, high yielding.

8. SEED PRODUCTION

The morphological characters and their expressions used for verification of genetic purity are listed in Table 3.

Nucleus seed

1. Select 1000 representative plants of parental variety and harvest them separately.
2. Examine seed colour, shape and size etc. and reject plants which are not true to the type.

3. Make two parts of seed each of the selected plants. One part for making nucleus bulk and other for evaluation.
4. Grow single plant progenies (SPP) in 20 blocks. Each block with 50 progenies (one row each of 4 m length) along with 5 row of parental variety for comparison with SPP from seeding to maturity stage. Adopt full package of practices.
5. Compare the characters of SPP with parental variety critically for important varietal characteristics.
6. Remove all the off-types as and when detected and reject the SPPs showing off-types.
7. The single plant progenies, which meet all the standards mentioned in the descriptor (true to type), are harvested separately. The remnant seed of single plant progenies yielding higher than general mean + 2 SE is bulked to get nucleus seed.
8. This nucleus seed is further multiplied and used for production of breeder seed.
9. Since in cross-pollinated crops like niger, a breeder has to selected a large number of plants so that the gene pool in properly represented, the above procedure becomes very laborious and difficult to practise. For all practical purposes the breeders generally select a large number of true to the type plants from the central part of the breeder seed production plot and bulk the seed of the selected plants to use as nucleus seed.

Table 3 Morphological characters and their expressions for verification of varieties

	Characters	Variations
1	Flower size	Very small / Small /Medium / Large / Very large.
2	Branching habit	Erect / Horizontal / Nearly horizontal / Dropping
3	Flower colour	Pale yellow / Yellow / Orange
4	Leaf colour	Pale green / Light green / Green / Dark green.
5	Leaf width	Very narrow/Narrow/Medium/Broad/Very broad
6	Leaf margin	Entire / Serrate / Dentate.
7	Stem colour	Green / Purplish green / Purple
8	Capitulum size	Very small / Small/ Medium / Large / Very large.
9	Lodging tendency	Absent / Present
10	Synchrony of maturity heads	Synchronous / Asynchronous
11	Seed shattering	Absent / Present
12	Seed colour	Brown /dark brown/ black

Breeder seed

The important safety measures to maintain the genetic purity during breeder seed production are : selection of field with no preceding crop of niger, proper isolation distance (400m), rigorous rouging in the seed plot, seed crop management, protective irrigation monitoring of seed plot by a team of experts, weed free plot and control of mechanical mixture.

REFERENCES

Abebe, M., Bingham, F.T., and Yermanos, D.M. 1978. The eco-physiology of noog (*Guizotia abyssinica* Cass). III. Growth, yield and mineral composition under saline conditions. Af. J. Agric. Sci. 5(2) 49-59.

Baagoe, J. 1974. The genus *Guizotia* (Compositae). A taxonomic revision. Bot. Tidsskrift. 69: 1-39.

Chavan, V. M. 1961. Niger and sunflower, Indian Central Oilseeds Committee, Hyderabad.

Dagne, K.1944a. Meiosis in inter specific hybrids and genome interrelationship in *Guizotia* Cass.(Compositae) Heriditas 121:119-129.

Fick, G. N. 1978. Breeding and genetics, pp. 279-338. In J.F. Carter (ed), Sunflower Science and Technology. No.19. Agronomy Series. Am. Soc. of Agron. Crop Sci. Soc. Madison, Wis.

Getinet, A. and Sharma S.M. 1996. Niger (*Guizotia abyssinica* (L.f.) Cass. Promoting the conservation and use of underutilized and neglected crops.5. IPGRI, Rome.p60.

Harlan, J . R. 1975. Crops and Man. Am. Soc. Of Agron. Crop Sci. Soc. Madison. Wis.

Hiramath, S.C. and Murthy, H.N. 1992. Cytological studies in *Guizotia* (Asteraceae). Caryologia 45:69-82.

Murthy, H.N. 1990. Cytological studies in *Guizotia bidentoides* Oliver and Hiern [*Bidens pinnatipartita* (Hoffm.) wild]. Cytologia 55:27-29.

Murthy, H. N., Hiramath, S.C. and Pyati, A. N. 1995. Genome classification in *Guizotia* (Asteraceae). Cytologia 60:67-73.

Naik, S. S. and Panda, B. S. 1968. Time of bud pollination in increasing fertility in self-incompatible niger (*Guizotia abyssinica* Cass). Indian J. Sci. and Ind. 2: 177-180.

Patel, O. P., Singh, C. B. Mishra, R.K. and Gour, V. K. (1983). Karyological studies in *Guizotia abyssinica* Cass. Cytologia 48: 221-230.

Prinz, K.D.1976. Untersuchangen zur "Okophysiologic won Niger-saat (*Guizotia abyssinica*)" thiopishcer and indischer. Herkunft. Ph.D. Thesis, Agr. Wiss. F.B. Univ. of Gottingen. 204 pp.

Purseglove, J.W. 1979. Tropical Crops. Dicoryledons. Longman. London.

Ramachandran T. K. and Madhava Menon, P. 1979. Pollination mechanism and inbreeding depression in niger (*Guizotia abyssinica* Cass). Madras Agric. J. 66:449-454.

Richharia, R.H. and Kamalkar, W.J. 1968. Somatic and meiotic chromosomes in *Guizotia abyssinica* Cass. Cytologia 9:249-253.

Seegler, C. J. P. 1983. Oil Plants in Ethiopia. Their taxonomy and agricultural significance. Center for Agricultural Publication and Documentation, PUDOC, Wageningen.

Sharma, S.M. 1990b. New potential areas of niger in India. Proceeding of three meetings held at Pantnager and Hyderabad, India, Jan. 4-7,1989.

Shrivastava, P.S. and Shamwanshi, K.P.S. 1974. Investigation on the extent of cross pollination and selfing and crossing technique in niger (*Guizotia abyssinica* Cass). JNKVV Res. J. 8: 110-112.

Vaughan, J. G. 1970. The structure and utilization of oilseeds. Chapman and Hall, London.

CHAPTER - 14

SAFFLOWER

D.M. Hegde, Vrijendra Singh and N. Nimbkar

1. INTRODUCTION

Safflower (*Carthamus tinctorius* L.), commonly known as *Kusum* (Hindi), *Kardi* (Marathi), *Kusube* (Kannada), *Kusuma* (Telugu), *Jafran* (Assamese), *Kusumbo* (Gujarathi), *Hubulkhurtum* (Kashmiri), *Kasumba* (Punjabi), *Pavari* (Sindhi) and *Sendurakam* (Tamil), is an important *rabi* oilseed crop of the country. For centuries, it has been under cultivation in India either for its orange red dye (carthamin) extracted from its brilliantly coloured florets and/or for its much-valued oil. Apart from its superior adaptability to scanty moisture conditions, safflower produces oil rich in poly-unsaturated fatty acids (linoleic acid 78%) which play an important role in reducing the blood cholesterol level.

2. TAXONOMY, ORIGIN AND CHROMOSOME NUMBER

Safflower belongs to division - Phanerogam, sub-division - Angiosperm, class - Dicotyledon, sub-class - Gamopetalae, series - Inferae, order - Asterales, family - Compositae and genus - *Carthamus*.

The genus *Carthamus* has 25 species with 10, 11, 12, 22 and 32 pairs of chromosomes. Only *C. tinctorius* is cultivated and all other species are wild and weedy in nature. The following is the list of species of safflower with their chromosome number:

Chromosome number [2n]	Number of Species	Species
20	4	*tenuis, alexandrinus, glaucus / anatolicus, syriacus*
22	1	*divaricatus*
2n=24	9	*tinctorius (cultivated species), palaestinus, nitidus, arborescens, caerulens, oxyacantha, flavescens / persicus, gypsicolus, curdicus*
2n=44	1	*lanatus*
2n=64	2	*baeticus, turkestanicus*
not reported	8	*leucocaulos, rhiphaeus, dentatus, ambigus, bossier, rechingeri, ruber, sartori*

Modern assessments, based on occurrence of closely related wild species, consider the probable origin of cultivated safflower in an area bounded by the eastern Mediterranean and Persian Gulf (Knowles, 1969). It encompasses southern parts of former USSR, Western Iran, Iraq, Syria, Southern Turkey, Jordan and Israel where closely related wild species are found. At an early stage in its evolution, safflower spread to Egypt, Ethiopia, Southern Europe, South Asia and Far East where distinctive types have evolved.

3. CLIMATIC REQUIREMENT AREA AND DISTRIBUTION

In India, safflower is cultivated between 14 and 22° N and 73° and 79° E during *rabi* season (September/October to March/April). It is normally grown below 900 to 1000 M. Safflower is considered to be day-neutral but varieties may show adaptation to a specific photo-period when grown at the limits of their normal range. Safflower is thermosensitive and does not favour extremes of either heat or cold. Tolerance to low temperature depends on variety and stage of development. Seedling stage is most tolerant of low temperatures while stem elongation, flowering and post-flowering stages are very sensitive. Safflower can tolerate temperature as low as -12 to -10°C at seedling stage. High temperature during early growth reduces the period spent in the rosette stage. The transition can be rapid when there is no cool period after emergence. A temperature of 15°C is highly favourable for rapid and even germination. A mean temperature of 20-21°C favour optimum vegetative growth. At flowering and day temperature ranging from 24-32°C is congenial for obtaining higher yields.

A rainfall of 650-1000 mm is considered optimum for obtaining maximum yields, depending on the soil type. In areas where there are no hot dry winds, and with adequate pre-planting soil moisture, 300 mm of rainfall will produce excellent yields if this falls mainly prior to flowering.

Excessive humidity during any stage of growth increases damage from fungal diseases. Persistent periods of high humidity when heads are mature encourage seed to germinate in the head, affecting yield and quality.

Traditionally cultivation of safflower has been in a band from the Mediterranean to the Pacific Ocean at latitudes between 20°S and 40°N, wherever a hot, dry climate suits the crop. Safflower is cultivated on a commercial scale mostly in India, USA, Mexico, Ethiopia, Australia, China, Argentina, Russia, etc., and to a limited extent in Pakistan, Italy, Spain, Portugal, Iran. During 1998, safflower area and production in the world were 1.174 million hectares and 0.947 million tonnes, respectively.

In India, safflower is mostly cultivated in Maharashtra, Karnataka and Andhra Pradesh and to a limited extent in Madhya Pradesh, Orissa, Bihar, Gujarat, etc. During 1998-1999, safflower area and production in India were 5.29 lakh hectares and 3.24 lakh tonnes, respectively. Maharashtra and Karnataka account for 61 and 35% of area and 64 and 33% of production of safflower in the country.

4. SALIENT FEATURES OF CROP PRODUCTION

Safflower prefers moderate to highly fertile, deep, moisture retentive and well-drained soils with neutral reaction. It requires clod free seedbed with a firm sub-soil and adequate moisture for good germination and plant stand establishment. In mono-cropped black soils of *rabi* areas, harrowing 3 to 4 times during monsoon is as effective as deep ploughing or subsoiling to keep fields weed free. Optimum time of planting safflower ranges from second fortnight of September to first fortnight of October. One must take advantage of first soaking rains during this period for planting. Under irrigated conditions, sowing date could be extended by 2 to 3 weeks beyond the suggested optimum period. Plant safflower with a spacing of 45 cm x 20 cm. A seed rate of 7 to 8 kg/ha is required. Place the seed at a depth of about 5 cm in the moist zone to facilitate good germination and emergence. Thin excess seedlings within 10-15 days after emergence and maintain the desired plant to plant spacing.

A fertilizer dose of 40-50 kg N, 20-30 kg P_2O_5 and 20 kg K_2O/ha is recommended under rainfed conditions. Under irrigated conditions, 60-80 kg N, 40-50 kg P_2O_5 and 30 kg K_2O/ha is recommended. It is advisable to decide P and K application based on soil test values. Under rainfed conditions, apply the entire quantity of fertilizer as basal dose by drilling. Under irrigated condition, apply 50% of N and 100% of P and K fertilizers at the time of planting and top dress the remaining 50% N after 5 weeks at the time of first irrigation.

Give one or two hand weedings and hoeings/harrowings at 25 to 30 and 45 to 50 days after planting depending on the length of rosette period and the severity of weed infestation. In black soil areas, to delay cracking, close them superficially with dust mulch as and when they appear and thereby minimise moisture losses. One additional interculture or hoeing using hoes/harrows/ sweeps in December before the crop canopy is closed and the spines become problematic is also advantageous.

Under scanty moisture conditions in dry-lands, yield can be boosted by 40 to 60% by providing one life saving irrigation at critical phase of crop growth (early stem elongation or flowering) or before soil moisture becomes limiting for crop growth. The crop is, however, very sensitive to excessive moisture.

Safflower is affected by a number of insect pests and diseases although only a few of them are of economic importance. Timely and appropriate measures are to be taken to manage them to minimize yield losses.

The crop is ready for harvest when the leaves and most of bracteoles except a few on last formed flower heads become dry and brown. Harvest the crop preferably in the early hours when shattering would be minimum and spines relatively soft. Cut plants with the help of sickles and stack them in the field in the form of small and well-pressed heaps until they are fully dried. Thresh either by beating with sticks or with the help of bullock drawn stone rollers or tractor and the resulting material is winnowed to clear seeds. The threshing and cleaning operations can be made with power operated threshers used for other crops such as wheat. Combine harvesters used in wheat could also be used for harvesting and threshing of safflower.

5. BOTANICAL DESCRIPTION

Safflower is a highly branched, herbaceous thistle like annual varying in height from 30 to 150 cm, generally with yellow flowers. The plant has a well-defined and frequently fleshy tap-root and normally produces numerous thin horizontal laterals. The tap root commonly penetrates upto 1.5 to 3 m depth. The stem is stiff, cylindrical, fairly thick at the base, becoming thinner as height increases, quite smooth, glabrous with light grey to green to white with fine longitudinal grooves and becoming brittle when mature. The central stem branches from 15-20 cm to form secondary stems, which themselves branch, each branch terminating in a flower head. The angle branches form to the stem varies from 30 to 70□, is a varietal characteristic which can be considerably affected by environment.

Leaf size and shape varies considerably between variety and on individual plants, from 2.5 to 5 cm broad and 10 to 15 cm long. They are usually large and deeply serrated on the lower stem, short, stiff, ovate to obovate at the inflorescence, when they become superimposed as the involucral bracts.

The inflorescence consists of numerous florets collected closely together on a circular, somewhat flattened receptacle. The number of florets varies with the variety, which is also influenced by environment, in the range of 20-200. The receptacle is surrounded by several layers of involucral bracts, outer ring being heavily spined. The individual flowers are themselves also provided with bracts in the form of small hairs. The involucre is conical with a small apical opening through which the corolla tubes of the flowers protrude. Flower colour can vary from a whitish yellow to red orange, with deep yellow the most common.

The capitula are borne at the end of stems or branches that is on the main branch blooms before terminating secondary branches. Flower heads can vary in diameter from 1.25 to 4 cm and the number per plant from 5 to 60 or more.

The fruit, achene, resembles a small, slightly rectangular, sunflower seed, but with a thicker, more fibrous hull. Pappus is normally absent, but may occur in some seeds and these are normally from the centre of the flower head. The testa is generally cream or off-white, but grey and mottled types occur.

Seed size is a varietal characteristic, is variable, and is affected by climatic conditions. The seeds are 4-9 mm long, and about half as broad with 100 seed weight of 2.5 to 8 g. Depending on the cultivar and head position, cultivated safflower produces 20 to over 100 developed seeds in each head. The proportion of hull can vary from 30 to 64%.

Stages of Development

Seeds after planting may take 6 or more days to germinate and emerge depending on the soil temperature. Emergence is followed by a slow-growing rosette stage, during which numerous leaves are produced near ground level, strong tap roots develop and begin to penetrate deep in to the soil, but no long stems form. The rosette period lasts from 20 to 50 days or more depending on cultivars and environment. Subsequently stem elongates quickly and branches extensively. Flowering normally begins in the primary capitulum, then the secondary capitula and so forth. Within a capitulum, flowering begins in the outer circle of florets and progresses centripetally towards the centre of the capitulum over several days, up to a week. The total bloom stage may last for 4 weeks or more, greatly influenced by growing environment. Time to flowering in a variety is genetically controlled, but can be greatly influenced by environment. Seed mass increases rapidly during the first 15 days after flowering. Seeds mature in 4-5 weeks from flowering.

Reproductive and Floral Biology

The flowers are borne terminally on the main stem, primary branches, secondary branches and tertiary branches (if present). In botanical sense, these flowers consist of a group of small individual flowers called florets. All these florets are collectively known as inflorescence.

Capitulum: The peduncle is expanded, flattened forming a receptacle. The inflorescence is of racemose type called as capitulum or anthopodium or simply as a head, which is characteristic of the family Compositae. The whole capitulum is surrounded by number of overlapping, green, leaf like serrated structures called bracts. Collectively this structure is known as involucre. These bracts serve as a protection to the florets developing inside. Since they are green also, carry on photosynthesis and supply food material to the developing seed. On the upper surface of the receptacle, sessile florets are present. The opening sequence of the florets is centripetal. Individual floret arises in the axil of densely, bristly receptacle. Unlike other members of this family, ray florets are absent. The disc florets situated on receptacle are small and inconspicious. They are bisexual, making head homogamous and carry on reproduction. In addition to the perfect florets, a few sterile florets may be seen in the outer whorls. Such sterile (staminate) florets were observed to the extent of 21 to 49%. The floral structure of staminate floret is similar to that of perfect floret with the exception of empty ovary present in staminate floret.

Disc floret: Each disc floret rises in the axil of scaly bracts, filaments structure encircling the ovary. It is epigyrnous. The calyx is reduced to a low ring or rudimentary. Five petals unite and form a tube, termed as tubular corolla. Since this claw or unguis structure is united forming a tube, the condition is gamopetalous or sympetalous. The upper extended portion of the petal is the limb, which remains free. It is oblong having five clift and basally dialated with acute apex. Corolla is regular in its symmetry and superior i.e., epigyrous condition. The corolla limbs are variously coloured as yellow, orange, red or white depending on the variety. Their main function is to attract insects and bees, when fully opened and to protect the anthers and stigma when they are in bud stage.

The andorcium consists of five stamens i.e., peatandrous. The anthers are united with each other forming an anther lobe but their filaments are separate, a condition known as syngenerious and superior in position. Each anther is vertically grooved, side faces the gynoecium and this condition is known as introse. They are basally attached called basifixed by the filaments. Each of the filaments shows cohesion i.e., union with the corolla tube internally exhibiting epipetalous nature. They are short and hairy in the middle. Mature anthers protrude out to form the corolla. The pollens are round, yellow, sticky and exine.

Gynoecium in the beginning is free but later coalesce and form a style which is bicarpellacy and syncarpous. A carpel has three distinct parts *viz.*, ovary, style and stigma. Ovary is inferior, unilocular with anatrapous or reversed type, single and basal ovule. The ovary is narrow at the base. From the top of the ovary starts single, free, hairy, long slender style that passes through the column of syngenerious stamens. The top portion of the carpel is stigma, which is long, bifid, and broader at the tip and covered with unicellular hairs, which shows a receptive surface where the pollens alight and germinate. The stigma portion of the carpel is markedly separated from style. When style elongates, corolla limbs open and stigma protrude through the corolla tube and anther lobe and becomes ready for pollination. Stigma remains receptive for 72 hours.

Flowering: All branches are terminated by individual flowering heads composed of 20 to 100 florets. Flowering begins in the head that terminates the main axis of the plant, proceeding basipetally, the upper branches flowering first and the lower branches at last. The secondary and tertiary branches continue the process in regular order. In an inflorescence, florets at the periphery of the capitula open first and flowering proceeds centripetally, the whole process taking about 3 to 5 days. The total flowering period of a whole plant varies from 3 to 4 weeks and usually takes about 25 days. Deokar *et al.*, (1974) observed the variety N-62-8 grown on vertisols at Jalgaon and reported that flowering of primary branches completed within 10 to 12 days, secondary branches in 24 days and tertiary branches within 30 to 33 days. They observed that flowering in individual plant was completed within about 40 days.

Pollination mechanism: The disc florets usually begin to open in the morning and the process goes till mid-day. A few, however, open in the afternoon. They remain open for one or two days before fading. In the bud stage, the style elongates and stigma usually emerges from the anther tubes before the pollen is liberated. The continued elongation of the style through the corolla tube pushes the brush like stigma through the anther tube until all the stigmatic surface of the pistil has grown well beyond the tip of the anthers. By this time this process of elongation is completed, the stigma is usually well covered with the floret's own pollen (Claassen, 1950).

An examination of the florets at this period shows that numerous pollen adhering all around the hairy surface of the stigma. In some cases, anthesis takes place before extension of the stigma thus rendering self-pollination possible. Besides this, it will be evident that the slightest movement of the capitula by wind or by the operation of the bees might easily carry pollen from one flower to its neighbour and geitonogamy is thus easily possible.

Bees visit the capitula in large numbers in the morning both for honey and pollen and pay particular attention to the newly opened florets in which anthesis has already taken place. They clear away the pollen adhering to the style and to accomplish this they have to strike it in the direction of the stigmatic surface. In this way, a few pollen grains are also carried to the stigma and pollination takes place. By this method of transfer of pollen from the anther to stigma, both self and cross-pollination are found to take place as the bees work daily on all the types of plants with open flowers. Therefore, both self and cross-pollination take place in safflower. A consideration of all the facts points to the probability that self-pollination is much more common than cross-pollination. Deokar and Patil (1976) observed about 99% self-pollination and 1% cross-pollination in Nagpur-7 variety. Claassen (1950) reported that natural crossing under Nebraska condition varies from 0 to 100% and that most plants were found to be outcrossed from 5 to 40%. In California, *Apis mellifera* was found the main agent while in India, *Apis dorsata* was found efficient forager.

6. SELFING AND CROSSING PROCEDURE

Although, the safflower is regarded as a self-pollinated crop, natural cross-pollination occurs to some extent. In order to prevent natural crossing from insects and wind to maintain purity selfing of the plants is necessary. For selfing, bags prepared from fine muslin cloth, butter paper and newspaper can be used. There are two methods of crossing in safflower which are described below:

1. Hand emasculation: Chavan (1961) described this process of crossing in safflower. Healthy capitula of the parental plants to be emasculated are chosen and these capitula are bagged before any florets are opened and labelled on their peduncle. A day before the anthers are likely to burst, the head is cleared by removing upper envelope of the involucral bracts in order to expose the young florets. The marginal whorl of the floret, which is likely to open on one day, are saved on the receptacle and all other younger florets are then off with the help of pointed forceps. The anther tube surrounding the stigma and style is carefully punctured at the bottom and a long narrow cut is given vertically, the entire anther column along with the surrounding corolla is removed. During the operation, it is likely that some pollen grains may adhere on the styler region, which are immediately brushed off and washed by a jet of water. A convenient number of florets are emasculated, washed and examined with a lens for any sticking pollen and then the capitulum is bagged. These emasculated florets are hand pollinated with the desired pollen parent next morning, when fresh pollen is available and the head is again bagged. Emasculation operation may be done in the morning or in the evening according to the convenience of the operator. Bags made of newspaper are quite suitable. This method of artificial crossing is commonly used in Maharashtra.

Knowles (1958) described a simplified technique which is more quick and without the use of rinse. In this method, the floret is squeezed with the sharp tweezers just below the attachment of the anthers. If the corolla is gently bent, it will break at the point where it was squeezed but the style will remain intact. Then the corolla and anther tube together slipped off gently and slowly from the style. While a number of styles do break in this operation, it does minimize the amount of escaped pollen. This method of artificial crossing is being used at various safflower research stations in the country at present and is found most effective in producing crossed seeds.

This hand emasculation method is highly time consuming. It also causes injury to the florets because of which seed setting is poor.

2. Mass pollination: Karve *et al.*, (1977) developed new technique of artificial crossing for production of crossed seeds. In this method, the parental plants are planted at such a time that the flowering coincides with the cooler part of the year. The potential female plants are pruned just prior to flowering, whereby only five to six upper most capitula are left on the plants. These capitula are individually covered by low-density polythene bags of an adequate size (10 x 5 cm), after cutting off the free laminae of the involucral bracts. In case of open capitulum, the anther dehiscence coincides with the elongation of style, so that the stigmas are covered with the pollen of the same floret when anthesis takes place. But in the case of capitula covered by polythene bags, the increased humidity inside the bag prevents the anthers from dehiscing so that when anthesis occurs, the stigmas protrude out of the florets, without any pollen shedding. These stigmas then can be pollinated with the pollen of any desirable pollen donor. The polythene bag is to be removed only for pollination, after which it is replaced. This process is to be repeated for 3 to 4 consecutive days until all the florets have been pollinated. The bag can be removed and the capitulum can be exposed only after all the florets of that capitulum have started to wilt. The only disadvantage in this method is that it can be used only in the cold season.

7. BREEDING OBJECTIVES

The major breeding objectives in safflower improvement are : (1) To increase seed yield productivity, (2) To improve oil content in seed, (3) To develop varieties resistant to insect pests and diseases and (4) To develop high yielding and early maturing spiny and spineless varieties and hybrids.

8. BREEDING METHODS FOR CROP IMPROVEMENT

Safflower is usually a self-pollinated crop, however, the stigma extrusion through the fused anther column before its dehiscence and pollen liberation has put it in the category of often cross-pollinated crops. In view of often cross-pollinated nature of safflower, the procedure followed for crop improvement are discussed below

Introduction and pure line selection: Introduction is one of the most ancient and simplest methods of crop improvement. However, introduction of varieties in a new environment only occasionally results in direct development of varieties. Usually, introductions need a few cycles of selection before they can be released as varieties, as the plants of an introduced variety behave differently in the changed environment. Therefore, acclimatization of the introduced variety in the new environment is essential before subjecting it to selection for getting promising types for further evaluation and release.

Selection is also the oldest and extensively practiced method of crop improvement as far as safflower improvement is concerned. As a result, 15 out of the 23 varieties developed and released so far for commercial production in the country have been developed by carrying out selection in the local cultivars. The safflower varieties developed by this method are N-630, Nagpur-7, N-62-8, A-300, Manjira, S-144, JSF-1, K-1, CO-1, Type-65, APRR-3, Bhima, HUS-305, Sharda and JSI-7. Many of the safflower germplasm lines available as sources of various desirable traits have resulted due to pure line selections practiced in the local cultivars. Safflower germplasm consists of a wide range of diversity for different traits of economic importance; However, no systematic efforts have been made to exploit the genetic diversity existing in safflower germplasm for crop improvement due to it being a rainfed crop of minor importance. Hence, a vast pool of genetic diversity for different traits, which exists in safflower, is lying untapped. The existence of a wide range of variability for different traits in safflower is the reason that most safflower varieties in India have been developed by following pure line selection and even today, this method is considered to be the most effective method of breeding new varieties in safflower.

Hybridization: The main aim of hybridization is to combine the desirable attributes of two or more varieties into one. Apart from generating variation for different traits in F_2 and subsequent generations, hybridization has proved to be of immense use in knowing the genetic architecture of different traits, which in turn has helped in devising proper methodologies to improve the specific traits and consequently the crop.

The choice of the parents for hybridization is one of the important factors contributing to success or failure of the crop improvement programme. Therefore, the selection of the parents should be based upon their performance in multi-locational, multi-year yield evaluation trials. In this context, Joshi

(1979) suggested important guidelines for selecting the parents for hybridization in a self-pollinated crop, the details of which are as follows —

1. Selection of the parents on the basis of their *per se* performance in respect of yield and other desirable economic traits.
2. Consideration of magnitude of expression of yield-contributing characters.
3. Consideration of genetic diversity of the parents, as it would provide an opportunity to bring desirable genes of diverse origin together.
4. Application of diallel cross approach to enable a breeder to identify the best general combining parents as well as the best cross combination out of all the cross combinations possible in a set of parental lines.

Evaluation of F_1 : For selecting the promising cross combinations for generation advancement, it is essential to evaluate the F_1 along with both its parents and check cultivars. It is better to evaluate F_1s in a replicated trial if sufficient seed is available. If seed is a constraint, F_1s can be grown along with both the parents and checks each in a single row side by side to assess the performance of the F_1 as compared to parents and checks in respect of yield and other desirable traits including reaction to pests and diseases. It is desirable to self the selected plants of promising hybrids for generation advancement as a high degree of cross pollination to the extent of 22% as mentioned above is reported in safflower (Patil *et al.* 1987).

Evaluation of F_2s : Only promising F_1 cross combinations, selected most preferably after evaluation in a replicated trial should be advanced for evaluation to F_2 generation. Each F_2 cross combination may be grown with a fairly good population size, so that it can accommodate individual plants representing all gene combinations possible in a cross to provide an opportunity to select the transgressive segregants. It is desirable to include the concerned F_1 and a check cultivar for comparing the mean of F_2 s with them as it would give an indication of extent of inbreeding depression in the F_2. The lesser the inbreeding depression the better it is as it suggests the existence of good amount of additive x additive components of variance, which is likely to be fixable through subsequent inbreeding. The selection of the promising individual plants in the desirable crosses may be practiced in the F_2 itself and the selected plants harvested and threshed separately for plant to progeny screening. For making individual plant selections in F_2, the studies carried out by several workers in safflower as summarized below may be considered as guidelines for fruitful outcome of the breeding programme. The studies on gene action in safflower revealed the importance of non-additive gene action for seed yield and the number of capitula per plant (Ramachandram and Goud, 1981, 1982b; Ranga Rao, 1983; Narkhede *et al.* 1992; Patil *et al.*, 1992) and that of additive gene action for oil content, seed weight and seed number. Importance of both additive and non-additive gene actions for these traits has been reported by Prakash and Prakash (1993). The association studies indicated that seed yield and oil content are the most important traits of which seed yield is more complex and highly affected by genetic-environment interaction in safflower (Ranga Rao and Ramachandram,

1979). It has also been reported that direct selection for yield is ineffective (Nie *et al.*, 1988). The correlation and path coefficient analysis studies in safflower carried out by several workers showed that number of branches per plant, seed weight per capitulum and 100 seed weight contributed the most to the seed yield either directly or indirectly (Nie *et al.*, 1988; Patil *et al.*, 1990, Nie *et al.*, 1993). Number of capitula/plant and seed yield/plant has been reported as the direct contributors to oil yield per plant (Prakash and Prakash, 1993). Seed weight per capitulum, which was a function of seed weight and number of seed was the direct component of seed yield. (Ramachandram and Goud, 1982a, Ramachandram, 1985; Ghongade *et al.* 1993; Nie *et al.*, 1993). Thus several studies suggested the importance of total number of capitula per plant, yield per capitulum, and seed weight for seed yield improvement in safflower. It is desirable to make individual plant selections at maturity of the crop as all the traits mentioned above can be taken into considerations while selecting the individual plants. The high oil containing individual plants may be selected by following the thumb-nail technique. In this method, the dry seeds from the main capitulum of the plant are taken out and pressed hard by using thumb-nail pressure, the entries having high oil are generally have thin hull and are thus pressed easily. On the contrary, seed with thick hull is not pressed at all by thumbnail pressure. Thus, a rough selection for high oil types is possible in the field itself.

The segregating material in F_2 and in subsequent generations, depending upon the trait under improvement is handled by one of the methods described below:

Pedigree method: In this method, individual plants having desirable attributes of the trait, which needs to be improved are selected in F_2 population. A very large number of such plants comprising of about 5 to 10% of the total population of a cross are selected and harvested and threshed separately to grow plant to progeny rows in F_3 generation. It is recommended to grow F_3 progenies in a replicated trial for identification of the promising progenies. These selected progenies are further advanced to F_4, F_5 and F_6 generation in subsequent years. Both inter-and intra-progeny selections of the promising types are practised in all these generations. It is desirable to self the selected plants at every stage of selection process, as this helps in making the progenies almost homozygous by the time they reach the F_6 generation. The preliminary yield trials of fairly uniform and homozygous progenies may be taken up at this stage and most outstanding ones of these may be further subjected to individual plant selections. These can be further evaluated in replicated trials to select the best of them for multi-locational testing before release. The pedigree method of breeding has been used most widely in improving seed yield, oil content and other desirable traits in safflower. The usefulness of this method in safflower improvement in India and abroad is evident from the summary of research achievements in safflower as furnished in Tables 1 and 2.

Bulk method: In the bulk method, the F_2 and subsequent generations are harvested in bulk to grow the next generation. This practice of generation advancement is followed up to F_6 generation. In F_7 generation the promising

Table 1. Summary of improved safflower varieties in India

S.N.	Variety	Breeding Method	Releasing center	Year of release	Seed yield (q/ha)	Days to maturity	Oil content (%)	Area of adaption	Remarks	Present status
1	N-630	Selection from local	Niphad	1942	10-12	130-135	30	Nagpur and Nasik	20% increase in yield over local	Not cultivated
2	Nagpur-7	Selection from local	Nagpur	1953	12-15	135-140	29-30	Vidarbha region	-	Not cultivated
3	A-300	Selection from local	Annigeri	1957	8-10	115-130	31	Karnataka and Bihar	Susceptible to rust	Cultivated
4	N-62-8	Selection from local	Niphad	1959	12-14	130-135	30-31	Marathwada	17% increase in yield over N-630, Not cultivated	
5	A-1	Hybridization (A-482-1 X A-300)	Annigeri	1969	10	115-130	29.5	Karnataka, Bihar, Orrisa & Rajasthan	Suitable under scanty as well as assured moisture regions	Cultivated
6	K-1	Selection from N-852	Kovilpatti	1969	8-10	100-120	30-31	Tamil Nadu	-	Cultivated
7	Tara	Hybridization (N-62-8 x *C. palaestinus*)	Jalgaon	1976	13-15	120-125	32-33	Western Maharashtra	18% increase in seed & oil yield over N-62-8	Denotified
8	Manjira	Pure line selection from SF-65	Hyderabad	1976	8-10	115-130	31	Andhra Pradesh	Medium-statured (75-80 cm)	Cultivated
9	S-144	Pure line selection from Local	Raichur	1976	8-10	115-130	32	Karnataka	Suitable for dry tracts of Raichur, Bijapur, Bellary & Gulbarga dist.	Cultivated
10	Type-65	Local selection	Kanpur	1977	12-14	180-190	32-33	Uttar Pradesh	Semi-spreading, bushy & tall	Cultivated
11	JSF-1	Selection	Indore	1979	10	180-190	30	Madhya Pradesh & Rajasthan	Suitable for early & late sowings	Cultivated
12	CO-1	Selection from PI 250528	Coimbatore	1979	10-12	100-125	32-33	Tamil Nadu	Non-spiny	Cultivated
13	Bhima	Selection from A-300	Solapur	1982	17-20	130-135	30-31	Maharashtra	34% increase in yield over Tara	Cultivated
14	APRR-3	Selection from EC 27250	Hyderabad	1985	8-2	120-130	31	Andhra Pradesh	Rust resistant	Cultivated

S.N.	Variety	Breeding Method	Releasing center	Year of release	Seed yield (q/ha)	Days to maturity	Oil content (%)	Area of adaption	Remarks	Present status
15	Nira	Hybridization (NS 1572 X EC 32012)	Phattan	1986	12-16R 20-24I	125-130	30-31	Western Maharashtra	Specially released for irrigated conditions	Cultivated
16	HUS-305	Selection from germplasm	Varanasi	1986	14-16	160-170	35-36	Uttar Pradesh	High oil, salt & wilt tolerant	Cultivated
17	Girna	Hybridization (A-1 X G 1254)	Jalgaon	1990	16-20	125-130	30-31	Maharashtra	Moderately wilt tolerant	Cultivated
18	Sharda	Selection from No. 168	Badnapur	1990	16-20	125-130	30-31	Marathwada	—	Cultivated
19	JSI-7	Selection from JSF 1909	Indore	1990	12-14	175-185	30-31	Madhya Pradesh	Non-spiny	Cultivated
20	DSH-129	MS 9 (0) X A-1	Hyderabad	1997	20-25	125-130	29	All safflower growing areas	20-30% higher seed yield than the A-1	Cultivated
21	A-2	Selection from A-1	Annigeri	1998	10-12	115-125	34	Karnataka	High oil	Cultivated
22	JSI-73	Hybridization (JSI-42 X JSI-7)	Indore	1998	10-15	145-165	30-33	Madhya Pradesh	Non-spiny	Cultivated
23	NARI-6	Hybridization (CO-1 X JL-8)	Phaltan	2000	10-15	117-137	35	All rainfed safflower growing areas	Non-spiny and high oil	Cultivated

R = Rainfed, I = Irrigated

Table 2. Details of safflower varieties developed in USA and other countries

Sl. No.	Variety and Ref. No.	Breeding method adopted and pedigree	Releasing Institution	Year of release	Seed yield (Kg/ha)	Maturity	Oil content (%)	Area of adoption	Specific feature of the variety
1	Nebraska-10 (N-10) [44]	Selection from N-852	Nebraska (USA)	1950	–	Early	–	California & Nebraska, USA	High seed yield and high oil content
2	US-10 [41]	Back cross (Nebraska-10 X Western oilseeds-14) X Nebraska-10)	USDA, California (USA)	1959	–	–	–	California, USA	resistant to root rot caused by *P. drechslri*
3	UC-1 [16]	Back cross (57-147 X N10) X US-10) x US-10	California (USA)	1966	–	–	–	California, USA	high oleic acid (78.3%)
4	Leed [45]	Pedigree U-1421 X Gila	Utah and USDA (USA)	1968	–	–	–	California, USA	Resistant to rust and root rot
5	Oleic Leed [43]	Backcross (UC-1 X Leed) X Leed	USDA, California (USA)	1976	–	–	–	California, USA	High oleic content
6	Sidwill [6]	Pedigree (F_3 composite of 12 F_2 plants) 87-42-3 X AC-1	Montana & North Dacota (USA)	1977	1754	–	32.8	Eastern Montana & Eastern North Dakota, USA	Resistant to Alternaria and bacterial blight
7	Hartman [2]	Pedigree 87-42-3 X AC-1	Montana, Sidney (USA)	1980	1801	Late	41.1	Montana, North & South Dakota, USA	Resistant to Alternaria leaf blight and pseudo-monas bacterial blight
8	Rehbein [1]	Pedigree 87-42-3 X AC-1	Montana, Sidney (USA)	1980	1743	Early	39.1	Montana, North & South Dacota, USA	Resistant to Pseudo-monas bacterial blight
9	Oker [3]	Pedigree 87-42-3 X AC-1	Montana, Sidney (USA)	1984	1755	Early	45	Eastern Montana, Western North, Dakota, USA	Resistant to Alternaria leaf spots and bacterial blight
10	Saffire [17]	Selection from bulk population of S-65-219	Lethbridge, Alberta, Canada	1985	–	Early	31.9	Alberta, Canada	Resistant to head rot
11	Girard [4]	Pedigree Cargill 1653 X Sidney selection 87-42-3 X Sidney selection, 88-74-2 X Boswell 671 X S-541	Montana, Sidney (USA)	1986	1167 (R) 3033 (I)	–	42.2 (R) 42.4 (i)	Montana, North Dakota & South Dakota, USA	Resistant to Alternaria leaf spot, Pseudomonas bacterial blight and Sclerotinia head rot

Sl. No.	Variety and Ref. No.	Breeding method adopted and pedigree	Releasing Institution	Year of release	Seed yield (Kg/ha)	Maturity	Oil content (%)	Area of adoption	Specific feature of the variety
12	Finch [5]	Pedigree (Rio X Royal) X Sidney selection 87-42-3	Montana, North Dakota, USDA - ARS (U.S.A.)	1986	1164 (R) 2811 (i)	–	41.3 (R) 41.2 (I)	Montana & Western North Dakota, USA	Resistant to Alternaria leaf spot and Pseudo-monas bacterial blight
13	Sironaria [9]	Backcross followed by multiple crossing (PI 311-737 X Gila) X Gila X CPI 69039	Griffith, Australia	1986	–	–	–	Australia	Resistant to Alternaria leaf spot and Phytophthora root rot
14	Sirothora [10]	Double cross followed by back crossing (PI 286-385 X Gila) X (PI 237-538 X Gila) X Gila	Griffith, Australia	1986	–	–	–	Australia	High seed yield, tolerant to Phytophthora root rot
15	Tomejil [7]	Backcross followed by single seed descent method (SH-202 X Oleic Leed) X SH-202	Cordoba, Spain		2786	–	43.8	Spain	High linoleic acid
16	Rancho [7]	-do-	-do-	-do-	2719	–	43.46	Spain	High linoleic acid
17	Merced [7]	-do-	-do-	-do-	2583	–	43.60	Spain	High linoleic acid
18	Alameda [7]	-do-	-do-	-do-	2455	–	45.3	Spain	High oleic acid
19	Rinconada [7]	-do-	-do-	-do-	2618	–	46.76	Spain	High oleic acid
20	Sahuaripa 88 [20]	Pedigree S1 CEN 1230 X Gila	C.d. Obregon, Mexico	1989	2810	Interme-diate	38.5	North west Mexico	Wide adaptability
21	Quiriego 88 [19]	Pedigree CM-106-S1 (I) X PCOY X Gila	Sonora, Mexico	1989	3020	Interme-diate	38.1	Sonoro Mexico	High seed yield
22	San Jose 89 [21]	Pedigree S1-CEN 1368 X S1-CEN 1178 X PO1-5	C.d. Obregon, Mexico	1990	2934	Interme-diate	37.6	Sonoro Mexico	High seed yield
23	AC Stirling [18]	Modified pedigree TH-5/3 X Gila X Mexican dwarf X RH-4	Lethbridge, Alberta, Canada	1991	2360-2730		35.0	Saskatchewan, Manitoba, Alberta, Canada	Suitable for both oil and bird feed.

individual plants possessing desirable traits are selected, harvested and threshed separately. The individual plant progenies are evaluated in a replicated trial with the standard check to identify the most desirable ones for multilocation evaluation. The advantage of bulk method is the selection of the promising genotypes naturally as the poor genotypes are eliminated due to their reduced frequency in each cycle of generation advancement. Finally, promising homozygous genotypes are retained at the end of F_7 generation. Another advantage of bulk method is that the breeder can handle several bulk populations of different crosses together which is not possible in any other breeding method. It is advantageous to self each bulk population of safflower during flowering. Otherwise the high rate of cross pollination in safflower would render the bulk population method ineffective due to the presence of a large number of heterozygous plants at the end of F_7 generation.

Single seed descent method : In this method, the advancement of each plant in F_2 and subsequent generations is carried out by a single seed selected randomly from each plant till the plants attain homozygocity. Generally, it is practiced till F_5 and F_6 generations. In F_7, large number of individual plants are selected to raise individual plant progenies. The promising progenies out of them are evaluated for yield and other desirable attributes in a replicated trial.

Back-cross method **:** Back cross method of breeding is used when a trait controlled by oligogenes is to be transferred from a genotype into a widely adapted variety. For transferring a trait from donor parent to the recurrent parent which is a widely adapted variety, a series of back-crosses are made between the hybrid and the recurrent parent. In each cycle of back-crossing, back-cross progenies possessing requisite characters are crossed with the recurrent parent. After 6 to 7 back-crosses, the resultant progeny will be homozygous for the alleles under transfer and for the alleles of all other traits. Selfing of the selected plants of last back-cross generation having desired traits will produce progenies homozygous for the gene under transfer and would be identical to the widely adapted variety (recurrent parent).

The back cross method of breeding has been used in safflower to develop phytophthora resistant variety US-10 (Thomas, 1964). This method has also been successfully used to develop high oleic cultivars UC-1 (Knowles, 1968) and Oleic Leed (Urie, *et al.*, 1979).

Mutation breeding: Mutation breeding is another tool to create variations in a crop. The handling of irradiated material is similar to the pedigree method. Irradiated material is space planted in M_1 generation and individual plants are harvested separately at maturity. In M_2, seeds of individual plants are raised separately and the mutants having desired traits are harvested separately. In M_3, selected individual plants are planted separately for selection of promising plants having desired attributes. The M_4 generation grown from promising M_3 plants is subjected to screening for identification of promising homozygous lines. Each line is harvested in bulk for evaluation along with the checks for seed yield and other desirable traits in F_s generation. The promising ones out of these are selected further for evaluation in multilocation trials to select the most promising one for release.

Sahu and Kumar (1977) studied the effect of chemical mutagen EMS on different characteristics of safflower and observed significant variation for number of effective capitula and seed yield/plant, but low variation for oil content and iodine value of oil.

Ramachandram (1984) indicated the importance of mutation breeding in achieving simultaneous improvement in seed yield and oil content in safflower as the mutagenesis shifts the mean of the trait under improvement in desired direction.

Heterosis breeding : Commercial exploitation of heterosis has led to breakthroughs in productivity of several crops. Reports of the existence of high heterosis for seed yield and other component traits in safflower has attracted several workers to find the ways and means to produce hybrid seed commercially in safflower (Urie and Zimmer, 1970, Karve *et al.*, 1979).

The identification of genetic male sterility sources in safflower followed by development of agronomically desirable genetic male sterile lines in the country has resulted in development and release of safflower hybrids DSH-129 and MKH-11 during 1997. These hybrids gave an average increase of 22% in seed yield and an increase of 30% in oil yield over the national check A-1.

The details of the male sterility systems being exploited for heterosis breeding in safflower have been discussed separately under "head hybrid development in safflower" in this chapter.

9. INTERNATIONAL AND NATIONAL ORGANIZATIONS INVOLVED IN SAFFLOWER IMPROVEMENT

Safflower research is carried out in India, USA, Mexico, Ethiopia, Australia, China, Argentina, Russia, Canada, etc. However, the major research work is carried out only in India and USA. In India, safflower research is carried out at Directorate of Oilseeds Research, Hyderabad and in different Agricultural Universities, *viz.*, UAS, Dharwad; MAU, Parbhani; MPAU, Rahuri, PDKV, Akola; JNKVV, Jabalpur; ANGRAU, Hyderabad; Osmania University, Hyderabad and in private organizations like, NARI, Phaltan; Nath Seeds, Aurangabad and Mahyco, Jalna.

10. REQUIREMENTS FOR SEED PRODUCTION (OF OPEN-POLLINATED VARIETIES)

Seed is one of the most important and the cheapest input of a crop. To obtain optimum production from the crop of any variety it is essential to use genetically pure seed. Therefore, it is utmost important to maintain the purity of the seed, otherwise the superiority of the variety is likely to be lost.

The often cross-pollinated nature of safflower and existence of high cros-pollination through honey bees warrants its consideration as a cross-pollinated crop for quality seed production. To avoid any chance of genetic contamination during seed production process, it is necessary to follow the field standards and cultural practices The field selected for safflower seed production should not have had safflower as the previous crop. The field should be well drained with medium deep to deep soil. The Isolation requirement of safflower for production of breeder and foundation seed is at least 400 m from other safflower fields. However, in case of certified seed, the minimum requirement of isolation is about 200 m from the neighbouring safflower fields. For roguing of off-types, a minimum of three visits to seed plots are necessary. The first visit should be carried out before flowering to rogue out the plants having different leaf shape, branch angle, spine index and colour of the plant etc. The second visit is required to be carried out during flowering of the crop to remove the plants with distinctly different flower colour, capitulum size, bract shape and size and spine index. It is essential to have the third visit during maturity to rogue out plants with different capitulum shape and size and maturity period etc.

11. MAJOR ACHIEVEMENTS IN THE CROP THROUGH BREEDING IN INDIA AND ABROAD

The major achievements in safflower improvement through breeding have resulted due to development of varieties and hybrids giving higher seed and oil yield, in addition to possessing greater resistance to diseases and pests and better adaptability to abiotic stress factors. The details of these achievements in relation to the breeding approaches adopted to carry out desired improvement in the specific traits of the cultivars of safflower both in India and abroad are given below :

In India : The major achievements in safflower improvement in India are furnished in Table 1.

Abroad : The achievements made by following different breeding strategies in safflower improvement in different countries are summarised in Table 2.

12. DEVELOPMENT OF HYBRID

The often cross-pollinated nature of the crop, existence of high heterosis for yield and other traits of economic importance and availability of genetic and cytoplasmic-genetic male sterility systems make safflower a prime candidate for hybrid development. The development of promising genetic male sterile lines in safflower has resulted in the development and release of safflower hybrids DSH-129 by DOR, Hyderabad and MKH-11 by Mahyco, Jalna for commercial cultivation in the country during 1997. The details of the male sterility systems being exploited for hybrid development in safflower are furnished below

Male sterility systems : In safflower, both cytoplasmic-genetic male sterility (CGMS) and genetic male sterility (GMS) systems are being exploited for hybrid development. However, the most prevalent and the only system which has been harnessed for development of safflower hybrids in India is

GMS system. The GMS systems being exploited for hybrid development in safflower in the country are of both monogenic recessive and dominant nature. The details of which are as follows —

1 *Single recessive genic male sterility:* The GMS sources governed by single recessive genes being exploited for hybrid development are :

(a) UC-148 and UC-149 GMS lines developed by Heaton and Knowles (1980).

(b) GMS male sterile lines developed by Ramachandram and Sujatha (1991).

(c) MSN and MSV male sterile lines developed by Singh (1996).

(d) DMS male sterile lines associated with dwarfness developed by Singh (1997).

Each of these male sterility sources segregates in the ratio of 1 male sterile (MS) : 1 male fertile (MF) plants. The earliest identification of MS and MF plants in all these GMS sources except the DMS, is possible at the time of flowering by observing pinched capitulum opening of male sterile plants and normal opening in case of male fertile plants. However, due to linkage of sterility and dwarfing genes in DMS lines, the identification of sterile and fertile plants is possible at 30-40 days after sowing. At the age of 30-40 days, male sterile plants remain dwarfs of 5-10 cm height. However, MF plants in contrast attain a normal height of 20-25 cm. The height difference between the two types facilitates the roguing of MF plants at this stage itself leaving behind 100% pure plant stand of dwarf MS plants.

2. *Dominant genic male sterility :* In safflower, the only source of male sterility governed by a single dominant gene is reported by Joshi *et al.* (1983). Owing to dominant nature of the male sterility causing gene, the MS lines as well as hybrids based on this source segregate in the ratio of 1 MS : 1 MF plants. The identification of MS and MF plants in this type too is possible only at flowering stage. The success of hybrids from this source depends upon the optimum honey bee activity in the crop at flowering so that 100% seed setting can be obtained in the 50% MS plants appearing in the hybrid population.

3 *Cytoplasmic-genetic male sterility :* The exploitation of CGMS for hybrid development in safflower has been carried out in the United States of America (Hill, 1989). However, a lack of suitable restorer for the sterile cytoplasm seems to be a major hurdle in commercial scale exploitation of cytoplasmic-genetic male sterility for hybrid development in safflower, as all the hybrids based on this system which were evaluated during 1998-99 at this Institute as well as at other locations in India exhibited a lack of fertility restoration in them (Singh *et al.,* 2000). Therefore, it appears that the CGMS system which is available in United States is facing some basic problems which need to be tackled before it can be exploited commercially in safflower.

In India also efforts are being made to develop CGMS system in safflower and the workers at Nimbkar Agricultural Research Institute, Phaltan have been able to identify a male sterile cytoplasm in F_3 population of an interspecific cross in safflower. Efforts are being made to identify a suitable restorer and a maintainer of the sterile cytoplasm.

As stated earlier, the hybrids released for commercial cultivation in safflower are based on a GMS system. Therefore, only this system is being discussed in relation to different aspects of seed production of safflower hybrids.

13. METHOD OF SEED PRODUCTION IN SAFFLOWER HYBRID

The different aspects of seed production of safflower hybrid are discussed below :

1. *Seed production of genetic male sterile (GMS) line or female parent* : The first requirement of seed production plot of GMS line of safflower is the isolation distance of at least one km from the neighbouring safflower plots.

Recommended agronomic practices are followed to raise a good crop of genetic male sterile line. The GMS line in safflower as mentioned earlier segregates into 50% MS and 50% MF plants. However, the identification of MS and MF plants in GMS lines of safflower is only possible at flowering, from the pinched brush like capitulum opening in MS plants and normal capitulum opening in case of fertile plants. Besides, no pollen production is observed in sterile plants whereas, abundant pollen is produced in the fertile plants. The difference in opening of the two types diminishes as the crop attains maturity, hence the marking of MS plants is necessary and is done by bagging one of the capitula of each MS plant at the time of flowering so that seed from MS plants (bagged ones) only can be harvested at maturity of the crop. Daily visit to the seed plot is required during flowering to mark the MS plants which flower on a particular day as well as to rogue out the off-types from the field. Only confirmed male sterile plants are bagged to maintain purity of seed. Any bagging of fertile plants by mistake would not only affect the purity of the seed but also the segregation ratio of sterile to fertile plants in subsequent generations. Optimum honey bee activity in the seed plot during flowering of the crop is desirable to obtain maximum seed production. Therefore, the spraying of insecticides is limited up to flower initiation only, as the later sprays may affect the honey bee activity in the field which in turn would adversely affect the pollination in MS plants and consequently result in realization of low seed yield. The harvesting of MS plants is carried out first in order to avoid any physical mixing of this seed with seed from fertile plants.

2. *Hybrid seed production :* Similar to the seed production of GMS line, the fast and foremost requirement of hybrid seed production plot in safflower is the isolation distance of at least 1 km from other safflower plots. Generally a row proportion having six rows of GMS line and two rows of male parent is followed in a hybrid seed production plot. Standard agronomic practices are followed to raise a good crop. As stated earlier, GMS line (female parent) segregates in 1 : 1 ratio of MS and MF plants and the MS plants are identified during flowering of the crop from the pinched brush type of opening and the presence of white to pale yellow anthers due to lack of pollen grains in them.

On the other hand the MF plants exhibit normal, wide flower opening and yellow anthers full of pollen grains. In hybrid seed production, the roguing of MF plants appearing in the GMS line is started about 5 to 6 days prior to the flower initiation in the crop by forced opening of the main capitulum of each plant with the help of a pair of scissors to check the anthers for the presence or absence of the pollen grains in it. The plants having flowers with anthers full of pollen grains are removed immediately and plants with flowers lacking pollen grains in the anthers are retained in the field. By following this technique, the roguing of all the MF plants in the female line is completed before the start of flowering in the crop. Any off-type appearing in either female or male parent rows is removed immediately. A daily visit to the seed production plot for 4-5 days during initial flowering is necessary to rogue out any leftover MF plants in the female line to ensure the purity of the hybrid seed. The harvesting and threshing of MS plants are carried out first to prevent any mechanical mixing of seed from pollinator parent. Every care is to be taken to maintain the purity of the hybrid seed while handling the material.

REFERENCES

Bergman, J. W., D. E. Baldridge, P. L. Brown, A. L. Dubbs, G. D. Kushnak and N. R. Riveland. 1987. Registration of Rehbein safflower. Crop Sci., 27 (3) : 610.

Bergman, J. W., D. E. Baldridge, P. L. Brown, A. L. Dubbs, G. D. Kushnak and N. R. Riveland. 1987. Registration of Hartman safflower. Crop Sci., 27 (5) : 1090-1091.

Bergman, J. W., G. Carlson, G. Kushnak, N. R. Riveland and G. Stallknecht. 1985. Registration of Oker safflower. Crop Sci., 25 (6) : 1127-1128.

Bergman, J. W., G. Carlson, G. Kushnak, N. R. Riveland, G. Stallknecht, L. E. Welty and D. Wichman. 1989. Registration of Girard safflower. Crop Sci., 29 (3) : 828-829.

Bergman, J. W., G. Carlson, G. Kushnak, N. R. Riveland, G. Stallknecht, L. E. Welty and D. Wichman. 1989. Registration of Finch safflower. Crop Sci., 29 (3) : 829.

Bergman, J. W. and N. R. Riveland. 1983. Registration of Sidwill safflower. Crop Sci., 23 (5) : 1012-1013.

Chavan, V.M. 1961. Niger and safflower - Central Oilseed Committee, Hyderabad.

Claassen, C. 1950. Natural and controlled crossing in safflower, Carthamus tinctorius L. Agron. J., 42:381-384.

Deokar, A.B. et al. 1974. Pattern of growth, flowering and yield attributes in safflower. Maharashtra Vidnyan Mandir Patrika, 9:98-103.

Deokar, A.B. and F.B. Patil. 1976. Vicinism in safflower. J. Maharashtra Agric. Univ., 1:231-234.

Fernandez-Martinez, J. and J. Dominguez-Gimenez. 1986. Release of five new safflower varieties. Sesame and Safflower Newsletter, 2 : 89-90.

Ghongade, R. A., B. P. Joshi and P. A. Navale. 1993. Correlations and path analysis of some yield components in safflower. J. Maharashtra agric. Univ., 18 (2) : 240-243.

Harrigan, E. K. S., 1987. Safflower registration of CV. Sironaria. Sesame and

Safflower Newsl., 3 : 47-49.

Harrigan, E.K.S., 1987. Safflower registration of CV. Sirothora. Sesame and Safflower Newsl., 3 : 49-50.Heaton, T. C. and P. F. Knowles. 1980. Registration of UC-148 and UC-149 male-sterile safflower germplasm. Crop Sci., 20 (4) : 554.

Hill, A. B. 1989. Hybrid safflower breeding, In : Proceedings : Second International Safflower Conference, Hyderabad, India, 9-13 January, 1989. Indian Society of Oilseeds Research, DOR, Hyderabad, pp. 169-170.

Joshi, A. B. 1979. Breeding methodology for autogamous crops. Indian J. Genet. 39 (3): 567-577.

Joshi, B. M., Y. S. Nerkar and N. D. Jambhale. 1983. Induced male sterility in safflower. J. Maharashtra agric. Univ., 8 (2) : 194-196.

Karve, A. D., A. K. Deshmukh and D. V. Nagvekar. 1979. Hybrid Safflower In : Abstracts of technical papers : International Congress on Oilseeds and Oils P. 13-14.

Karve, A.D., B.P. Singh and D.V. Nagvekar. 1977. New method for manual cross pollination in safflower. Oilseeds J., 7(1):22.

Knowles, P.F. 1958. Safflower. Adv. Agron., 10:294-323.

Knowles, P. F. 1968. Registration of UC-1 safflower. Crop Sci., 8 (5) : 641.

Knowles, P.F. 1969. Centres of plant diversity and conservation of crop germplasm : Safflower. Econ. Bot., 23(4):324-329.

Mundel, H-H., H. C. Huang, L. D. Burch and F. Kiehn. 1987. Registration of Saffire safflower. Crop Sci., 27 (2) : 364-365.

Mundel, H-H., R. J. Morrison, H. C. Huang and F. Kiehn. 1993. Registration of AC Stirling safflower. Crop Sci., 33 (1) : 201.

Musa, G.L.C. and V. S. Munoz. 1990. Registration of Quiriego 88 safflower. Crop Sci., 30 (4) : 961- 962.

Musa, G.L.C., S. Munoz-Valenzuela and R. D. Garcia-Perez. 1993. Registration of San Jose 89 safflower. Crop Sci., 33 (2) : 356.

Narkhede, B. N., A. M. Patil and A. B. Deokar, 1992. Gene action of some characters in safflower. J. Maharashtra agric. Univ., 17 (1) : 4-6.

Nie, Z., F. F. Chen and X. C. Shi. 1988. A study on selection indices of seed yield in safflower. Oil Crops of China, 1 : 15-19.

Nie, Z., F. F. Chen and X. C. Shi. 1993. Path analysis of characters related to seed yield in safflower. Oil Crops of China, 3 : 26-29.

Patil, B. R., S. G. Deshmukh and M. P. Deshmukh. 1990. Studies on correlation and path analysis in safflower. Annals of Plant Physiology, 4 (1) : 86-91.

Patil, P. S., A. B. Deokar and B. K. Katule. 1987. Breeding. In : P. S. Patil (ed). Safflower. All India Coordinated Research Project on Oilseeds (Safflower). Mahatma Phule Agricultural University. Zonal Research Station, Solapur. pp. 55.

Patil, P. S., A. M. Patil and A. B. Deokar. 1992. Line x tester analysis for combining ability in safflower. J. Maharashtra agric. Univ., 17 (1) : 64-66.

Prakash, K. S. and B. G. Prakash, 1993. Yield structure analysis of oil yield in safflower. Oleagineeux, 48 (2) : 83-89.

Ramachandram, M. 1984. Genetic improvement of seed yield and oil content in safflower. Paper presented at safflower worker's group meeting held at Sukhadia Univ., Durgapur, 6-10 Aug, 1984.

Ramachandram, M. 1985. Genetic improvement of oil yield in safflower. Problems and prospects. Oilseeds J., 2 : 1-9.

Ramachandram, M. and J. V. Goud. 1981. Genetic analysis of seed yield and oil content and their components in safflower. Theor. Appl. Genet., 60 : 191-195.

Ramachandram, M. and J. V. Goud. 1982a. Components of seed yield in safflower (*Carthamus tinctorius* L.), Genetica Agraria, 36 (3/4) : 211-221.

Ramachandram, M. and J. V. Goud. 1982b. Gene action for seed yield and its components in safflower. Indian J. agric. Sci., 42 (2) : 213-220.

Ramachandram, M. and M. Sujatha. 1991. Development of genetic male sterile lines in safflower. Indian J. Genet., 51 (2) : 268-269.

Ranga Rao, V., 1983. Combining ability for yield, percent oil and related components in safflower. Indian J. Genet., 43 (1) : 68-75.

Ranga Rao, V. and M. Ramachandram. 1979. Stability parameters for yield and its components in safflower. Mysore J. agric. Sci., 13 : 297-308.

Sahu, G. R. and H. Kumar. 1977. EMS induced genetic variability in safflower (*Carthamus tinctorius* L.) Malaysian Appl. Biol. 6 (1) : 25-32.

Singh, Vrijendra. 1996. Inheritance of genetic male sterility in safflower. Indian J. Genet., 56 (4) : 490-494

Singh, Vrijendra. 1997. Identification of genetic linkage between male sterility and dwarfness in safflower. Indian J. Genet., 57 (3) : 327-332.

Singh, V., M. B. Deshpande, M. K. Galande, S. R. Deshmukh and N. Nimbkar. 2000. Current status of research and development in safflower hybrid in India. Extended Summaries. National Seminar on Oilseeds and Oils Research and Development : Needs in the Millennium. Feb 2-4, 2000. Indian Society of Oilseeds Research, DOR, Hyderabad, pp. 62.

Thomas, C. A. 1964. Registration of US-10 safflower. Crop Sci., 4 (3) : 446-447.

Urie, A. L. and D. E. Zimmer. 1970. Yield reduction in safflower hybrids caused by female selfs. Crop Sci., 10 (4) : 419-422.

Urie, A. L., W. F. Peterson and P. F. Knowles. 1979. Registration of Oleic Leed safflower. Crop Sci., 19 (5) : 747.

Williams, J. H. 1964. Registration of N-10 safflower. Crop Sci., 4 (3) : 446.

Zimmer, D. E. and A. L. Urie. 1968. Registration of Leed safflower. Crop Sci., 8 : 511.

CHAPTER - 15

SESAMÉ

S.M. Sharma

1. INTRODUCTION

Sesame, *Sésamum indicum* (Syn. *Sesamum orientale* L.) also known as benniseed, gingelly, sim-sim and is variously known as *til* (Hindi, Assamese, Bengali, Marathi), *tal,* (Gujrati), *Tilli* (Punjabi), *nuivvulu* (Telugu), *ellu* (Tamil) and *rasi* (Oriya) in different parts of India.

2. TAXONOMICAL CLASSIFICATION, CHROMOSOME NUMBER AND ORIGIN

Taxonomy: Phylum – Angiospermae; order – Bignoniales; subphylum – Dicotyledons; Division – Lignosae; Family – Pedaliaceae;

The genus *Sesamum*, Pedaliaceae, contains over 30 species, of which only *S. indicum* L., is cultivated. Sometimes, the wild species *S. angustifolium* and *S. radiatum* are also cultivated in Africa. The taxonomy and cytogenetics of this genus require investigations that are more definite. Kobayashi (1991) listed 37 species including the new species such as *S. mulayanum* Nair (Mehetre *et al.,* 1993) and *S. indicum* var. *sencottai* and var. *yanamalami* (Devarathinum and Sundaresan, 1990) added after that. However, many of the wild species, their distribution areas and chromosome numbers are not known. According to Kobayashi (1991), 22 of the species were found only in Africa, 5 only in Asia (India, Sri Lanka, East Indies or in all areas). Seven species were reported from both Africa and Asia and one each was reported in Crete and in Brazil.

Classification below the species level is difficult because the limits of sub-species and cultivars are ill-defined and often controversial. Identification of genotype complexes can be achieved by grouping accessions having similar appearance and performance, but the variability of cultivars creates special problem in classification because there may not be clearly definable discontinuity within a species. The existence of many transitional forms, the fuzzy boundaries of cultivars with intergrading variation and the susceptibility of cultivars to environmental influences further complicate the classification. The major diagnostic characters used in type classification of cultivated sesame are: (i) Phyllotaxis (ii) Nectary number per axis (iii) Carpel number per capsule and (iv) Capsule number per axil (node).

The chromosome numbers are known for about only one third of the species. The analysis of the chromosome number of 12 *Sesamum* species (Kobayashi 1991) showed that there are three groups: 2n= 26 *(S. indicum* and the wild *S. alatum, S capense, S. schenckii, S. malabaricum (*initially classified as *S. orientale ssp. malabaricum);* 2n=32 *(S. prostratum, S. laciniatum, S. angolense, S. angustifolium);* 2n = 64 *(S. radiatum, S. occidentale, S. schinzianum).* The basic chromosome numbers for sesamum are x = 8 and 13. The basic chromosome number in *Ceratotheca* and *Pedalium*, two related genera, is x = 8. The x = 13 probably resulted from ancient polyploidy. The chromosome number, distribution and plant habit of wild, cultivated and partly cultivated species of the genus *Sesamum* are given in Table 1 (Kabayashi, 1981; Thangavelu, 1994).

A variety of cytogenetic, botanical, archaeological, biochemical and historical-linguistic criteria have been used to know the ancestry, origin and evolution of sesame. Two alternative centres viz., East Africa (Ethiopia) or Asia (Indian Subcontinent or further east or central Asia) have been proposed for its origin. The African origin of sesame has been supported by the fact that most wild sesame species are endemic there. A centre of origin in the Indian subcontinent is supported by finding charged sesame seeds, in Harappa (Pakistan) which may be 5000 years old. Bedigian and Harlan (1986) concluded that India was the centre of origin because *S. malabaricum,* which is found only in certain parts of India, is very similar to sesame and is cross-fertile with it. Bedigian *et. al.* (1985) cited in support of this hypothesis that the lignans sesamin and sesamolin occurred in the oil of the cultivated *(S. indicum),* and four wild species viz., *S. alatum, S. malabaricum, S. radiatum* and *S. angustifolum,* of which only *S. malabaricum* found only in certain areas in India, is close to *S. indicum.* It seems that the origin of cultivated sesame is still in doubt. The origin and evolution require clarification with more detailed cytogeneitc studies and suitable DNA comparisons.

Table 1. Distribution and salient features of Sesamum species (Kabayashi, 1981; Thangavelu, 1994)

S. No.	Species	2n	Occurrence	Salient features	Crossability with *S. indicum*
1	*S. indicum*	26	Cultivated, wide-spread from tropical to temperate zone	Erect, simple or branched, leaves variable, corolla compa-nulate-whitish to pink, nectary; sessile, seed coloured to white, smooth, edible	F_1s highly fertile
2	*S. malabaricum*	26	Wild, India, (Malabar coast), Africa	Fertile in crosses with *S. indicum*	Some sterility observed
3	*S. mulayanum*	26	Wild Indian subcontinent	Similar to *S. indicum* in plant morphology, wild, corolla pink, fertile in crosses with *S. indicum*	F_1s highly fertile
4	*S. laciniatum*	32	Wild, India (Kerala, Tamil Nadu), Africa	Prostrate, occurs in rocky habitat, dark pink corolla, simple, serrate leaves	Shrivelled seeds
5	*S. prostratum*	32	Wild, India (Kerala, Tamil Nadu, Mahara-shtra), Africa	Prostrate, occurs in rocky habitat, dark pink corolla, simple, serrate leaves	Viable seeds, F_1s sterile amphidiploid
6	*S. alatum*	26	Wild, Nigeria, Sudan, Mozambique, India	Erect, branched, leaves foliate, corolla pink to carmine, nectary sessile, seed winged, blackish.	Fruit normal. Seeds healthy but non-viable
7	*S. radiatum*	64	Partially cultiva-ted, India (Mah-arashthra), Africa, Upper Gu inea, Sri Lanka	Erect, simple or branched, leaves entire, corolla violet purple, nectar sessile, long capsule, seed blackish, rugose	Shrivelled seeds
8	*S. schencki* (S. grandiflorum)	26	Wild, S.W. Africa, India (Travancore), East Indies	Erect, intermediate, leaves 5-7, partite, corolla purple, seed winged	F_1s with end period fertility, F_2s reported
9	*S. capense*	26	Wild, South Africa	Erect, branched, leaves foliate, corolla violet-purple, nectary sessile, seed winged, blackish	Shrivelled non-viable seeds
10	*S. angolense*	32	Wild, Angola	Erect, simple or branched, leaves entire, corolla violet purple, nectary sessile, long capsule, seed blackish, rugose	
11	*S.angustifolium*	32	Partially culti-vated, Congo, Mozambique, Uganda	Erect, simple or branched, leaves entire, corolla rose, nectary sessile, long capsule	
12	*S. occidentale*	64	Partially culti-vated Africa, Sri Lanka, India		
13	*S. mombazense*	?	Wild, Congo	Erect, branched, leaves, entire, corolla purple, long capsule	

S. No.	Species	2n	Occurrence	Salient features	Crossability with *S. indicum*
14	*S. antirrhinoides*	?	Wild, Angola	Erect, branched, leaves entire, corolla purple, nectary stipulate, seed narrowly winged.	
15	*S. schnizianun*	?	Wild, S.W. Africa (Damaraland) East Indies	Erect, branched, leaves oblong-lanceolate, corolla pale-rose, long capsule, nectary stipulate, seed large	
16	*S. pedalioides*	?	Wild, Angola	Erect, branched, leaves entire sessile, seed narrowly winged	
17	*S. calycinum*	?	Partially cultivated Angola, Mozambique	Erect, branched, leaves lanceolate entire, corolla rose, nectary sessile, small capsule	
18	*S. baumii*	?	Partially cultivated, Angola	Erect, slender, leaves entire, corolla pale rose, nectary sub sessile, seed rugose	
19	*S. heudelotii*	?	Wild, Upper Guinea	Prostrate, leaves lanceolate oblong, fruit unknown	
20	*S. repens*	?	Wild, Lower Guinea	Prostrate, leaves sessile, entire, corolla rose, long capsule, seed flat	
21	*S. marlothii*		Wild, S.W. Africa, East Indies	Erect, lower 3- foliolate, nectary sessile, seed blackish	
22	*S. rigidum*		Wild, Angola	Erect, branched, leaves entire, corolla pale-rose, seed oblique, flat or convex	
23	*S. biapiculatum*	?	Wild, Congo		
24	*S. caillei*	?	Wild, New Guinea		
25	*S. digitaloides*	?	Wild, Africa		
26	*S. denterii*	?	Wild, Africa		
27	*S. latifolium*	?	Wild, East Africa		
28	*S. lepidotum*	?	Wild, Africa		
29	*S. microcarpum*	?	Wild, Africa		
30	*S. sabulosum*	?	Wild, Sudan		
31	*S. somalense*	?	Wild, Somalia		
32	*S. talbotii*	?	Wild, Nigeria		
33	*S. thonnerii*	?	Wild, Africa		
34	*S. triphyllum*	?	Wild, Africa, East Indies		
35	*S. trifoliatum*	?	Wild, India		

3. CLIMATIC REQUIREMENT, AREA AND DISTRIBUTION

The main distribution of the crop is in the counties between 25°N and 25°S latitudes but it is found growing up to 40°N (China, Russia and the USA) and 30°S (Australia) to 35°S (South America). It is normally found below 1250m altitude, however, some varieties are adapted up to 1500m. It is grown in Nepal up to 2000 m. The high altitude types are usually small, quick growing and relatively unbranched and low seed yielders. In the same variety, the oil content is also high at low altitudes.

Environmental factors : Climate and soil are the two main components of the physical environments of the plant. Sesame has great adaptability with regard to rainfall. It will produce an excellent crop with a rainfall of 500-650 mm but down to 300mm (under irrigation) and up to 1000 mm rainfall will already provide a crop under certain conditions, particularly with newer varieties. For maximum yields, the precipitation should be distributed over the period of plant growth as follows: germination to first bud formation 35%; bud formation to main flowering 45%; flowering to maturity not more than 20%. Heavy rain at flowering will drastically reduce yield. Rainfall when plants are ready for harvest reduces yield by increasing susceptibility to diseases and prolonging the period required for drying the capsules. Sesame is extremely susceptible to water-logging, and continuous rains at any time during growth period will greatly increase the incidence of diseases.

Sesame normally requires fairly high temperature during its life cycle. Temperature for optimum growth from seedling emergence to flowering and fruiting ranges between 27-33°C. A temperature of 25 -27°C encourages rapid germination, initial growth and flower formation. If temperature falls below 20°C for a length of time, germination and seedling growth are delayed and these processes are inhibited at temperatures below 10°C. Temperatures above optimum (i.e. 40°C or above) at flowering can seriously affect fertilization and the number of capsules set. However, the highest yields are obtained under irrigation in arid regions.

Since the rate of net total dry matter production per unit of ground area is related to the daily amount of photosynthetically active radiation (PAR) intercepted by the crop, low yields in *Kharif* season (summer rainfall areas) could result from low radiation levels caused by heavy cloud cover resulting in reduced radiation input or from low plant density rendering sub-optimal interception by crop canopy. Reduced solar radiation depresses dry matter production and seed yield.

In the intercropped sesame, yields can be drastically reduced by shading by the companion crops. The stage at which shading occurs has great influence on the level of yield reduction. Recent advances of plant breeding have reduced plant heights of the traditional companion crops e.g., sorghum, millets, pigeonpea etc. putting sesame in a more equitable position in the competition for space and light improving thereby the potential of sesame as an intercrop with such crops.

Sesame is a quantitative short day plant and with a 10-hour day will normally flower in 40-50 days, but many varieties have become locally adapted to various photoperiods. Early cultivars are less sensitive to day length than later types.

Soil requirement: Sesame grows well on wide range of soils from highly sandy soil to the black cotton and clay soil, but it thrives best on well-drained moderately fertile soils of medium texture. Shallow soils with impervious subsoil are not suitable. Soils with neutral reaction are preferred, although good results are obtained on slightly acidic and slightly alkaline soils (pH 5.5 to 8.0) but at higher pH, soil structure becomes of increasing importance. However, many soils on which sesame is grown are saline.

Each of the above mentioned factors are important in the adaptation of sesame to a particular region. Studies with a wide range of genotypes indicate that there is considerable diversity in response to these factors. The world germplasm should be classified on the basis of climatic response, adaptation to soil type, soil reaction, fertility etc.

During the past three decades, there has not been much shift in the area under sesame in the country, which has been around 2 million hectares with minor fluctuations from year to year due to seasonal conditions in individual years. However, there has been much increase in production during the same period due to increase in productivity from about 231 kg / ha in 1970-71 to about 332 kg/ha in 1998-99 from an area of 1673 thousand hectares.

Statewise, Rajasthan, Orissa, Maharashtra, Madhya Pradesh, Gujarat, Uttar Pradesh, Andhra Pradesh, Karnataka, Tamil Nadu and West Bengal are the major producers of sesame. However, sesame crop is grown in all the states of the country.

4. SALIENT FEATURES OF CROP CULTIVATION

Growing seasons and conditions : The sesame crop is grown in different regions of the country is different seasons and also in more than one seasons in the same region. Thus, the crop has different varieties and crop production practices for different regions and growing conditions (seasons). In northern India, the crop is grown as rainfed *Kharif* crop. The crop is also grown as a semi-*Rabi* crop (September-January) in parts of Madhya Pradesh, Maharashtra, Gujarat and Andhra Pradesh. Sesame crop has also gained popularity as a summer crop (February-May) after potato or late paddy in Bengal, Bihar, Orissa etc. However, in southern India, the crop is taken in three seasons (i) *Kharif* / semi *Rabi*, (July / August – November / December) under rainfed condition, (ii) *Rabi*, (November- February) and (iii) Summer irrigated crop (March-June). A large proportion of *Kharif* season crop is taken as a mixed crop with various millets, oilseeds, pulses etc.

Sowing : In order to provide an ideal condition for germination, the soil should be well pulverized and brought to fine tilth by ploughing deep in summer, and with the onset of monsoon 2 or 3 harrowing in crosswise direction are done for *Kharif* season crop. Two or three harrowing are enough for semi-*Rabi* or *Rabi* season crop. The field should be perfectly levelled and kept free from weeds.

Seed rate, spacing and sowing methods : Sesame is usually broadcasted. For easy intercultural operations and higher yields, line sowing is adopted. Spacing between and within the rows depends on the specific variety chosen, its plant type (branching potential, close or open-branching etc.), specific season etc. Commonly, 30-45 cm row to row and 10 - 15 cm within the row, plant to plants distance is kept after thinning, depending on above factors.

A seed rate of 4-7 kg / ha is adequate to get the required plant stand. Wherever drills are used, the seed rate may be reduced to 2.5 to 3.0 kg / ha. In order to facilitate easy seeding and for even distribution of seed, the bulk is increased by mixing the seed with either sand or dry earth or well sieved farm yard manure.

Deep sowing is avoided as sowing deep (more than 3 cm) effects germination and plant stand. For the prevention of seed borne diseases, seed is treated with Thiram 3 g / kg seed.

Manners and Fertilizers: It is essential that the crop is adequately fertilized. It is suggested to incorporate about 10 tonnes of farm yard manure per hectare about a month before sowing, Fertilizer dose may vary according to the variety and available nutrient status of the soil. It is generally desirable to apply half of the nitrogen and full doses of P_2O_5 and K_2O at the time of sowing, and the remaining half quantity of nitrogen after 30–35 days of sowing. The top dressed nitrogen is incorporated into the soil by hoeing with hand hoe or bullock-drawn blade hoe. The recommended doses of fertilizer for the major rainfed sesame growing states are 20-40 kg N, 15-30 kg P_2O_5 and 10-20 kg K_2O per hectare, and for irrigated growing situations are 60-90 kg N, 30 to 40 kg P_2O_5 and 20-30 kg K_2O / ha.

Weed Management : A weed-free field is necessary for maximum production. The crop requires first weeding at 15-20 days after sowing. Along with thinning, one more weeding will be necessary at 30-35 days after sowing. The interculturing should be done with hand-hoe or bullock- drawn blade-hoe. Pre-sowing soil incorporation of Fluchloralin (1.0 kg a.i./ha) or Alachlor (1.5 to 2.0 kg a.i./ha) as pre-emergence weedicide were found effective to control the weeds.

Crop protection: Sesame crop is affected by a number of insect pests viz., leaf roller/capsule borer (*Antigastra catalaunals* Dup), Gall fly (*Asphondilia sesami Folt*), Bihar hairy caterpillar (*Spilosoma abliqua* W.) etc. However, leaf roller/capsule borer is the key pest.

The important diseases are: seedling blight *(Phytophthora parasitica var. sesami)*, bacterial blight *(Xanthomonas campestirs pv. sesami)*, Alternaria leaf spot (Al*ternaria sesami), Stem and root rot (Rhizoctonia bataticola, Macrophomina phaseolina),* phyllody (mycoplasma), bacterial leaf spot (*Pseudomonas syringae pv. sesami), powdery mildew (Oidium sp, sphaerotheca sp.), Cercospora* leaf spot *(Cercospora sesami)* etc.

Effective management measures to control these insect pests and diseases are known and recommended.

Harvesting and threshing: The crop is preferably harvested at physiological maturity without waiting for drying, to avoid seed losses due to capsule dehiscence. Usually, the crop is threshed by gentle beating of well-dried plants with sticks.

5. BOTANICAL DESCRIPTION

Plant description

Root : Usually tap and branched

Leaves: Lower leaves opposite, lobed with serrated margins and

	upper mostly alternate, narrow, oblong, simple, exstipulate, muclagenous, glandular; hairs present on the lower surface, upper suface glabrous.
Inflorescence :	Cymose
Flower:	Pedicellate, hermophrodite, zygomorphic, complete, hypogynous, ebracteate
Calyx:	5 sepals, gamosepalous, sepals lanceolate, acute, hairy; aestivation imbricate or valvate.
Corolla :	5 petals, gamopetalous, 2- lipped, corolla 5-lobed, the upper lip consists of 2 equal sized loves while the lower lip includes 3 petals, out of which central one is the largest.
Androecium:	4 stamens, didynamous, epipetalous, instead of 5th stamen usually a staminode is present. Anthers 2-celled (dithecous), sagitate, introrse and dehisce by longitudinal slit. They are 1 to 2 mm long, white to yellow or purplish in colour with yellowish pollen grains.
Gynaecium:	Two carpels (bicarpellary) or four carpels, syncarpous; ovary superior, placentation axile, style filiform, stigmas 2 and lobed.
Fruit:	Loculicidal capsule, about 2.5 cm long, erect, hispid, beak short, valves separating half way down, valved.
Seeds:	Small and many, smooth, albuminous with straight embryo.

Floral biology

Sesame is an indeterminate plant, that is, it continues to grow and produce flowers and capsules as long as conditions permit. As a result, later in the growing season, the plants have fully mature capsules, immature capsules, flowers and buds at any one time. The plants have acropetal flowering and the flowers are borne on short pedicels in the leaf axils. A single dominant allele (T-) determines if there is one flower / axil or 2 or more flowers / axil (tt). When the flowers are solitary, the two lateral primordia produce nectaries, when there are 2 or more flowers (usually 3), the laterals bloom a few days later. The flowers vary in size, colour and markings. The 5 stamens (4 functional and one sterile) are inserted at the base of the corolla, which is tubular and campunulate.

Sesame is considered a self–pollinated crop. However, if insect activity is high, out-crossing can reach high levels. Yermonas (1989) reported that when sesame plots were surrounded by another flowering crop, cross-pollination was very low (under 1%), whereas in an isolated plot in a dry area with minimal other vegetation, out crossing reached 68%. Varying rates of cross-pollination were reported from various countries depending on insect activity. Therefore, in critical studies it is essential to isolate or bag the plants to ascertain self-pollination. Weiss (1971) considered an isolation distance of 180-360m as sufficient. Sesame is classified as a short-day plant. However, plants with varying photoperiod requirements for flowering have developed due to long term selection in regions with varying day lengths. In areas where sesame is grown in 2 or 3 seasons during the year, cultivars with different photoperiod responses have been developed. There are significant interactions of temperature and day length with flowering rate. Genotype and environment variations have been found for many characters including anther, pollen grain and pistil. These differences could lead to differential transmission of alleles by male and female gametes.

6. SELFING AND CROSSING TECHNIQUES

Selfing is achieved by covering the plant with brown paper bag before opening of flowers. All open flowers or fertilized flowers are removed. The bags are closed tightly to prevent insects from entering. Selfed seeds can also be obtained by growing the plant in greenhouse or by enclosing them in large, insect proof cage-like structures.

For crossing, emasculation is easily performed by pulling off the corolla from the buds on the day before opening. The gamopetalous corolla are smoothly slipped out by a gentle pull. As the stamens are epipetalous, they get themselves simultaneously removed along with corolla.

Flowers open at dawn. The pollen grains are shed shortly thereafter, remain viable for about 24 hours. The stigma are usually receptive from hours before to 24 hours after flower opening. Most reports indicated that pollen grains stored at room temperature (20-25°C) remain viable upto 24 hours after flower opening.

Although, the best times for emasculation and pollination are the afternoons and mornings respectively, these processes can be performed at the same time. For pollination, healthy and robust flowers of male parent are collected by smooth slipping of the corolla column. The anther with fresh pollen are exposed by splitting open the tip of the corolla, and brushed on stigma of the emasculated flowers. The technique of emasculation and pollination has been discussed in detail by Yermanos (1980).

Cytogenetics : The basic chromosome numbers for *Sesamum* are x = 8 and 13. The basic chromosome number in *Ceratotheca* and *Pedalium*, two related genera is x = 8. The x = 13 might have resulted from ancient polyploidy. The chromosomes of *S. indicum* are small (0.66 m to 1.80 m), and subterminal centromed (Kobayashi, 1991). Zhan *et al.* (1987) reported that 5 chromosomes were median and 8 submedian, one with a satellite, whereas earlier studies suggested that all the chromosomes had terminal centromeres. Giesma banding facilitating identification of each of the 13 chromosomes of the genome was achieved (Zhan *et al.,* 1990). Interrelationships of the species within and between the chromosome number groups (2n = 26, 32 and 64), and their evolution have been studied by Prabakaran (1996). A summery of his cross-compatibility findings is presented in Table 2.

In the 2n = 26 chromosome species, *S. indicum* and *S. malabaricum* are closely related, their F_1, hybrids have normal meiosis and are fertile. In crosses of *S. malabaricum* as female and *S. indicum* as male, the F_1 hybrids are male sterile (Thangavelu, 1994, Prabakaran *et al. 1995,* Prabakaran, 1996) whereas the reciprocal F_1 hybrids are fully fertile. This indicates some nuclear-cytoplasm interactions and differences between these species. The *S. indicum* and *S. mulayanum* crosses yield partly fetrile F_1 hybrids, while the reciprocal crosses set no seeds (Biswas and Mitro, 1990). In crosses of *S. indicum x S. Schenckii (S. grandiflorum),* the F_1 hybrids are partly fertile and F_2 progenies are obtained indicating that the two species are related (Nayar and Mehra, 1970, kobayashi, 1991). Reciprocal crosses of the cultivated species with *S. capense* (as *S. alatum)* produced capsules but the seeds were inviable. *S. capense* seems more distantly related. Embryo rescue may be needed to produce viable F_1 hybrids between the two species. In the 2n = 32 chromosome group, *S. prostratum* x *S. laciniatum* produced fertle F_1 hybrids and F_2. In reciprocal crosses of *S. laciniatum* and *S. angolense,* a few but inviable seeds are obtained as also in reciprocal crosses of *S. radiatum* with *S. laciniatum.* Embryo rescue may be helplful in both of these combinations. Of the 2 n = 64 chromosome group, fertile reciprocal F_1 inter-specific hybrids were obtained between *S. radiatum* and *S. occidentale* and between *S. radiatum* and *S. schinzianum.* They may in fact, represent one biological species.

Table 2. Cross compatibility (C) and incompatibility (I) between cultivated sesame and wild *Sesamum* species collected in India[Z].

Female Parent	Male Parent						
	S. indicum 2n=26	*S. alatum* 2n=26	S. malabaricum 2n=26	*S. laciniatum* 2n=32	*S. prostratum* 2n=32	*S. occidentale* 2n=64	*S. radiatum* 2n=64
S. indicum	V	I	C	C	C	I	I
S. alatum	C^Y	V	I	I	I	I	I
S. malabaricum	C	I	V	C	C	I	I
S. laciniatum	C.S.	I	C	V	C	I	I
S. prostratum	C.S.	I	C	C	V	I	I
S. occidentale	I	I	I	I	I	V	C
S. radiatum	I	I	I	I	I	C	V

[Z] Modified from Prabakaran 1996; C= viable hybrids, I = no seeds, S = sterile hybrids, V = intraspecific.
[Y] One viable hybrid only.

The relationship across the three chromosome number groups is little known. The introgression of desirable genes for various wild species into cultivated one is of particular interest. Crosses should be attempted between several unrelated cultivars of the cultivated species and divergent accesssion of the wild species for every desired combination. Several wild *Sesamum* species are reported to be resistant to *Antigastra catalaunalis (Shoot webber)* viz., *S. malabaricum S. alatum, S. laciniatum, S. radiatum;* resistant to powdery mildew viz *S. malabaricum, S, alatum, S. occidentale* and resistant to phyllody viz., *S. alatum, S. mulayanum* and *S. prostratum* and to "wilt" in *S. mulayanum* and *S. prostratum.* Resistance to Phytophthora blight. *Fusarium* wilt, leaf blight and seedling blight were reported in *S. alatum* and *S. radiatum* (Thaugavelu, 1994, Lee *et al.,* 1991).

Genetics: Most of the published reports on inheritance dealt with monogenic traits and major genes (Brar and Ahuja 1979). Only a few polygenic control was reported. The mode of genetic control of some characters affecting yield are as follows (dominant italicized):

Branched vs. unbranched, one gene (Nb, nb)

Normal vs. fascinated, single or duplicate genes (F_1 f_1; F_2 f_2)

Single vs. multiple (2-3) flowers and capsules / leaf axil, one gene (T, t)

Fertility vs. Complete male and female sterility, one gene (SC, sc)

Fertility vs. male sterility, one gene (MS, ms)

Bicarpellate vs. quadricarpellate capsules, one gene (TC, tc)

Dehiscent vs. indehiscent capsules, one gene with peiotropic effect (ID, id) or two complementary genes

Normal vs. paper shell capsule, two complementary genes (P_1 p_1; P_2 p_2) Capsule length, polygenic

Photoperiod response, multigenic

Only one study on linkage (made by Dhillon in 1974) was reported by Brar and Ahuja (1979). In plants that are *id id* the cotyledons and leaves are cupped, apparently either a pleiotropic effect or a close linkage. Seed colour is controlled by several loci, one of which is epistatic. The homozygous recessive *(xx)* prevents pigment formation resulting in white seeds.

7. BREEDING OBJECTIVES

The breeding objectives in sesame include higher yields, improved plants architecture, appropriate length of growing season, and resistance to diseases and pests. The other objectives vary with the level of technology. Sesame is grown with low or no inputs in developing countries, while in developed countries such as USA and South Korea, it is grown with reasonable or high inputs. Sesame fits into cropping systems either as a major crop or a second crop, and also grown as a mixed/intercrop with peanuts, pigeon peas, sunflower, millets etc. Non-shattering types are needed for combine harvesting, but shattering ones are preferred when traditional systems prevail and manual labour is readily available.

Breeding objectives also differ with marketing and quality aspects. Seed size, shape and coat structure and colour are not important for the confectionery market. Higher oil content is an objective but potential genetic increase would be slight. Increasing seed yield and thereby increasing oil yield per unit area is a better option. Modification of fatty acids or amino acids composition may create additional marketing niches. High contents of lignans e.g. sesamin and sesamolin and their derivatives such as sesamol and sesaminol as well as other related compounds which prevent oxidation could expand the use of sesame oil as an anti-oxidant additive to oil containing products like peanut butter.

Specific breeding objectives, based on the recommendations of three FAO Expert Consultations, discussion and recommendations of many authors are summarized below, whereas diseases and pest resistance objectives are presented in Table 3.

Specific breeding objectives in sesame

SEED CHARACTERS

Large or medium – large, well filled, shape and color to satisfy market demands, especially for confectionery use.

Seed coat rough or brittle, easily removed by dry decortication

Short-term (a few months) dormancy, appropriate for local cropping systems.

Higher oil content

Higher lignans content for longer shelf-life

SEEDLING CHARACTERS

Fast, vigorous germination and emergence, with strong hypocotyl elongation to assure better emergence.

Rapid growth in the early stages to overcome weeds and to give good stand establishment.

Ability to germinate and withstand lower temperatures in early stages of growth in temperate production areas. Some variability was found in a germplasm survey in Korea (Lee and Kang 1982).

PLANT CHARACTERS

Roots

Rapid root growth.

Deep tap root penetration with a well-distributed secondary root system for maximum exploitation of soil moisture.

Leaves

Medium to broad at the base and narrow-lanceolate toward the apex, to permit maximum sunlight penetration

Short petioles

Abscission early and complete at maturity.

Stems

Uniculm if inputs medium to high; or under low input conditions, moderately branched at lower nodes to compensate for poor stands.

Internodes short with corresponding adjustment of capsule angle.

Height varying with conditions and seasons

Growth habit

Determinate with uniform and short flowering and capsule periods, or indeterminate (with prolonged ripening period), according to local needs and farming systems; under low input conditions, either may be desirable depending on the conditions and the growing season. Under medium or high input conditions, the first would be preferable.

Flowers

To start 20-30 cm from soil surface, to reduce biomass and facilitate harvest.

Number of flowers and capsules per leaf axil; one or three.

Capsules

To start forming about 20-30 cm above soil surface, as above

Bicarpellate (4) locules), for larger seed size.

Long or short, as desired in area; capsule length appears neutral in total yield in the indeterminate varieties studied.

One or 3 per leaf axil; number of capsule/leaf axil appears neutral in total yield in the indeterminate varieties studied.

Upright angle, appressed.

Full seed-set, without aborted ovules.

Seed retention (see also above)

Effective seed retention (non-shattering) mechanisms suitable for machine harvesting or for traditional harvest methods, through better capsule indehiscence genotypes or strong placentation or still other mechanisms to be elucidated.

PHYSIOLOGICAL CHARACTERS

Neutral day-length response and thermo-insensitivity.

Improved harvest index, more favorable reproductive to vegetative ratio

Maturity early, medium or late, according to local requirements.

Higher efficiency in nutrient uptake under low and high fertility conditions.

Resistance to water logging, drought, salinity, and other abiotic stresses.

Potential to respond with higher seed yields to applications of fertilizers and/or irrigation, without concomitant undesirable effects (e.g., lodging).

Rapid natural desiccation at physiological maturity

YIELDS

High and stable seed yields of good quality under a wide range of environmental conditions.

Optimal number of capsules per unit area.

Table 3. Diseases and pest resistance objectives

Pathogen	Disease	Resistance Report
FUNGAL		
Alternaria sesami (Kawamura), Mohanty & Behera	Alternaria leaf spot	In cultivated and wild species
Cercospora sesami Zimmerman	White spot	In cultivated species
Corynespora cassiicola (Berk & Curt.) Wei	Corynespora	In cultivated species
Fusarium oxysporum (Schelt.), *sesami* Jacz	Fusarium	In cultivated and wild species
Macrophomina phaseolina (Tassi) Goid. Ssp. *Sesamica*	Characoal rot	In cultivated and wild species
Oidium sp. & others	Powdery mildew	In cultivated and wild species
Phytophthora parasitica (Dastur)var. sesami Prasad [Z]	Phytophthora blight	In cultivated and wild species
BACTERIAL		
Pseudomonas syringae Van Hall Pv. *sesami* (Malkoff) Young, Dye & Wilkie	Bacterial leaf spot	Genetic variation reported
Xanthomonas campestris (Pamel) Dowson pv. *sesami* (Sabet & Dowson) Dye [Z]	Bacterial blight	Genetic variation reported
VIRUS	Leaf curl	In cultivated species
MYCOPLASMA	Phyllody	Genetic variation in cultivated and wild species
INSECT PESTS		
Acherontia styx	Sphingid moth	Genetic variation reported
Antigastra catalaunalis	Webworm	Genetic variation in cultivated and wild species

[Z] Pathogenic races were reported.

8. BREEDING METHODS

The role of different breeding methods is summarized here —

A. *Plant Introduction* : In the earlier phases of breeding the use of introductions directly or after some selection, subsequent breeding efforts are required for improved performance or specific need. It is advisable to introduce promising and other cultivars and landraces from diverse geographical areas. In some cases, introductions may be adopted directly or as a source of valuable genes. It is difficult to get genetic progress directly from introduction. A large number of introductions have been made into India (Sharma, 1994, Rana *et al.* 1994), but no cultivar has directly originated from introductions.

B. *Selection* : Both, mass selection and single plant selection have been very useful in developing improved cultivars because the local landraces are often very heterogeneous. Mass selection aimed initially at elimination of the less desirable materials can provide better adapted and higher yielding material. Single plant selection has also produced successful cultivars but it provides greater opportunity when new variability is created by controlled crosses (Brar and Ahuja, 1979, Sharma. 1985, Zhan *et al.* 1990; Osman and Yermanos, 1982).

C. *Hybridization* : Breakthrough in productivity has to come from controlled crosses aimed at creating new and wider variability because the genetic variability within the regions has been exhausted. The main hybridization approach in sesame is the pedigree method, less common are backcross, composite crosses with bulk populations and selection, and population improvements with or without male sterility. The crosses have been mainly between two cultivars, but sometimes more complex crosses are made. Crosses of F_1 x F_1 and 3-way crosses followed by pedigree selection have given more rapid progress in certain cases. Modifications of the classical pedigree methods such as single seed descent have been initiated in sesame. Pathirana (1995) concluded that the bulk method which is less expensive, can be used with the same amount of success as the other methods he compared viz., classical pedigree, single seed descent, wherein the later 3 procedures selections were made in F_4 and tests for yield and other agronomic traits were conducted in F_6 and F_7.

Srinivas *et al.,* (1992) compared the effective use of bulk and single capsule descent procedures and concluded that bulk method was superior to single capsule descent in four crosses but was less successful in two crosses, where the parents were closely related. Early generation testing was found ineffective. The composite cross approach has been successful and several cultivars have been evolved (Kinman and Martin, 1954, Weiss, 1983)

The backcross method has bean used to transfer desirable traits such as indehiscence, male sterility etc. into adapted cultivars. Its utility will no doubt enhance as better and more productive cultivars are developed and more sources of specific and useful loci are identified.

D. *Induced Mutations* : Although sesame has a wide range of genetic variability in its germplasm collections, certain highly desirable trials, such as good seed retention and resistance to certain diseases and insect pests, have not bean found so far. Three FAO Expert

Consultations (Anon 1981, 1985, and Ashri 1987) recommended that induced mutations be used to enhance the genetic variability of sesame for such characters that can be easily identified in large segregating population e.g. seed retention, modified plant architecture, modified growing period and resistance to diseases and pests. Two varieties viz. Usha and Uma evolved as a result of mutation breeding cover sizeable area in Orissa State.

Sesame seeds are generally less sensitive physiologically to radiations (mainly gamma-rays) and chemical mutagens (mainly ethyl methane sulfonate (EMS). Therefore, higher doses can be used (Ashri, 1982), but they may cause additional undesirable mutations masking the desired mutants. Genetic differences in sensitivity to mutagens have been reported (Ashri, 1982, Kamla, 1980, Pathirana 1992). Single plant selection from M_2 rows has been reported to be the best (Murty and Oropeza, 1989). The other alternative may be harvesting the M_1 plants in bulk and growing the M_2 in bulk by cultivars and treatments. In this approach, it is not possible to determine the number of mutation events that occurred resulting in 2 identical mutant plants which appeared in a particular M_2 bulk. Selection in M_3 generation can also be made for quantitative traits.

A wide range of mutation have been reported (Kobayashi, 1958, 1965, Murty *et al.* 1985). Induced mutation may have agronomic value, some can be useful as markers and still others can be useful in genetic and molecular studies. Induced mutants with non-shattering trait (Cagirgan, 1996, 1997) in Turkey and determinate growth habit (Ashri, 1988, 1995) in Israel have been reported.

A naturally occurring male sterile mutant found in Venezuela more than 30 years ago is still used in breeding. More recently male sterile mutants have bean obtained in India (Rangaswamy and Rathinam, 1982, Ramnathan *et al.*, 1992, Murty 1997) and some experimental male sterile mutants in Turkey (Cagirgan, 1997). Some of the male sterile mutants induced by Murty (1997) were female sterile also. Polypetalous mutant (star flower), with split corolla tube (Murty and Oropeza, 1989; Murty and Bhatia, 1990) greatly reduced the opportunity for self pollination.

Induced disease resistant mutants have been reported in several countries including 'Ahnsan' in Korea (Lee *et al.,* 1985), for resistance to *Fusarium* and *Rhizoctonia, Corynespora* and *Phytophthora*, 'ANK-82' in Sri Lanka (Pathirana, 1992) for resistance to *Phytophthora.*

E. *Hybrid cultivars* : F_1 could potentially be used in sesame. Some of the necessary conditions for the production of F_1 hybrids have bean met. High levels of heterosis have been demonstrated in various hybrid combinations. The research in the area of F_1 hybrid production was intensified with the discovery of a stable source of male sterility (Osman and Yermanos 1982). Cross pollination of male sterile plants by insects presents no problem, and each female plant (msms) can produce thousand of seeds. F_1 hybrids can be produced economically by using the available genic male sterility and rouging out the male fertile plants (Ms ms) from the intended female rows.

F_1 hybrid seed can be produced economically by hand emasculation and pollination if adapted male sterile lines are not developed. However, large scale availability of F_1 hybrid seed and its widespread use on commercial scale will depend on the discovery of a cytoplasmic-genic male sterility system.

An alternative approach for the production of F_1 hybrid seed is the use of split corolla mutant as a female parent. If the crossing blocks are in isolated areas, without other flowering plants in the vicinity, insect pollinators produce a high percentage of out-crossed progenies. With the parents having high combining ability, the boost in yield could be considerable even if the cross-pollination rate is only to the extant of 60 to 70%.

F. *In vitro Techniques*

1. *Tissue culture:* Several tissue culture experiments have been done in sesame. Plants were regenerated from shoot apical meristem and from hypocotyl segments (Ram *et al.*, 1990). In this study, some regenerates were found deviating from the normal source phenotype, although there are no further reports on somaclonal variation in sesame. Somatic embryos were produced directly from zygotic embryos in culture and from callus culture, although at a low frequency, originating from cotyledon and hypocotyl segments. Embryo-like structures were produced from callus tissues by George *et al.,* (1987). Somatic embryos were induced from hypocotyl – derived calli (Jeya Mery and Jayabalan, 1997). For this, out of the 7 auxins and 4 cytokinins tested, 2,4-D was the most effective. Micro-propagation from shoot tip cultures has been reported by George *et al.*, (1987) and Lee *et al.*, (1985). Only root formation was obtained from hypocotyl and cotyledon explants (Batra *et al.*, 1991).
2. *Embryo Culture:* Procedures for embryo-culture have been developed with a view to rescue inter specific F_1 hybrids. A three – stage embryo culture has been described by Qu *et al.*, (1994). In their studies, the embryos of *S. schinzianum* x *S. indicum* were cultured 9 to 13 days after pollination, on MS medium (Murashige and Skoog) containing Nitsch organic salts, 0.2 mg BA and 3-0 mg GA/l and 5 % sucrose. The embryo that germinated were transferred to MS medium supplemented with 0.2 mg BA/l for 2 weeks and then to MS medium supplemented with 0.5 mg NAA/l.
3. *Anther culture:* Anther culture in order to obtain haploid plants were reported (Ranweera and Pathinana, 1992, Kang, 1994), although no haploid plants have been obtained so far.
4. *Protoplast culture:* Several investigations on protoplasm isolation and fusion protocols for sesame have been reported (Shoji *et al.,* 1988; Bapat *et al.,* 1989; Dhingra and Batra, 1990). There are no selective media or suitable experimental lines to isolate the fused cells available.
5. *Transgenic cells and plants:* Transformation of sesame has not been reported so far.

9. ORGANIZATIONS INVOLVED IN CROP IMPROVEMENT

Sesame is not a mandate crop of any international agricultural research centres. There is limited research in its main areas of adaptation, and in developed countries it is a minor crop with little, if any, research attention paid. There is limited international cooperation, limited exchange of know-how, materials and trained personnel.

The international cooperation to a limited extent is sponsored by IDRC (International Development Research Centre of Canada), IPGRI (International Plant Genetic Resource Institute) which succeeded IBPGR, FAO, GIRRID (German: Israel fund for Research and International Development) etc. Research programmes of national interest have been taken up in countries of Africa, Central America, Europe etc.

10. LIST OF VARIETIES UNDER CULTIVATION IN INDIA

The varieties of Sesame recommended for cultivation are given in Table 4, whereas the details of the important varieties are given in Table 5.

Table 4. The sesame varieties recommended for cultivation

Sr	State	Varieties
1	Andhra Pradesh	Gauri, Madhavi, Rajeshwari, Varaha, Gautama, Sweta, RT 54, RT 103
2	Assam	Gauri, Madhavi, Rama
3	Bihar	krishna, Kanke white, JT 21, TKG22
4	Gujarat	Gujarat Til 1, Gujarat Til 2, RT 54, RT103
5	Haryana	Haryana Til 1, RT 46, RT 125
6	Karnataka	E 8, DS 1
7	Kerala	Kayamkulam 1, Thilothama, Soma, Surya, Thilak, Thilthara
8	Maharashtra	Phule Til 1, N 8, Tapi, Padma, RT 54, RT103
9	Madhya Pradesh	N-32, JT 7, TKG-21, TKG-22, TKG-55 JTS 8, RT 46, and Rama
10	Chattisgarh	RT-46, and Rama
11	Orissa	Viayank, Kalika, Kanak, Uma, Usha, JT 21, TKG 22, TKG 55
12	Punjab and HP	Punjab Til 1, TC 25, TC 289
13	Rajasthan	RT 46, RT 54, RT 125, RT 103, RT 127
14	Tamil Nadu	TMV 3, TMV 4, TMV 5, CO 1, TSS 6, Pauyur 1, Vri 1
15	Uttar Pradesh	T 12, T 13, Sekhar (T 78), RT 125, TKG 21, TKG 55, Uma
16	West Bengal	Tilottama (B 67), Rama (Imp Sel. 5)

Table 5. Details of the important varieties grown in Central part of India

No	Varieties	Maturity (days)	Oil %	Yield kg/ha	Special features
A	For *Kharif* season				
1	N. 32	95	53.6	6-7	White seed
2	JT-7	85	54.0	6-7	White, medium seeds
3	TKG-21	78	55.9	7-8	White seed resistant to bacterial leaf spot and Cercospora leaf spot.
4	TKG-22	80	53.6	7-8	White seed tolerant to Phytophthora blight
5	TKG-55	77	52.5	7-8	White seed; tolerant to Phytophthora blight and resistant to Macrophomina stem and root rot.
6	JTS-8	85	46.5	7-8	White seed, tolerant to phyllody, Phytophthora blight, Alternaria leaf spot and bacterial blight.
B	For semi-*Rabi* season				
1	RT-46	85	49-2	6-8	White seed, resistant to oozing and Macrophomina
2	Rama	90	45.0	10-15	Reddish brown seed

11. REQUIREMENTS FOR SEED PRODUCTION

For production of seed the general seed certification standards are basic, and together with the following specific standards constitute the standards for certification of sesame seed.

(i) Land requirements: Land to be used for seed production of sesame should be free of volunteer plants.

(ii) Field inspection: A minimum of three inspections are made, the first before flowering, the second during flowering and third at maturity and prior to harvesting.

(iii) Field standards

A. General requirements

1. Isolation – An isolation of 100m for foundation and 50m for certified seed production programme is recommended to maintain the genetic purity.

B. Specific requirements – Offtype percent should not be more than 0.1 for foundation and 0.2 for certified seed production programme, whereas at foundation seed stage the infection of Cercospora leaf spot should not be more than 0.5 percent and for certified seed 1.00 percent.

(iv) Seed Standards

Factor	Standard for each class cropped	
	Foundation	Certified
Pure seed (minimum)	97%	97%
Inert matter (maximum)	3.0%	3.0%
Other crop seed (maximum)	10 kg	20 kg
Weed seeds (maximum)	10 kg.	20 kg.
Other distinguishable varieties (maximum)	10 kg.	20 kg
Germination (minimum)	80%	80%
Moisture (maximum)	9.0%	9.0%
For vapour-proof containers (maximum)	5.0%	5.0%

12. MAJOR ACHIEVEMENTS THROUGH BREEDING

Systematic and scientific efforts to improve the crop were started at Pusa in 1925 with the collection of sesame including cultivars from areas now in Burma, Shri Lanka, Pakistan, Bangladesh etc. Controlled self-pollination in individual plant and selection resulted in the isolation of thirty distinct types, among which NP 3, NP6, NP 7, and NP 29 were found to be suitable for cultivation in sesame growing states in North India. Inter-group crosses between *S. indicum* (2n = 26), *S. prostratum* (2n = 32) followed by colchicine treatment of the hybrid derived from it resulted in the amphidiploid, named *S. indicatum* (2n = 58) which showed variable pollen and seed fertility but was true breeding. Parthsarthy and Kedarnath (1949) reported some 26-chromosome derivatives which showed resistance to the phyllody disease presumably coming from *prostratum* parent. From this material, two lines 'S' and 'R' were finally selected. These were as good as the parent NP6 in seed yield with less incidence of phyllody under the field conditions, and line 's' was selected for cultivation.

The research work on sesame was subsequently taken up by the State Departments of Agriculture of major sesame growing states. The work was strengthened by the Indian Central Oilseeds Committee (Hyderabad) and PIRRCOM (Project for the Intensification of Regional Research on Cotton, Oilseeds and Millets).

In Uttar Pradesh, the work on sesame started in 1932 at Kanpur. Types 49 and 10 were evolved by selection and Type 4 and Type 10 were recommended for cultivation. Later on T-12 and T-13 were evolved by selection from the local material collected from Almora and Hamirpur districts respectively in 1960. Locally adapted exotic germplasm lines viz., Giza 23, Giza25, EC-10042 and EC 10045 were utilized in hybridization programme which resulted in development of strains CST-781, CST-782, CST-784, and CST-785 and out of them CST-785 was selected for general cultivation. Variety T 78 was selected in 1999.

In the erstwhile Punjab state, research on sesame started at Gurdaspur in 1930 and T-5, T-22 were evolved by selection. Later on Punjab –1 and TC-25 in 1978, were released for general cultivation. Variety TC-289 was released in 1986. Similarly, selection at Hissar resulted in the release of variety Haryana Til 51.

In Rajasthan, Pratap (C-50) a selection from local material was released in 1968 at Kota. The research work was shifted to Sumerpur (Palti) and later to Mandore (Jodhpur) where intensified research efforts resulted in release of varieties RT-46, RT-54, RT-103, RT-125, RT127, which are under cultivation in different parts of the country.

In Madhya Pradesh, before the inception of All India Coordinated Research Project in 1967, the research work was started in Gwalior which resulted in evolution of sesame varieties G-5 and G-35. At Nowgaon (Chhatarpur) the sesame research resulted in identification of lines N-62-10, N-62-130, N 62-39, N 99-135, N 62-32 etc. Out of these N 62-32 was released as N-32 for general cultivation in the state in 1971 and N 62-39 was recommended for cultivation in Telangana region of Andhra Pradesh. The research work from Jabalpur was shifted to Tikamgarh which, resulted in release of sesame varieties TKG-21, TKG-22, TKG-55 and JTS 8 which are at present recommended for general cultivation.

In Maharashtra, in the erstwhile Bombay state, the research work on sesame was started at Niphad and at Nagpur in 1936-37. Variety N 58-2 was developed at Niphad in 1960 and D-7-11-1 was evolved at Dhule in 1962. At Nagpur, No. 41 and No. 128 for *Kharif* and No. 8 for semi-*Rabi* cultivation were evolved. T–85 (selection from Kopergaon 5-9) was recommended for Telangana region of Andhra Pradesh. The research work at Jalgaon, after 1967 resulted in development of cultivars Phule til –1, JLT-7 (Tapi), JLT-26 (Padma) etc. which are in cultivation.

The earlier research work in Gujarat resulted in development of suitable varieties of sesame viz. Patan 64, Patan 65, Astrung, Mrug-1, Purva-1, Tejpur-5, Surnagar-4, etc. which have been under cultivation in the past. Now the higher yielding varieties Gujarat Til-1, Gujarat Til-2 are being widely cultivated.

Sesame improvement work in Tamil Nadu started at Tindivanum and released varieties TMV-1 (selection from local material), TMV-2, TMV-3, TMV-4, and Kerur research station released varieties KRR-1 and KRR-2. The research work at Coimbatore resulted in development of varieties TMV-5 (1978) and TMV-6 (1979), variety Co-1 (1983). More recently TSS-6 has been released for Southern Tamil Nadu. The Vridhachalam centre has released variety Vri 1.

In Kerala, research on sesame was intensified as Kayankulam, which resulted in the development of Kayamkalum-1, in 1980, Thilothama in 1982 and, Soma and Surya in 1985. These are suitable for cultivation in paddy fallows of the state. More recently, Thilak (19980 and Thilthara (1999) have been evolved and recommended for cultivation.

In Karnataka, the sesame improvement work was started at Raichur in 1951 and later on, at Dharwad, which resulted in development of variety E-8, suitable for northern Karnataka. Recently variety DS 1 has been released for cultivation.

The sesame improvement research in Yellamachalli and Jagtial (Karimnagar) in Andhra Pradesh has resulted in Yillamachalli –1, and Yellamachalli-2 for central region and Rajeshwari for Telangana region. Earlier, the selections from the local material from Vishakhaptnam had resulted in the development of cultivars Gauri in 1974 and Madhavi in 1978. These varieties were suitable for growing in coastal districts of Andhra Pradesh. Most recently, variety Sweta has been released from Jagtial for Telangana region.

In Orrisa the research on sesame was initiated during the late fifties and pure line selection (Set-14) was released as Vinayak during sixties. Hybridization work was started in 1969 which resulted in the evolution of varieties BS-6-1 (Kanak) and BM 3-7 (Kalika), a mutant of Vinayak both in 1979. Later on varieties OMT 11-6-3 and OMT 11-6-5 were released in 1990 and are the predominant varieties there.

The research on sesame improvement in Bengal started with the selections from the local material from Dhaka which resulted in isolation of lines B-1, B-4, B-9, B-67 etc, and B-67 in 1974 was found suitable for general cultivation in Bengal and Bihar. Later on, research at Berhampore (Murshidabad District) resulted in development of variety Imp. Sel-5 (Rama) which is suitable especially for summer pre- *Kharif* season.

Selection work in Bihar resulted in isolation of lines M2-3, M3-2, M-3-3 and Kanke white. Kanke white has been under cultivation in this region. Another cultivar developed in Bihar viz. Krishna is very popular since 1986. A list of vareities under cultivation at present is given in Table 4.

REFERENCES

Anon, 1981, Conclusions and recommendations, p. 192-195- In – A Ashri (ed.), Sesame status and improvement. FAO Plant Production and protection paper 66, Rome.

Anon, 1985, Summary of recommendations of the oil crops network-p 249-251, In: A Dmran (ed-), Oil Crops: Sesame and safflower, IDRC- MR 105 E. IDRC, Ottawa.

Ashri, A. 1981. Increased genetic variability for sesame improvement by hybridization and induced mutations. P 141-145. In : A Ashri (ed-), Sesame: status and improvement. FAO Plant production and Protection paper 29, Rome.

Ashri, A. 1982. Status of breeding and projects for mutation breeding in peanuts, sesame, and castor-beans, p-65-80. In: Improvement of oil seed and industrial crops by induced mutations, IAEA, Vienna.

Ashri, A. 1987. Report on FAO/IAEA expert consultations on breeding improved sesame cultivars, Vienna, September 1987, FAO/AGP, Rome.

Ashri, A. 1995. Sesame research overview: current status, perspectives and priorities. P1-17. In:M.R. Bennett and I.M. Wood (eds.), Proc. 1st Australian sesame Workshop, NT Dept. Primary Industry and Fisheries, Darwin.

Ashri, A-1988. Sesame breeding objectives and approaches, p.152-164. In: A. Omaran (ed.), oil crops; sunflower, linseed and sesame- IDRC – MR 505 E, IDRC, Ottawa.

Bapat, V.A., L. George, and P.S. Rao 1989. Isolation, culture and callus formation of sesame (*Sesamum indicum L.*) protoplasts. Indian J. Expt. Biol. 27 : 182-184.

Batra, A., M. Dhingra, and R. George. 1991. Induction of diversified root system in *in vitro* culture of *Sesamum indicum L.* J. Phytol. Res. 4:93-97.

Bedigian, D. and J.R. Harlan 1986. Evidence of cultivation of sesame in the ancient world. Econ. Bot. 40:137-154.

Bedigian, D., D.S. Seigler and J.R. Harlan, 1985, Sesamum, sesamolin and the origin of sesame, Beochem, Systematic Ecol. 13:133-139.

Biswas, A.K. and A.K. Mitra. 1990 Interspectific hybridization in three species of *sesamum.* Indian J. Genet. Plant Breed. 50:307-309.

Brar, G.S. and K.L. Ahuja. 1979. Sesame, its culture, genetics, breeding and biochemistry Ann. Rev. Plant Sci. 1:245-313, Kalyani Publ.. New Delhi.

Cagirgan, M.L. 1997. Mutation breeding of sesame for intensive management-p.85-95. In:2nd FAO/IAEA Res. Coord. Mtg. Induced mutations for sesame improvement IAEA, Vienna.

Devarathinam, A.A., and N. Sundaresan 1990. A new wild variety of *Sesamum indicum* L., var. sencottai ADR & MS. Compared with *S. indicum* (L) ver. Yanamalai ADR & MS and *S. indicum* (L). J. Oilseeds Res. 7:121-123.

Dhingra, M, and A. Batra, 1990- High yielding preparation of viable protoplasts from hypocotyl of *Sesamum indicum* L. Current Sci. 59:325-326.

George, L., V.A. Bapat and P.S. Rao.1987. *In vitro* multiplication of sesame (Sesamum indicum L.) through tissue culture. Ann- Bot. 60:17-21.

Jeya Mary, R., and N. Jayabalan. 1997, Influence of growth regulators on somatic embryogenesis in sesame. Plant cells, Tissue organ culture 49:67-70.

Kamala, T. 1990. Gamma ray effects on polygenically controlled characters in sesame. (*Sesamum indicum L.*) J. Nucl. Biol. 19-179-183.

Kang, C.W. 1994. Sesame breeding and agronomy in Korea. P.65-90. In: R.K. Arora and K.W. Riley (eds.). Sesame biodiversity in Asia: Conservation, evaluation and improvement. IPGRI, New Delhi.

Kang, C.W. 1997. Breeding for disease and shatter resistant high yielding varieties using induced mutations for sesame improvement, IPGRI, New Delhi.

Kinman, M.L. and A.J. Martin. 1954. Present status of sesame breeding in the United States. Agron. J. 46:24-32.

Kobayashi, T. 1958. Radiation genetics of sesame III Morphological changes and mutants induced by ionizing radiation. Japan J. Genet. 33:239-261.

Kobayashi, T. 1965. Radiation induced beneficial mutants. P. 399-403. In: The use of induced mutations in Plant breeding. Peregamon Press, Oxford.

Kobayashi,T. 1981. The type classification of cultivated sesame based on genetic characters, p. 86-89. In: A Ashri (ed.). Sesame status and improvement. FAO Plant Production and Protection Paper. 29. Rome.

Kobayashi, T. 1991. Cytogenetics of sesame (*Sesamum indicum*). P. 581-592. In: T. Tsuchiya and P.K. Gupta (eds), chromosome engineering in plants: genetics, breeding, evaluation B. Elsenier, Amsterdam.

Lee J.I., Y.H. Park, Y.S. Park, and B.G. Im. 1985. Propagation of sesame (*Sesamum indicum* L.) through shoot tips culture (in Korean, English Abstr.). Korean J. Breed. 17: 367-372.

Lee, B.H., J.I. Lee, N.S. Seong, C.W. Kang and C.H. Park. 1991 studies on interspeicific hybridization in sesame. II. Development of hybrid plants from *S. indicum x S. radiation* (in Korean, English Abstract). Koran J. Breed. 23: 42-47.

Lee, J.I., C.W. Kang, and S.T. Lee. 1984. Breeding of sesames (*Sesamum indicum* L.) for oil quality improvement III. Variation of fatty acid composition in gamma – ray irradiated M_2 population (in Korean, English Abst.) Res. Rep. Office Rural Dev. Suweon 28*:* 134-143.

Murty, B.R. and F. Orapeza, 1989. Diversity pattern, in Sesamum mutants selected for a semiarid cropping systems. T. A G. 77: 275-286.

Murty, G.S.S. 1997. Induced mutants for the improvement of sesame and hybrid seed production. P. 31-38. In: Proc. 2nd FAO/IAEA Res. Coord. Mtd., Induced mutations for sesame improvement. IAEA, Vienna.

Murty, G.S.S. and C.R. Bhatia, 1990. Inheritance of polypetalous corolla mutation in sesame. Indian Acad. Sci. Plant Sci. 100:7-10.

Murty, G.S.S. D.C. Joshua, N.S. Rao, and C.R. Bhatia 1985. Induced mutants in sesame. P 188-190. In*:*A. Ashri (ed.) sesame and safflower*:* status and potential, FAO Plant production and protection paper 66, Rome. Nayar, N.M. and K.L. Mehra. 1970. Sesame- its uses, botany, cytogenetics and origin, Econ. Bot. 24:20-31.

Osman, H.E. and D.M. Yermanos, 1982. Genetic male sterility in sesame: reproductive characteristics and possible use in hybrid seed production. Crop Sci. 22*:*492-698.

Parthsarthy, N., and Kedarnath. 1949. The improvement of sesame crop in India. Indian J. Genet. 9:59-71.

Pathirana, R. 1992 Gamma ray induced field tolerance to Phytophtora blight in sesame. Plant Breeding. 112:167-170.

Pathirana, R. 1995. Comparison of selection procedures in breeding for seed yield in segregating sesame population. Euphytica 82:73-78.

Prabakaran, A.I. 1996. Genetic diversity of wild sesame from southern India. Plant Genet. Research New Delhi. 10:44-46.

Prabakaran, A.I. and S.R. Sree Rangaswamy. 1995. Observations on interspeicific hybrids between Sesamum *indicum* and *S. sehinziamum* (in Chinese, English Abst) oil crops of china. 16:33-35.

Qu, Z.. X.Y. Wu and F.J. Xia 1994. Embryo rescue and plant regeneration in interspecific hybridization between *S. indicum* and *S. schinzianum.* Oil Crops of China 16:33-35.

Ramnathan, T., M.S. Durairaj and A.J. Prabhkaran 1992. Induced male sterility in *Sesamum indicum* L. Sesame Safflower Newslet., 7: 34-36.

Ram, R.D., Catlin, J. Romero, and C. Cowley, 1990. Sesame :new approaches for crop improvement p. 225-228. In J. Janick and J.E. Semon (eds.). Advances in new crops Timber Press, Portland.

Ranaweera, K.K. D.S., and R. Pathirana. 1992 Optimization of media and conditions for callus induction from anthers of sesame cultivar M 13. J. Natl Sci. Council Sri Lanka 20:309-315.

Sharma, S.M. 1985. Sesamum research and its progress in India. P. 11-27. In A. Dmran (ed.). Oil crops. Sesame and Safflower. IDRC-MR105 e, IDRC, Ottawa.

Sharma, S.M. 1994, Utilization of national collections of sesame in India. P. 135-156. In: R.K. Arora and K.W. Riley (eds.) Sesame biodiversity in Asia. Conservation, evaluation and improvement. IPGRI, New Delhi.

Shoji, K., K. Masuda, M. Sugai, and T. Kobayashi. 1988. Relative DNA content of fused protoplasts and colony cells in *Sesamum indicum* Cytologia 53:205-211.

Srinivas, T., G. Raghunatham, and A.R.G. Ranganath. 1992. Evaluation of bulk and single capsule descent breeding procedures in sesame *Sesamum indicum L.)* Dleagneux 47 :645-648.

Thangarelu, S. 1994. Diversity in wild and cultivated species of sesame and its uses. P. 13-23- In; R.K. Arora and K.W. Riley (eds.) Sesame biodivesity in Asia: Conservation evaluation,. and improvement IPGRI, New Delhi.

Weiss, E.A. 1971. Castor, sesame and safflower, Leonard Hill, London, p. 311-525.

Weiss, E.A. 1983 Oil seed crops. Longman, London. P. 282-340.

Yermanos, D.M. 1980. Sesame. P.549-563. In: W.R. Fehr and H.H. Hadley (eds.) Hybridization of crop plants. Am. Soc. Agron. Crop Sci. Soc. Am. Madison, WI.

Zhan, Y.X., M. Cheng. and A.Z. Wee. 1987. Cytogenetic studies in sesame *(Sesamum indicum* L.) I. Karyotype analysis. Sesame. Safflower Newslet. 6: 17-18.

Zhan, Y.X., X.C. Zhou, S.K. Liu, and Y.R. Li. 1990. Studies on heterosis in sesame and its utilization: heterosis and genotypic correlation of characteristics (in Chinese, English Abst.) Scientia Agr. Sinica 23: 27-36.

CHAPTER - 16

COTTON

G.K. Koutu and K.C. Mandloi

1. TAXONOMICAL CLASSIFICATION, CHROMOSOMES NUMBER AND ORIGIN

Cotton primarily a fiber crop is also an important source of edible vegetable oil and protein belongs to the genus *Gossypium* in the family Malvaceae. It is a large, rich and economically important genus with 40 species, of which four are under cultivation. Among the linted species *G. arboreum* L. and *G. herbaceum* L are the old world Asiatic diploid cottons (2n = 26), whereas, American or Upland cottons species *G. hirsutum* L. and *G. barbadense* L. are the New World allo-tetraploid cottons (2n = 52).

The species of *Gossypium* are classified into genomic groups A to G in diploid and A to D in allo-tetraploids. The genus is distributed in Asia, Africa, America, Australia, Arabia and Pacific Islands. The tetraploid cultivated cotton is evolved from a diploid 'A' genome species *G. herbaceum* var. *africanum* and 'D' genome progenitor species *G. raimondii* by the evolutionary process of migration, interspecific hybridization, polyploidy and natural selection.

2. CLIMATIC REQUIREMENT, AREA AND DISTRIBUTION

In India cotton is grown in extremely diverse agro-climatic conditions varied from arid subtropical to humid tropical wet climates and from deep heavy vertisols to sandy loamy alluvial and red lateritic soils spread over three distinct zones viz., North, Central and South (Table 1). In India, all the four species are grown in 9.2 million hectares with 90 percent area under American cotton. One-third area of cotton is in Maharashtra followed by Gujarat (17%), A.P. (11%), Punjab (8%), Haryana, Rajasthan and Karnataka (7% each), M.P. (6%) and Tamil Nadu (3%).

3. SALIENT FEATURES OF THE CROP CULTIVATION

Popular varieties of *G. hirsutum, G. arboreum, G. herbaceum* and *G. barbadense* and hybrids cultivated in different zones of India are presented in Table 1.

Table 1. Popular varieties and hybrids under cultivation in India

Varieties and hybrids	North zone (Punjab, Haryana and Rajasthan)	Central zone (Gujarat, M.P. and Maharashtra)	South zone (Karnataka, Andhra Pradesh and Tamil Nadu)
G. hirsutum	F-414, F-1054, LH-1134, HS-6, H-777	LRA-5166, PKV-081, Khandwa-2, Vikram, Rajat G.Cot-10	MCU-5, MCU-7, Surabhi
G. arboreum	LD-327, LD-941 HD-107	AKH-4, AKA-5 Jawahar Tapti	K-12
G. herbaceum	—	Digvijay, G.Cot.-11, 13, 17 and 21	Jayadhar
G. barbadense	—	—	Suvin
Hybrid	Ankur-651	JKHy-1, H-6, H-8, H-10, NHH-44, JKHy-2.	DCH-32, TCHB-213, RCH-2, DHH-11.

Optimum time of sowing for irrigated cotton is in May and June whereas, in rainfed conditions sowing with commencement of monsoon is recommended. Spacing of 60 x 30 cms for compact plant type; 60 x 60 cms for *hirsutum* varieties while, 60 x 90 cms, 90 x 90 cms and 120 x 120 cms for hybrids are recommended.

Nutrients are supplied through chemical fertilizers based on soil test. Recommended per hectare rate of Nitrogen is 40 - 60 kg for *arboreum* and *herbaceums*; 60 - 90 kg for *hirsutum* and 120 - 150 kg for hybrids. Whereas, 30-40 kg Phosphorus; 40 kg Potassium and 10-15 t organic manure per hectare is recommended for all the species. 25 kg zinc/ha is applied once in three years at the time of sowing. Normally 2-3 irrigation is required for hybrid cotton.

Cotton is grown as mono crop and two-year rotation as cotton-Jowar and cotton-wheat. It is inter-cropped with hybrid sorghum, black gram, soybean, pigeonpea, maize, sunflower and ground nut.

4. BOTANICAL DESCRIPTION

Cotton plant has long tap-root and monopodial growth of main stem. Spirally arranged leaves have long petiole, two small stipules and broad palmately, 3-5 lobed, velvety to glabrous leaf blade with three to nine palmately arranged veins. Small nectar as a gland like cavity is present on the main vein of the lower surface of the leaf. Out of two buds present in each leaf axil normally only one develops. Lower branches of the plant, which do not bear flowers, are termed as vegetative branches or monopodial branches. Upper branches, which bear flowers, are called fruiting branches or sympodial branches. Flower is surrounded by three leaf bracts (epicalyx). Green coloured calyx is reduced to a cup shaped structure at the base of the five lobed corolla tube. Five large petals of white to yellow colour with or without definite red or purple spot at the base are closely convoluted in the bud with twisted aestivation. Petal are slightly united to each other at the base and fused with the staminal tube (epipetalous) A tube like structure formed by the fusion of 100-150 stamens surrounds the style (monoadhelphous condition). Single celled kidney shaped anthers varies in colour from white to gold. Superior ovary is composed of 3-5 carpels with several ovules in axil placentation. Flower opens in the morning with receptive stigma. Pollen grains remain viable up to twelve hours.

The boll (fruit) is a spherical to ovoid leathery capsule that splits at maturity along the carpel edges marked as furrows on the surface of the boll. Number of locules in a boll varies from three to five with six to nine seeds in each locule.

5. SELFING AND CROSSING

It is a self-pollinated crop with bisexual flowers. However, depending upon the climatic conditions 0-60 per cent natural cross-pollination is possible.

Flower buds which are likely to open on the next day are emasculated by removal of anthers gently with the help of nail between 3 to 6 PM and covered with a tissue paper bag of red colour for identification and to prevent the chance of natural crossing. Emasculated buds are pollinated next day between 8-12 A.M. By one flower of male parent, 4-5 buds can be pollinated. After pollination, the stigma of emasculated buds is covered with a straw tube and the red tissue paper bags are replaced by white tissue paper bags. A thread is tied on the pedicel for identification. The fertilization occurs after 12-13 hours of pollination.

For selfing, flower bud, which is likely to open next day, is covered with tissue paper bag and tied with cotton thread to avoid, cross-pollination.

6. BREEDING OBJECTIVES

Higher yield : The prime objective is to improve the yield of economic produce by evolving high yielding varieties and hybrids.

Earliness : It permits double cropping and reduces the cost of production.

Fiber quality : In cotton, fiber qualities like length, strength, micronaire fineness, uniformity and maturity coefficient are essential in textile industries.

Wider adaptation : It helps in stabilizing cotton production in fluctuating environments.

Resistance to biotic stress : Development of resistant varieties against insect pest viz., jassids, aphids, thrips, whitefly, boll-worm and red cotton bug and diseases like bacterial blight, fusarium wilt, root rot, grey mildew and alternaria leaf spot are the cheapest way of minimizing the cost of production and stabilizing the yield.

Resistance to abiotic stresses : Resistant varieties/hybrids suitable for abiotic factors such as drought, soil salinity and water-logged conditions are the requirement of the day.

Synchronous maturity : It will help in harvesting the crop in one or two picking.

Quality of oil : Cotton seed is an important source of edible oil, but it contains gossypol an alkaloid, which is harmful for human health. Hence, there is need to develop cotton varieties/hybrids with least gossypol in the seed.

Naturally coloured cotton : Development of naturally coloured cotton varieties and hybrids for eco-friendly requirements and export.

7. BREEDING METHODS FOR CROP IMPROVEMENT

Various breeding methods which are used for genetic improvement of cotton are introduction, pure-line selection, mass selection, pedigree/bulk method, back cross method, heterosis breeding, mutation breeding, interspecific hybridization and biotechnology. Four breeding approaches viz., recurrent selection, disruptive selection, diallel selective mating and biparental mating are used for population improvement. The varieties developed by using these breeding methods are presented in Table 2.

Table 2. Development of varieties by different breeding methods

Breeding method	Variety developed
Introduction	PRS-72, G-67, Savitri
Pure line	Cocanada-2, Lohit
Mass selection	F-414, Bikaneri Narma, Narmada, H-777 and SRT-1
Pedigree method	LH-900, LH-1134, F-281, RST-9, Khandwa-3, G.Cot-14, DHY-286, PKV-081, Sahana, MCU-6, 8 and 9, LRA-5166, CNH-36, AKA-8401, Digvijay, G.Cot.-17, G.Cot. -13 and JK-4
Mass Pedigree	Maljari, Jawahar-Tapti and Khandwa-2
Back cross method	V-797 & Digvijay
Heterosis	H-4, H-6, H-8, H-10, JKHy-1, JKHy-2, NHH-44, PKV Hy-1, DCH-32
Mutation breeding	Indore-2, MCU-5, 7 and 10.

8. ORGANIZATIONS INVOLVED IN COTTON IMPROVEMENT

International Cotton Advisory Committee (ICAC) located at Washington in USA advises on various matters of cotton research and marketing. In India, cotton breeding work is carried out by three main organizations i.e., (i) Central

Institute for Cotton Research (CICR), Nagpur, (ii) All India Coordinated Cotton Improvement Project (AICCIP), through its 11 main research centres and 16 sub-centres located in different cotton growing states and (iii) Private registered seed companies. There are about nine prominent companies (several others) which are engaged in development of superior hybrids. Apart from this Directorate of Cotton Development, Cotton Corporation of India and East India Cotton Association are engaged in cotton development and marketing.

9. IMPORTANT VARIETIES AND HYBRIDS

Characteristics of Important varieties and hybrids are presented in Table 3.1 and 3.2.

Table 3.1. Important cotton hybrids and their characteristics

SNo.	Hybrid	Yield (q/ha)	Ginning (%)	M.F.L.	Area of adaptation
hirsutum x *hirsutum*					
1	JKHy-1	30	34	28	M.P. & A.P.
2	H-4	30	34	28	Gujarat & M.P.
3	NHH-44	25	35	26	M.S. & M.P.
4	Savita	30	30	32	T.N. & A.P.
5	MECH-4	35	28	28	M.S.
6	Fateh	34	34	26	Punjab
hirsutum x *barbadense*					
7	Varlaxmi	25	30	32	Karnataka, A.P. &T.N.
8	DCH-32	30	28	32	T.N., A.P. & M.P.
9	DHB-105	30	28	32	Karnataka
arboreum x *hirsutum*					
10	DH-7	15	36	20	Gujarat
11	DH-9	15	36	24	

Table 3.2. List of cotton varieties under cultivation in India and their description

S No.	Variety	Yield (q/ha)	Ginning (%)	M.F.L.	Area of adoption
hirsutum					
1	Khandwa-2	8-10	34.0	23.0	M.P.
2	Khandwa-3	10-12	34.0	23.0	M.P.
3	DHY-286	14.0	36.0	28.0	M.S.
4	Vikram	10-12	36.0	23.0	M.P
5	AKA-8401	18-20	36.0	24.5	M.S.
6	NH-452	15-16	37.0	25.0	M.S.
7	Rajat	16.0	39.0	25.0	M.S.
8	G. Cot-16	16.0	38.0	26.0	Gujarat
9.	B.N.	16.0	33.0	22.0	Rajasthan

10	F.-414	18.0	33.0	22.5	Punjab
11	H.777	18.0	34.0	22.6	Haryana
12	LH-900	27.0	34.0	22.0	Punjab
13	MCU-7	14.0	33.0	23.5	T.N.
14	MCU-5	20.0	34.0	29.5	T.N. & A.P.
15	LRA-5166	14-16	35.0	24.5	M.P., M.S. & A.P.
arboreums					
16	Lohit	10.0	39.0	18.5	U.P.
17	G.-27	12.0	36.0	17.0	Punjab
18	AKH-4	7-10	39.0	24.0	M.S.
19	G.Cot.-15	10.0	34.0	21.8	Gujarat
20	Maljari	8-10	39.0	24.0	M.S.
21	J.Tapti	15-18	35.0	24.0	M.P.
Herbaceum					
22	G.Cot-17	14.0	37.0	25.5	Gujarat
23	Digvijay	7-8	34.0	20.5	Gujarat
24	G.Cot-11	9-10	39.0	23.5	Gujarat
Barbadence					
25	Suvin	7-8	30.0	38.0	T.N.

10. MAJOR ACHIEVEMENTS IN THE CROP THROUGH BREEDING IN INDIA AND ABROAD

In India and abroad, remarkable progress has been made in cotton breeding. The major achievements of cotton breeding includes : (i) Yield improvement (ii) Quality improvement and (iii) Resistance to insect.

(i) Yield improvement : In India significant improvement has been achieved in yield of cotton. The present average (1996-97) lint yield of cotton is 327 kg/ha, which was only 81 kg/ha during 1947-48. Moreover, the present average seed cotton yield of best varieties/hybrids is 40 q/ha, which was only 15 q/ha during 1947-48. This increase in cotton yield could be achieved through development of high yielding varieties and hybrids. India is the pioneer country in the world for successful exploitation of heterosis (hybrid vigour) in cotton on commercial scale. The first hybrid H-4 was released in 1970 from Cotton Research Station, Surat of Gujarat by late Dr. C.T. Patel. Since then several cotton hybrids have been released for cultivation in South and Central Zone. First intra-*hirsutum* hybrid JKHy-1 released by Jawaharlal Nehru Krishi Vishwa Vidyalaya, Jabalpur is still doing extremely well under rainfed and irrigated condition. The first Interspecific hybrid [*hirsutum* x *barbadense*] Varlaxmi was released by Karnataka in 1972.

(ii) Quality Improvement : Significant achievements have been made in improvement of fiber quality characters viz., length, strength, fineness, maturity coefficient, uniformity and spinning capacity. The MCU-5 was the first extra long staple variety of *G. hirsutum* released in India. Suvin a variety of *G. barbadense* release is capable of spinning 120 counts which is a distinct landmark in the quality improvement in India. At present the staple length of the best variety is 37 mm and spinning ability is 120 counts as compared to varieties of 24 mm and 28 counts in 1947-48. Significant improvement has been made in deploid cotton [K-8, K-9, K-11, Jawahar-Tapti PA-183 and AKA-8401] for increasing fibre length and fineness.

(iii) Insect pest and disease resistance : In India and abroad extensive work has been done on insect pest resistance. In upland cotton, variety MCU-5 VT, resistant to Verticillium Wilt was released from Coimbatore. All the varieties of genera *arboreum* and *herbaceum* released after 1967, are resistant to fusarium wilt. Some jassid resistant varieties (Khandwa-2, SRT-1, Dhy-286 and PKV-081) and hybrids (PKV Hy-2, NHH-44 and JKHy-1) of *hirsutum* have been released. In USA Monsanto Seed Company has developed transgenic Bt cotton which is resistant to *Helicoverpa armigera*. In India C.I.C.R., Nagpur has also developed Bt cotton which are being evaluated.

11. METHOD AND REQUIREMENTS OF HYBRID SEED PRODUCTION

Conventional method : Hybrid seed production carried out by hand emasculation and pollination as suggested for hybridization is known as conventional method. Foundation seed of male and female parents are used for the production of certified hybrid seed. The female and male parents in 2 : 1 to 4 : 1 ratio are planted at wider spacing generally 120 x 150 cm in well prepared land with facilities of irrigation. The off-type plants are rogued out before initiation of crossing programme. Crossing work is started after one week of flower initiation.

Non conventional method : Genetic male sterility and Ctyloplasmic genetic male sterility systems are used in cotton. It reduces the cost of emasculation. However, pollination has to be done manually. The GMS / CMS “A” lines are developed and the hybrid seed is produced by crossing it with identified “R” lines. While, the “A” lines are maintained by corresponding “B” lines.

CHAPTER - 17

MALE-STERILITY AND ITS EXPLOITATION

Dr. Mridula Billore

1. INTRODUCTION

The inability of a plant to produce functional gametes thereby causing unfruitfulness can be termed as sterility. Sterility characterized by inability of a plant to produce viable pollen grains is known as male–sterility. Even the production of non-functional pollen grains may be designated as male-sterility. Nevertheless, female gametes are functional and such plants are fruitful, when pollinated by functional pollen.

Male-sterility is a device for pollination control and has been characterized as a mechanism, which facilitates or enforces out breeding, consequently enhancing heterozygosity in the population. It is a means of emasculating the plants genetically and utilized as a tool in hybrid seed production of various crops, including cereals, pulses and vegetables; on commercial scale.

Male-sterility is reported in almost all the cultivated crop species (Duvick, 1966). Virtually, all the diploid species of plants, whether domesticated or wild, have at least one male-sterility locus (Beadle, 1932 and Rick, 1984).

2. OCCURRENCE AND CAUSES OF MALE-STERILITY

Koelreuter was the first to report male-sterility in flowering plants in 1763. Later on, several cases of male-sterility in angiosperms have been reported. After the report of male-sterility in flax by Bateson and Gairdner (1921) the phenomenon has been reported in many crop species. Allard (1960) and Duvick (1966) presented a good account of male-sterility in flowering plants. Frankel and Galun (1977) and Kaul (1988) have also reviewed work on male-sterility.

Sterility may be caused by chromosomal aberrations, gene action or cytoplasmic influences. These agencies cause abortion or modification of the entire flowers, stamens, pistils or may upset the normal development of pollen, embryo sac, embryo or endosperm. In higher plants, male-sterility is due to abnormal development of male gametophytes, resulting in pollen abortion (Hanson and Conde, 1985). Female fertility has been reported in some crops such as sorghum (Casady *et al.*, 1960). Two dominate nuclear genes viz., Fs_1 and Fs_2 with complementary effect in double heterozygote induce female-sterility.

According to Gableman (1956) male-sterility could be of three forms among plants viz., (i) A situation may exist where pollen grains are absent or extremely few which may be designated as the pollen sterility, (ii) There could be mal formation or absence of the stamens or the male flower, a situation called the staminal sterility and, (iii) The pollen is normal but fails to dehisce, a situation known as positional or structural sterility. The male-sterility is usually designated to pollen sterility. In Sudangrass, a male sterile line known as 'antherless' has been reported which does not bear anthers (Karper and Stephens, 1936). Among all these three forms, the pollen sterility is more frequent and usually referred to as male-sterility.

Sears (1947) suggested three types of male-sterility viz., cytoplasmic, genetic and cytoplasmic-genetic. Lasa and Bosemark (1993) proposed four additional types viz., non-genetic chemically induced male-sterility, functional male-sterility, environmental cytogenetic male-sterility and genetically engineered male sterile system.

For the convenience of understanding following terms need to be precisely explained which may be frequently encountered in the details of consequences of male-sterility in crops.

(i) *A-line:* The male sterile line of any kind is designated as A-line. By virtue of their nature, they are essentially used as female parent or seed parent in the production of hybrid seed.

(ii) *B-line (Maintainer):* This term is used to denote a line, which could be utilized for perpetuation and multiplication of A-line. In case of cytoplasmic-genetic male-sterility the maintainer line is isogenic (i.e. genetically identical with the A-line but differs in cytoplasm.

In case of genetic male-sterility the maintainer line is usually heterozygous at the fertility locus and hence, the F_1 of A x B results in 50% sterile plants and 50% fertile plants.

(iii) *R-line:* This is also known as restorer line as it is used for restoration of fertility (functional pollen production) in resultant (A x R) progenies. Restoration implies production of fertile plants in F_1 generation of (A x R) crosses.

3. GENETIC MALE-STERILITY

This type of male-sterility is also known as genic, Mendelian or nuclear male-sterility as the genes located on nuclear chromosomes solely govern it.

According to Beadle (1932) and Duvick (1966) almost all the diploid plant species, if investigated would reveal at least one sterility locus.

In most of the cases, a single gene determines the sterility, however, two or more genes (Duvick, 1966 and Driscoll, 1972) or polygenes (Athwal and Borloug, 1967) have also been reported to control male-sterility. Genetic male-sterility occurring spontaneously in nature has been observed in crops like cotton, barley, rice, sunflower, sweet pea, garden pea, soybean, tomato, tobacco, brinjal, lettuce etc. Mutations for this trait have been induced through, ionizing radiations and chemical mutagens (Kaul, 1988).

The inheritance of genetic male-sterility follows the simple Mendelian principles (Fig. 1) where the sterility is under the control of recessive genes.

In few crop plants viz., tomato, sorghum, barley, rice, wheat, pigeonpea, brassica, sunflower, watermelon and lettuce, male-sterility is governed by recessive gene(s). Male-sterility controlled by single dominant gene has been reported in safflower (Beadle, 1932; Rick, 1984 and Jain, 1959), cotton (Allison and Fisher, 1964; Weaver and Ashley, 1971 and Bowman and Weaver, 1979) and wheat (Sasakuma *et al.*, 1978). Role of cytoplasm is not involved in the determination of this kind of male-sterility.

Like any other morphological qualitative trait, which is governed by a few major genes, genetic male-sterility results in plants due to mutation (spontaneous or induced) of the fertility locus situated on the chromosome within the nucleus. Here, the cytoplasm is not involved.

Genetic male-sterility is a result of pollen abortion, as reported in sorghum (Singh and Hadley, 1961) or due to failure of anthers to dehisce as reported in tomato (Mital and Thomas, 1961) or anther abortion, pistillody of anther or several other causes.

4. MAINTENANCE OF GENETIC MALE STERILE LINE

In most of the cases, so far observed, genetic male-sterility results from the action of recessive genes. Hence, the male sterile line is essentially homozygom (msms). The maintainer line cannot be msms or MsMs for obvious reasons. The maintainer line is isogenic (genes identical with A line) at all the loci, except the sterility/fertility locus. It is Msms. It produces two kinds of pollen grains. The male sterile plants fertilized by ms pollen grains set seeds and are genotypically msms. On the other hand, those A-line plants, which are fertilized by Ms pollen grains, are heterozygous at sterility locus (i.e. Msms). When the F_1 seeds are sown, 50% plants are sterile and rest 50% are fertile. The later should be identified and removed during hybrid seed programme or maintenance programme, before the sterile plants start flowering. Early marker genes would facilitate such identification and rouging. The production of hybrid seed based on this kind of sterility is relatively costly because fertile plants are to be rouged out before sterile plants flower and only 50% plants are sterile. The hybrid seed is produced only on 50% plants of A-line (Fig. 1)

Inheritance of Male Sterility

Parents — ms ms (Male sterility) x ms ms (Male fertility)

↓

F_1 — ms ms (Male sterility)

↓

F_2 — 1 Ms Ms, 2 Ms ms, 1 ms ms — Ratio 3 : 1

(Male fertility) (Male sterile)

Maintenance of A Male Sterile Line

Male sterile strain A ms ms	x	• Male fertile strain B	Cross the male sterile line with the male fertile sib line
	↓		
ϕ ms ms Male sterile (line A)	x	ms ms • F_1 (Male fertile)	The F_1 (Ms ms) is backcrossed with the male sterile line A maintained by sib mating. Seed from the male sterile plants only is harvested
1 Ms ms Male fertile (Seed not harvested)	: x ↓	1 Ms ms ϕ Male sterile (Seed harvested)	
1 Ms ms Male fertile (Seed not harvested)	: x ↓	1 Ms ms ϕ Male sterile (Seed harvested)	Maintained by sib mating. Seed from the male sterile plants only is harvested.
Maintained indefinitely			

Fig. 1 Inheritance of genetic male sterility. Male sterility is produced by a recessive nuclear gene designed as ms; the dominant allele Ms produces male fertility. During the maintenance of male sterile lines, sib mating is achieved through natural pollination.

5. TRANSFER OF GENETIC MALE-STERILITY

Any desirable variety, to be converted into a male sterile line, be used as a pollen parent. The male sterile line is used as a seed or female parent. Because genetic male-sterility is usually recessive in nature, the pollen parent cannot be msms. When male sterile line is pollinated with pollen from the desired variety, the genetic constitution of F_1 shall be 50% genes from A line and 50% from desired variety (donor). With respect to sterility/fertility locus, the F_1 shall be Msms (i.e. heterozygous). Self-pollination of F_1 plants will yield plants in F_2 generation, which are genotypically MsMs, Msms, msms. Among the F_2 plants, about 25% should be genotypically male-sterile. Among these male-sterile plants, select the plants, which appear similar to the desired variety (donor or recurrent parent) and pollinate them with pollen from desired variety. Four to five such cycles will lead to a male sterile plant in the genotypic background of desired variety.

6. OCCURRENCE OF GENETIC MALE-STERILITY

In cotton, one recessive gene, ms_2 (Richmond and Kohel (1961) and a dominant gene for male-sterility, Ms_4 (Allison and Fisher, (1964) have been reported. In cucurbits, recessive genes for male-sterility have been reported in muskmelon (*Cucumis melo*), pumpkin (*C. pepo*) and watermelon (*Citrullus vulgaris*), but to date the male-sterility of this kind could not be used for hybrid seed production. Several genes cause male-sterility in tomato (Rick and Butler, 1956), but the critical factor for its utilization is incidence of cross-pollination. In maize and sorghum, numerous genes for male-sterility are known (Beadle, 1932; Stephens, 1937), but none is used for production of F_1 hybrid seed. Bhale and Bhat (1990, 1993), nevertheless, reported that genetic male-sterility based hybrids were superior to hybrids based on cytoplasmic genetic male-sterile lines, in case of cotton. Gilmore (1964) suggested that in sorghum with perfect flowers, genetic male-sterility could be used for schemes of recurrent selection.

Vranceanu (1974) evolved sunflower hybrids using genetic male-sterility. Male-fertility (Msms) was linked with red anthocyomin pigmentation enabling rouging of fertile plants in female lines during hybrid seed production.

7. DISADVANTAGES OF GENETIC MALE-STERILITY

As pointed out earlier, about 50% plants among the male-sterility line (seed parent for hybrid seed production) are fertile; such plants should be identified and rouged out in time. Otherwise, these plants shall cause contamination in hybrid seed programme. Moreover, in the absence of suitable marker gene/trait it would be difficult to identify sterile plants till flowering. Secondly, because only 50% plants of A line are available for seed production, cost of hybrid seed is higher. Thirdly, the male-sterility genes are not stable. Hence, mutation of male-sterile genes may impair the hybrid seed production.

8. EXPLOITATION OF GENETIC MALE-STERILITY

Although the genetic male-sterility has been reported in several crop species like wheat, barley, maize, sorghum, rice, lucern, sunflower, watermelon, muskmelon, pumpkin, cotton, tomato, tobacco, sugar beet, castor, pigeonpea etc., this phenomenon was utilized for hybrid seed production in few crops like pigeonpea (Reddy *et al.*, 1978), sunflower (Veranceanu, 1974), safflower (Prasad *et al.*, 1995) etc. However, its exploitation for commercial hybrid seed production has been done for crops like pigeonpea and sunflower.

9. CYTOPLASMIC MALE-STERILITY

When the male-sterility is determined by factors carried in the cytoplasm it is designated as cytoplasmic sterility and is maternally inherited. The plants, which possess sterile type of cytoplasm, remain unfruitful unless some pollinators are present in the vicinity. Nevertheless, the F_1 plants resulting from such pollinations are again male-sterile because the cytoplasm of the zygote is entirely that of the female gamete. Cytoplasmic genes are most often maternally transmitted in plants (Fig. 2).

CYTOPLASMIC MALE STERILITY

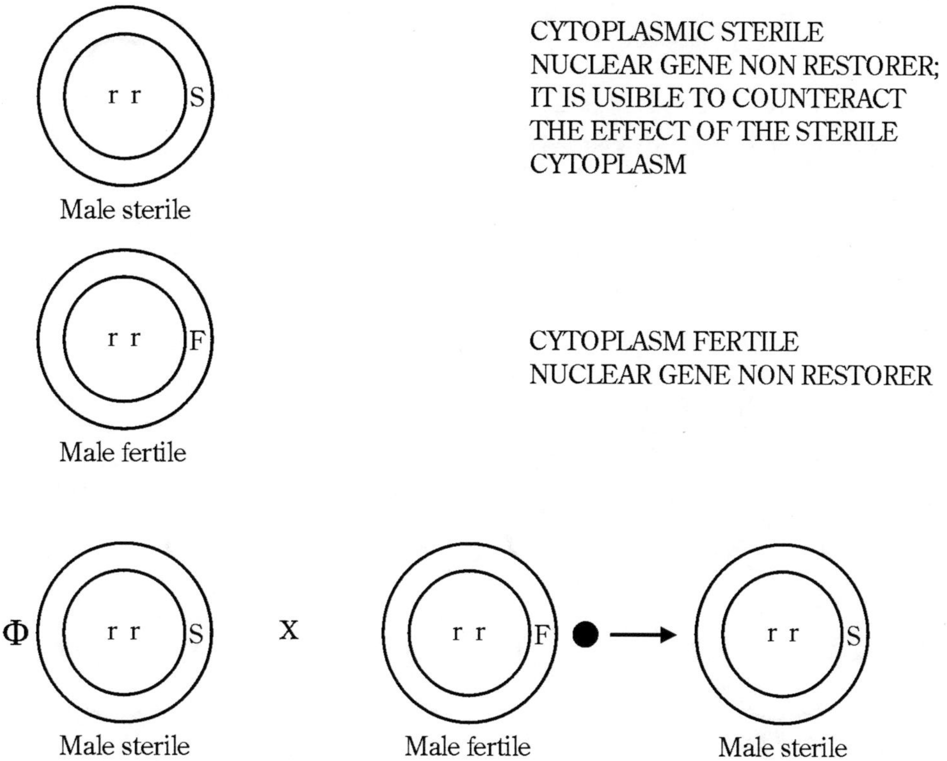

Fig. 2. Cytoplasmic male sterility. Male sterility is casued by the cytolasm (depicted by S) The F cytoplasm is the normal cytoplasm. If a restorer gene (the dominant allele of r) becomes availalbe, the male sterility will be inclused in the cytoplasmic genetic group. This type of male sterility shows strict cytoplasmic inheritance.

10. UTILIZATION OF CYTOPLASMIC MALE-STERILITY

Since, the hybrid produced by using cytoplasmic male-sterility remains, male sterile or fruitless regardless of the genotype of pollen parent it imposes a major limitation to its utilization. It can be used only for those crops where seed is not economically important cytoplasmic male-sterility could be utilized for production of hybrids which could be propagated vegetatively and whose flowers or vegetative parts are economically important. This male-sterility cannot be utilized for crops, which are obiligately sexually propagated and seed is an economically important produce. Duvick (1966) reported use of cytoplasmic male-sterility for production of hybrid petunia.

11. MAINTENANCE OF CYTOPLASMIC MALE-STERILE LINE

The cytoplasmic male-sterile line could be conveniently maintained by vegetative propagation, if the later is possible. Else there should be an isogenic line (genotype similar to the sterile line) differing only for cytoplasm.

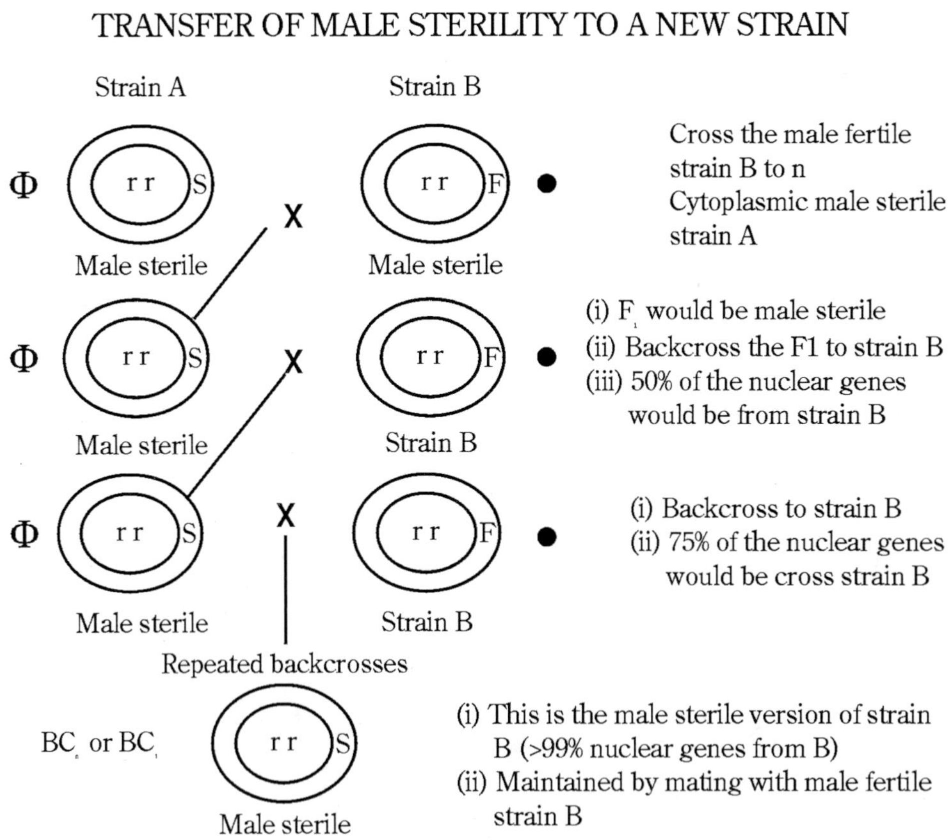

Fig. 3. Transfer of the male sterile cytoplasm from strain A to strain B. Strain A is used as the non-recurrent parent in the backcross programme with strain B. After 6-7 backcross the progeny will be similar to strain B in their nuclear genotype. the strain B serves as the maintainer of the new male sterile strain.

12. TRANSFER OF CYTOPLASMIC MALE-STERILITY

The procedure for incorporating cytoplasmic male-sterility in any desired line is straightforward. An existing cytoplasmic male-sterile line (A-line) used as donor or non-recurrent parent and crossed with the desired variety to be sterilized (used as recipient/recurrent parent). The F_1 is necessarily to be used as a female and is pollinated with pollen from (desired variety) the recurrent parent.

Five to six backcrosses, with selection for plants similar in appearance to the desired variety, will result in male-sterile line. This is A line and the original variety is its maintainer or B line (Fig. 3).

13. SOURCE OF CYTOPLASMIC MALE-STERILITY

The wide hybridization (interspecific or intergeneric) could be attempted to obtain cytoplasmic male-sterility. Besides, chemicals like ethidium bromide (EB) has been used for inducing cytoplasmic male-sterility in pearl millet (Burton and Hanna, 1976) and barley (Minocha *et al.*, 1983)

14. BASIS OF CYTOPLASMIC MALE-STERILITY

It was postulated by Shrivastava *et al.* (1993) that cytoplasmic inheritance of male-sterility is associated with mitochondrial DNA and is supposed to be the key cytoplasmic factor responsible for sterility. Recent evidences at molecular level suggested that the genetic determinants controlling male-sterility in maize, sorghum, brassica, rice and an array of other crops are attributable to mitochondrial genome.

15. CYTOPLASMIC-GENETIC MALE-STERILITY

In a plant cell, normal development requires co-ordination of gene expression between the three different genomes, viz., the nuclear, the chloroplast and the mitochondrial. A kind of male-sterility is known among 140 species of plants from 47 genera and 20 families (Laser and Larsten, 1972; Hanson and Cande, 1985), which results from interaction of nuclear genes with factors in cytoplasm. Among these more than 50% arose naturally and about 20% were discovered in intra-specific crosses and remainder in inter-specific crosses (Frank, 1989). There is now substantial evidence that cytoplasmic-genetic male-sterility and subsequently restoration of fertility is attributed to the co-ordinated action of nuclear and mitochondrial genome. Interaction of nuclear with mitochondrial genome governs the regulation of mitochondrial gene expression (Bailey-Serres *et al.*, 1986).

The sites of extra-nuclear factors conditioning male-sterility are located in mitochondrial genome (Levings and Pring, 1976). Thus, the combination of both nuclear gene(s) and extra nuclear factors determines the fertility or sterility in such plants. Based on these combinations, there can be maximum of six kinds of genotypes, only one of them is sterile. (Fig. 4).

1. RR genes with F cytoplasm Fertile (R-line)

2. RR genes with f cytoplasm Fertile (R-line)

3. Rr genes with F cytoplasm Fertile

4. Rr genes with f cytoplasm Fertile

5. rr genes with F cytoplasm Fertile (B-line)

6. rr genes with f cytoplasm Fertile (A-line)

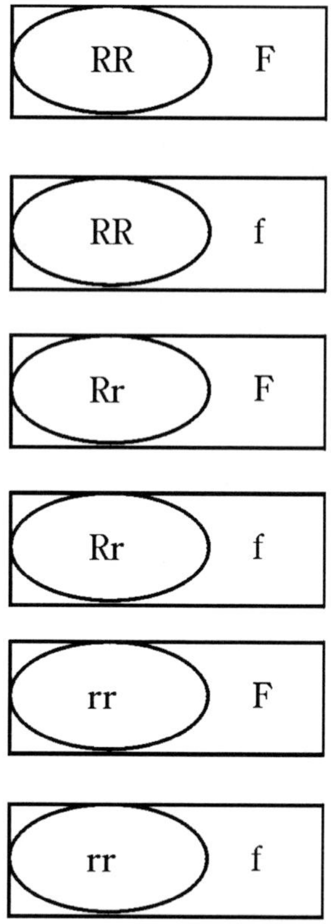

Fig. 4. Various combintaions of fertility and sterility

16. GENETICS OF CYTOPLASMIC GENETIC MALE-STERILITY

The cytoplasmic genetic male-sterility (CMS) is known is several crops. Recessive genes commonly determine the male-sterility, although dominant genes are also reported in some crops. In sorghum, male-sterility is determined by a single recessive gene (Maunder and Pickett, 1959). However, more than one gene pair has also been suggested. The first male-sterility to be reported and utilized commercially for hybrid seed production is attributed to interaction of nuclear genes from Kafir with factors in milo cytoplasm (Stephens and Holland, 1954). However, several different cytoplasms are now available, which induce male-sterility in sorghum (Quinby, 1980; Rao *et al.*, 1984). In maize, male-sterility was described as early as 1931 by Rhoades. It was discovered that certain maize lines which carry nuclear genes called restorer genes (Rf), which are able to suppress the male sterile phenotype (Jones, 1950). There are three types of CMS lines in maize, viz., cms T, cmsS and cms C, based on T, S, and C types of cytoplasm. Two dominant genes Rf_1, Rf_2, restore fertility of cms T, restorer gene Rf4 restores male-sterility of Cms C (Pour *et al.*, 1981) and Rf_3 restores male-sterility of Cms S (Laughman and Gabay-Laughman, 1983).

Burton (1958) developed a cytoplasmic genetic male sterile line in pearl millet, which was subsequently utilized for hybrid seed production in India, viz., Tift 23A (Burton, 1965). The fertility is restored by a single dominant gene (Govila, 1989). Two cytoplasmic genetic male sterile lines of rice, widely used in India, viz., 58025A and 682829A are based on WA cytoplasm (Yuan, 1996). However, some lines possessing CMS ARC and CMS-oryza perennis are also available (Virmani *et al.*, 1996). Single pair of dominant gene restores the fertility of male-steriline line based on WA cytoplasm. Jones and Clark (1943) showed that the male sterile condition of onion resulted from interaction of a recessive gene (ms) and cytoplasmic factor(s). For restoration of sterility, one dominant gene is responsible. In addition to the S system of Jones and Clark (1943), Schweisguth (1973) discovered a T cytoplasm and 3 recessive genes.

The cytoplasm of *Gossypium harknersii* and the genome of *H. hirsutum* interaction produce a stable and dependable cytoplasmic genetic male-sterility in cotton (Basu *et al.*, 1995). A single dominant gene 'Rf' from *G. harknessii* is essential for fertility restoration. *G. barbadense* possesses modifier gene or fertility enhancement factor 'E' for fertility in male-sterile restorer system involving *G. harknessii.* Cytoplasmic-genetic male-sterility was discovered in sunflower by Leclercque (1969), which was an important landmark in the development of commercial sunflower hybrids. Genes for genetic restoration of fertility were found in wild species by Kinman (1970).

17. HYBRID SEED PRODUCTION

For this purpose suitable restorer line or pollinator is crossed with the A-line so as to produce all fertile, normal, viable hybrid seeds. Out of the six types shown in Fig. 4 only first two are such restorers or R-line. They produce all fertile progenies as given in Fig. 5.

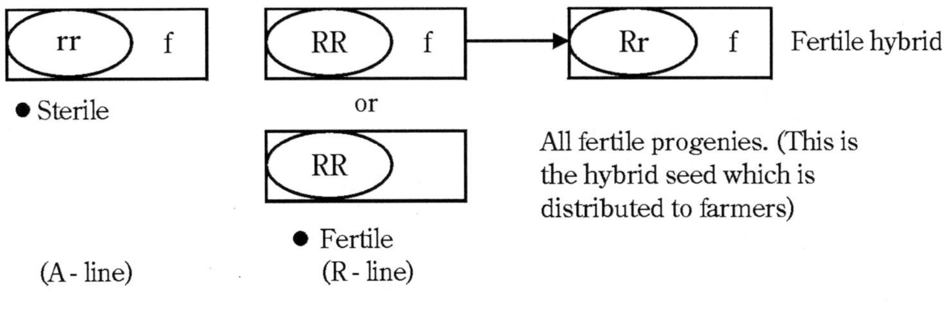

Fig. 5 Fertility restoration

18. OCCURRENCE OF CYTOPLASMIC GENETIC MALE-STERILITY

Several workers viz., Hanson and Conde (1985), Levings (1990) and Braun *et al.* (1992) have reported this type of male-sterility in various plant species. The crop species in which this sterility system has been found include maize, onion, sorghum, pearl millet, wheat, rice, barley, cotton, sugar beat, field bean, flax, tobacco, red pepper, tomato, carrot, radish, sunflower and brassica etc.

19. MAINTENANCE OF CYTOPLASMIC GENETIC MALE STERILE LINES

The maintenance of this kind of male sterile lines is a simple matter. The male-sterility in most of the known cases, where it has been employed as a tool for production of hybrid seed, the male-sterility is governed by the interaction of single recessive nuclear gene and genetic factors in mitochondria. A maintainer line is developed which is genotypically exactly identical (isogenic) with the genotype of male sterile line (A-line), nevertheless, it differs from the later in cytoplasm (Fig. 6 as shown below).

A line B line

Fig. 6 Maintenance of cytoplasmic genetic male sterile lines

Since the factor in cytoplasm does not interact with nuclear gene(s) for sterility to produce male-sterility, the plants of B line produce viable pollen grains in the anthers. These are genotypically similar to the A line (except that they are haploid). Pollen grains of B line, therefore, set seeds on A plants. The seeds are genotypically identical with A line.

On commercial scale, the A line is propagated by planting two rows of A line alternating with 1 row of B line in isolation (distance depends upon he crop). For example, in sorghum, isolation distance is 300 m from other sorghums and 400 m from *Sorghum halepense*.

20. SOURCES OF CYTOPLASMIC GENETIC MALE-STERILITY

Spontaneous or induced mutations or inter specific crosses are the major sources of this kind of male-sterility. Spontaneous cytoplasmic genetic male-sterility has been isolated in maize (C, S and T type), pearl millet (Tifton and Ludhiana types) and sunflower. Besides, wild species of crop plants are important sources of sterile cytoplasm. Such cytoplasm has been exploited by placing the nucleus from one species in to them, through inter specific hybridization and back crossing. Such transfer has been made in several crops as given in Table 1.

Table 1. List of some crop plants in which male-sterility is found

Plant family	Name of the crop plants		Type of male-sterility found
	English	Scientific	
Gramineae	Barley	*Hordeum vulgare*	GMS
	Wheat	*Triticum aestivum*	GMS,CGMS
	Maize	*Zea mays*	GMS,CGMS
	Sorghum	*Sorghum bicolour*	GMS,CGMS
	Pearl millet	*Pennisetum americanum*	CGMS
	Rice	*Oryza sativa*	GMS, CGMS
	Sugarcane	*Saccharum officinarum*	CGMS
Leguminosae	Luceme	*Medicago sativa*	GMS
Compositae	Sunflower	Helianthus annuus	GMS, CGMS
Cucurbitaceae	Watermelon	*Citrullus vulgaris*	GMS
	Muskmelon	*Cucurbita moschata*	GMS
	Pumpkin	*Cucurbita maxima*	GMS
Malvaceae	Cotton	*Gossypium hirsutum*	GMS, CGMS
Solanaceae	Tomato	*Lycopersicon esculentum*	GMS, CGMS
	Tobacco	*Nicotiana tabacum*	GMS, CGMS
Liliaceae	Onion	Alum cepa	GMS, CGMS
Chemopodiaceae	Sugarbeet	Beta vulgaris	GMS, CGMS
Linaceae	Linseed	*Linum usitatissimum*	CGMS
Umbelliferae	Carrot	*Daucus carrota*	CGMS

GMS = Genetic male-sterility; CMS = Cytoplasmic male-sterility;
CGMS = Cytoplasmic genetic male-sterility.

21. FORMS OF CYTOPLASMIC GENETIC MALE-STERILITY

On the basis of its origin Lacadena (1968) suggested two forms of cytoplasmic-genetic male-sterility viz., autoplasmic and alloplalsmic male-sterility. The former type refers to the cases where cytoplasmic male-sterility is originated within a species through spontaneous mutational changes in the cytoplasm, most likely in the mitochondrial genome, while the alloplasmic is the one, which has, arisen from inter-generic or inter-specific crosses and may result due to incompatibility or poor co-operation between the nuclear genome of one species and the organellar genome of another.

22. UTILIZATION OF CYTOPLASMIC GENETIC MALE-STERILITY

Cytoplasmic genetic male sterile lines are of immense importance in the exploitation of hybrid vigour. It can be utilized successfully for the development of commercial hybrid in both seed propagated and vegetatively propagated crop plants. Krug (1932) made the first proposal for the production of hybrid maize through the use of cytoplasmic genetic male-sterility. Since then, highly successful cytoplasmic-genetic male-sterility sources have been developed in onion and sugar beat. In onion, it has been reported to be exploited by Jones and Davis (1944) to produce first onion hybrid (California hybrid red No. 1). Workable CMS has been reported at low frequency in many crops and would probably be found, even though rarely, in all crops, if they were searched intensively.

23. HYBRID SEED PRODUCTION

The cytoplasmic-genetic male sterile lines are being employed in India to produce hybrid seed on commercial scale of several crop species, including sorghum, pearl millet, rice, cotton (on a limited scale) and vegetables.

The line that restores fertility of the CMS line possesses a restores nuclear gene but may possess a fertile or sterile cytoplasm. The restorer gene is completely dominant to the gene causing male-sterility in several crops. However, a set of modifier genes has been reported to govern fertility restoration in cotton. The pollen grains of restorer line (R line) are genotypically different from the A line. In most of the cases the genes complement those of A line to cause considerable heterosis. The genes interact with factors for sterility in cytoplasm (mt DNA) in such a way that F_1 between A and R line produce viable pollen grains. Consequently self-pollination/cross pollination of the F_1 plants set seed and produce a heterotic crop as given in Fig. 7.

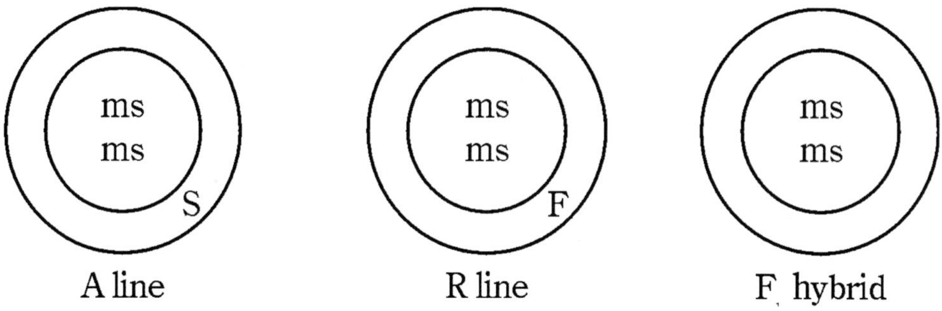

Fig. 7 Hybrid seed production

For hybrid seed production on commercial scale, usually 4 to 6 rows (up to 8 rows in case of rice) of A line are alternately planted with 2 rows of R line. The seed field is usually surrounded by at least 4 rows of R line (Border rows). For successful hybrid production, it is desirable that R line synchronously flowers with A line and produces ample pollen grains which are easily wind borne, otherwise, insect vectors or other means are required.

24. LIMITATIONS OF USING CYTOPLASMIC GENETIC MALE-STERILITY

Certain minor genes or modifier genes and some environmental factors have been reported to influence the degree of fertility restoration. A small amount of cytoplasm is also contributed by male gamete which may break down the male-sterility, although partially. It has also been found in sorghum that a temperature higher than 40°C breaks down the male-sterility producing a few seeds on the selfed panicle heads. Besides, maintenance of A, B and R lines need to be handled for a hybrid seed production programme. A brief comparison of three types of male sterility is presented in Table 2.

Table 2. Comparison of three types of male sterility in crop plants

S. No.	Particulars	Genetic male sterility	Cytoplasmic male sterility	Cytoplasmic genetic male sterility
1	Control	It is controlled by nuclear genes	Controlled by cytoplasmic genes	Controlled by both nuclear and cytoplasmic genes
2	Material included	Consists of A and B lines	Consists of A and B lines	Consists of A, B & R lines
3	Effect of temperature	May become fertile at low temperature	No effect of temperature	No effect of temperature
4	Maintenance	Maintained by mating sterile plants with heterozygous male fertile plants	Maintained by crossing of A line with B line	Maintained by crossing of A line with B line
5	Uses	Used for production of hybrids in both seed and vegetatively propagated crops	Used for development of hybrid in vegetatively propagated crop plants	Can be used in both seed propagated and vegetatively propagated species
6	Demerits	The male sterile and fertile plants are observed in 1:1 ratio. The fertile plants have to be removed in hybrid seed production plot	It can be used in seed propagated plants because the F_1 is sterile	It requires handling of three types of material i.e. A, B and R lines
7	Examples	Barley, wheat, Lucerne, watermelon, muskmelon, pumpkin, rice, cotton, tomato, tobacco, sunflower etc.	Onion, sugarcane, forage crops etc.	Pear millet, sorghum, cotton, wheat, maize, rice, sugar beet, sunflower, tobacco, tomato, etc.

25. VARIABILITY OF CYTOPLASM

The major limitation in the exploitation of the CMS lies in the fact that it can lead to a high degree of uniformity in cytoplasmic genomes, if only a single source of cytoplasm is used, which even in the presence of nuclear diversity can make a crop genetically vulnerable. One of the examples of use of single cytoplasm and genetic vulnerability is the outbreak in epidemic proportion of Helminthosporium leaf blight in maize in the 1970's in USA. This occurred due to the extreme susceptibility of most of the commercial hybrids to the leaf blight. This susceptibility was produced by the Texas male sterile cytoplasm present in these hybrids. A similar case in India is the susceptibility of early pearl millet hybrids to downy mildew and ergot and again this susceptibility was contributed by male sterile parent (female parent) of these hybrids or due to sterile cytoplasm.

Occurrence of disease epidemics due to the narrow cytoplasmic base of hybrids, under cultivation has alarmed breeder about the danger of maintaining a high degree of cytoplasmic genetic homogeneity and further induced them to look for alternative and diverse sources of CMS in important crops.

Various cytotypes have been identified in crops viz., sorghum, maize, brassica, pearl millet, rice, wheat, cotton etc. In sorghum, four cytotypes A_1, A_2, A_3 and A_4 (Worstell *et al.*, 1984) have been reported in USA. Out of these A_1 (milo cytoplasm) has been used extensively for commercial hybrid seed production. In maize, three distinct CMS cytoplasm have been designated as CMS-T, CMS-C and CMS-S. In pearl millet, some hybrids have been developed under HB series on male sterile tift 23 A (with A_1 cytoplasm) which was identified in 1965 (Burton, 1965). However, the pearl millet breeders have recognized three sources of cytoplasm i.e. A_1, A_2 and A_3, at present. Similarly, cytotypes have been reported in other crops also.

26. CHEMICALLY INDUCED MALE-STERILITY

In the chemical induction of male-sterility, physiological male-sterility is artificially induced by spraying the plants with gametocides to cause stamen sterility without any injury to the pistil. Chemicals, which sterilize the stamen, with little or no effect on normal functioning of the pistil are referred to as male gametocides or chemical hybridizing agents (CHA) or chemical pollen suppressants or chemosterilants. Here after we will refer it as male gametocides two genotypes, which exhibit good specific combining ability, are planted in alternate strips and the seed parent is chemically male–sterilized. This enables hybrid seed production on seed parent.

27. ADVANTAGES OF MALE GAMETOCIDES

Use of male gametocides obviates the need for genetic or cytoplasmic genetic male-sterility system. It simplifies the hybrid seed production programme as the development of male-sterile and restorer parent and the maintenance of A, B and R lines is not at all required for hybrid seed development. Also, it is regarded as a rapid method of developing male-sterile lines, as only one season is required to induce male-sterility. Further, it is less laborious and less expensive than transfer of male-sterility through backcross.

28. USE OF MALE GAMETOCIDE

A number of chemicals have been found to be effective on various crop plants (Table 3). However, these chemicals affect both male and female gametes and the plant growth in general. Hence, these should be used cautiously. In China; chemically induced male sterile lines have been used for the development of hybrids of rice and released for commercial cultivation (Shen, 1980). It has been used to produce rice hybrids on a small scale (Illyas-Ahmed, 1996). The hybrids based on chemically induced male-sterile line of rice had some difficulties viz., use of gametocide is not completely effective and efficient in a multi-tillered crop like rice where panicles emerge continuously over a period of 5-7 days. Moreover, unfavorable weather at critical period when the gametocide is to be sprayed also adversely influences the hybrid seed programme. For example, rains during effective emasculation period will reduce the effectiveness of male gametocide.

Table 3. List of gametocides found effective on some crop plants

Male gametocide	Crop species on which found effective
Sodium Methyl Arsenate	Rice (used in China)
Zinc Methyl Arsenate	Rice (used in China)
Naphthalene Acetic Acid (NAA)	Cucurbits
Gibberellins	Rice, Maize, Sunflower, Onion, Lettuce
Maleic Hydrazide	Wheat, Tomato, Cucurbit, Onion
FW-450	Cotton, Tomato, Groundnut, Sugarbeet
Ethrel	Wheat, Rice, Sugarbeet

29. NON-POPULARIZATION OF MALE GAMETOCIDES

As per Sneep and Hendriksen (1979) following are the reasons for non-popularization of male gametocides:

(i) Less effective and partial male sterilization.

(ii) Incomplete and erratic pollen abortion.

(iii) Partial female-sterility leading to reduced seed set.

(iv) Phyto-toxicity exhibiting harmful side effect on plants viz., leaf deformation, stunting of plant growth and wilting etc.

(v) Short-lived and growth stage specific effect make it necessary to repeat the application.

(vi) High genotype x environment x chemical interaction.

(vii) High cost of gametocide and their repeated application make their use uneconomical.

30. PROPERTIES OF AN IDEAL MALE GAMETOCIDE

Sneep and Hendriksen (1979) and Chopra (1985) have described the following features, which an ideal gametocide should possess:

(i) It should be selective in inducing male-sterility without affecting ovule fertility.

(ii) It should give consistent results i.e. the result should be reproducible.

(iii) It should be less expensive and its method of application should be simple and economical.

(iv) It should be safe for the use and should have the minimum side effect on plant growth.

(v) It should not have mutagenic effect on plants.

31. GAMETOCIDES v/s MUTAGENS

Several mutagens have also been reported to induce genetic as well as cytoplasmic male-sterility of heritable nature. However, the sterility induced by the gametocide is non-heritable, hence, it requires to be induced every year for hybrid seed production.

The prospects for the large-scale application of gametocide appears bright in India though extensive, efforts are needed to establish their use for hybrid seed production on commercial scale.

32. FUNCTIONAL MALE-STERILITY

Inability of anthers to dehisce and disperse otherwise normal and viable pollen grains is referred to as functional male-sterility. The possible reason of this restriction on pollen release could be due to the presence of tough or hard outer layer surrounding the microsporangium. This is also known as positional male-sterility. It is governed by a single recessive gene (Nuttal, 1963). The vegetable crop plants viz., tomato, brinjal and cabbage have been reported to posses this type of male-sterility (Sneep *et al.*, 1979). The former two crops have stigma hidden within the stamen core that prevents cross-pollination and precludes the use of such male sterile lines. If such functional male-sterility is found to be associated with exerted stigma, only then it could be utilized for hybrid seed production.

Besides, the unstability of male-sterility particularly under dry, hot and windy weather also restricted its use for large-scale hybrid development. However, identification of suitable genetic marker(s) may help to ensure the elimination of self-ones in the hybrid seed production plot.

It has been reported by Reover (1948) that a recessive gene in tomato produces normal viable pollen but due to non-dehiscence of anther, selfing is prevented. However, if anther dehiscence is induced manually, selfing could be observed in this mutant. The male sterile mutant can thus be easily maintained and used for hybrid seed production without emasculation. Similar mutants have been found in maize (Horowitz and Obregon, 1951), eggplant (Jasnum, 1954) and beans (Wyatt, 1983). Although, some of the mutants mentioned seem to have potential for hybrid seed production but have not been used commercially. However, Barabas (1991) described a method of hybrid seed production in tomato involving functional male-sterility.

33. ENVIRONMENTAL GENIC MALE-STERILITY

Environmental factors like the temperature, photoperiod etc. have been reported to influence the expression of plants. Photoperiod is known to induce flowering in crop plants, while the male-sterility mechanism is known to be influenced by temperatures. For example, the cytoplasmic-genetic male-sterility of sorghum based on milo-kafir system breaks down when temperatures exceed 40°F. Castor, *Ricinus communis*, is highly polymorphic. Normally, castor is monoccious with about 50 to 70% male flowers occurring towards the base of a racemose inflorescence and 30 to 50% pistillate flowers in the upper portions. But, the stability of sex expression varies with the environment, age of the plant and nutrition (Shifriss, 1956). Female tendency is highest in winters, while male tendency predominates in summer and rainy season.

The use of cytoplasmic-genetic male-sterility is cumbersome for exploitation of heterosis. Three lines, viz., A, B and R are to be maintained and a separate crossing block has to be established for hybrid seed production. An alternate approach is being now utilized in China for hybrid seed production of rice on commercial scale.

Alteration of fertility/sterility in response to environmental factors, especially to temperature or photoperiod, is utilized for, what is called, two line heterosis breeding is used in rice. This type of male-sterility is referred to as environmental sensitive genic male-sterility (EGMS). Prof. Shri Min Shong of China discovered a photoperiod-sensitive plant of rice NONGKEN 585 in Japonica variety (Shi, 1981; Shi and Deng, 1986). Similarly, some lines have been identified which are thermo-sensitive genic male-sterile (TGMS). Such lines have complete pollen sterility at 30-32°C but partial to normal fertility at 24-27°C. Among the indigenously identified TGMS lines in India, are ATG_1, ATG_4 and ATG_8 (Siddiqu, 1996), which appear promising for two-line heterosis breeding. ATG_8 is a reverse TGMS type, in that; it is sterile below 24°C and fertile above 28°C. The sensitivity is reported to be governed by a single recessive gene.

34. UTILIZATION OF TGMS LINE FOR HYBRID SEED PRODUCTION

The TGMS line does not require a maintainer or B line for its perpetuation because it produces viable pollen grains above 30-32°C. During the sterile phase the line is used as seed parent/female parent for production of hybrid seed. During fertile phase it is used for own seed production. Since there are wide temperature differences across latitudes and attitudes in India, TGMS lines are more useful. Some of the TGMS lines have been characterized with respect to their critical sterility point (CSP) and critical fertility point (CFP). Four TGMS lines received from IRRI were evaluated in India. The hybrid seed production on some of these, e.g. IR 68949, could be taken up in late dry season of *rabi* while during *kharif*, its seed could be multiplied. TGMS has several advantages in rice over CMS system for hybrid seed production. There is no need for maintainer line. Any genotype could be utilized for hybrid seed production as male parent. Thus, genic base is widened. Negative effects associated with cytoplasm, as in case of cotton hybrids based on CMS lines (Bhale and Bhat, 1990; 1993) are avoided. Seed production is simple and more efficient. However, it is desirable that TGMS lines should consist of 100 per cent sterile plants and the degree of male-sterility (percentage of abortive pollen gram) should be over 99.5%. Moreover, the change from fertility to sterility and vice-versa should be quite clear. The male-sterile period should last for more than 30 consecutive days.

35. MAINTENANCE OF GENETIC PURITY OF EGMS

The maintenance of genetic purity of TGMS/PGMS lines is essential. Isolation of the TGMS/PGMS lines from other undesirable lines should be at least 200 m.

36. GENETICALLY ENGINEERED MALE-STERILITY SYSTEM

The development of recombinant DNA and plant transformation techniques has created new possibilities for engineering genetic male sterility. It has broadened the use of male sterility by an understanding of which gene(s) and protein(s) confer male sterility and how they do so. Several genes and strategies for using these genes for inducing male sterility and for pollination control have emerged since the first genetically engineered nuclear male sterility gene was reported in 1990 by Mariani *et al.*

Recent evidence at molecular level suggested that genes controlling CMS in crops like maize, sorghum, rapeseed, mustard, rice, sugarbeet and lucern etc. are located on mitochondrial DNA and are deeply involved in the expression of fertility/sterility in higher plants. Substantial evidences suggested that the CMS trait be encoded by the mitochondrial (mt) rather than the chloroplast genome (Levings and Brown, 1989). In view of the unique feature of mt-DNAs emerging from the current observations on the biochemical, molecular and genetically characterizations of mitochondrial genetic systems, the role of plasmid like DNAs in plants is quite evident. In addition, the transfer of CMS is attainable in one step through mitochondrial manipulations involving protoplast fusion or microinjection, which otherwise requires 6-7 backcrossings in routine conventional transfer scheme. In future, such mitochondrial transformation will be an asset to introduce new CMS genes into species without CMS and for engineering improvements into existing CMS genes.

Other approaches for inducing male sterility through genic manipulations include modification of location and mode of action of chemical hybridizing agents (O'Keefe et *al.*, 1994), induction of male sterility by *in vitro* culturing and creation of CMS by protoplast fusion. Induction of CMS by *in vitro* culture was reported in *N. syslnestris*. Two CMS plants CMS I and CMS II were obtained by protoplast culture without somatic hybridization. Somatic hybrids from *Nicotiana, Petunia, Brassica* and *Oryza* genera had exhibited CMS trait. Analysis of CMS cultivar and their hybrids supported the hypothesis that CMS phenotypes are the result of incompatible interactions between the nucleus and several mitochondrial genes.

Thus, it is evident that the development of recombinant DNA and plant transformation techniques has created new possibilities for engineering genetic male sterility system. Also, combined with future advances in transformation technology the manipulation of male sterility could become an important application of plant biotechnology which will promise both improved CMS and nuclear male sterility systems for future hybrid seed production. Further, it is worth mentioning that the speed with which Indian breeders have developed high grain yielding hybrids using male sterility system is a most remarkable plant breeding success story, which suggests that the future for hybrid varieties in India is still very bright.

REFERENCES

Allard, R.W. (1960). Principles of Plant Breeding. John Wiley and Sons, Inc, New York.

Allison, D.C. and Fisher, W.D. (1964). *Crop Sci.*, 4:548-549.

Athwal, D.S. and Borlaug, N.E. (1967). *Indian J. Genet.*, 27: 136-142.

Bailey–Series, J.; Dixon, L.K.; Lid dell, A.D. and Learer, C.J., (1986). *Theor.Apple-Genet.*,73: 252-260.

Basu, A.R.; Narayanan, S.S. and Bhat, M.G. (1995). Hybrid Cotton: In "Hybrid Research and Development" Ed. M.Rai and S.Mauria. Proc. Of Indian Soc. Of Seed Tech: Seminar, New Delhi pp. 63-80.

Bateson, W. and Gairdner, A.E. (1921). *J. Genet.*, 11:269-275.

Beadle, G.W. (1932). *Genetics*, 17:413-431.

Bhale, N.L. and Bhat, M.G. (1990). *Indian J. Genet.*, 50(1):37-44.

Bhale, N.L. and Bhat, M.G. (1993). Proc. FAO-ICAR. Regional Exprt Consultation on hybrid cotton in India. Oct. 22-25, 1990.

Bowman, D.T. and Weaver, J.B.Jr. (1979). *Crop Sci.*, 19: 628-630.

Burton, G.W. (1958). *Agron. J.*, 50: 230-235.

Burton, G.W. and Hanna, W.W. (1976). *Crop Sci.*, 16:731-732.

Burton, G.W. and Powell, J.B. (1968). *Adv.Agron.*,20: 49-89.

Casady, A.J., Meyne, E.G. and Weibel, D.E. (1960). *J.Heledity*, 51(1): 35.

Chopra, V.L. (Ed.) (1989). Plant Breeding Theory and Practice. Oxford and IBH Publishing Co., New Delhi.

Driscoll, C.J. (1972). *Crop Sci.*, 12:516-517.

Duvick, D.N. (1966). In "Plant Breeding" Ed. K.J. Frey. The Iowa State University Press, Ames, Iowa pp. 85-138.

Frank, S.A. (1989). *Am. Nat.*, 133:345-376.

Frankel, R. and Galun, E. (1977). Pollination mechanisms, reproduction and plant breeding. Springer, Berlin, Heidelberg, New York.

Gableman, W.H. (1956). Brookhaven symposia in Biol., 9:113.

Gilmroe, E.C. (1964). *Crop Sci.*, 4:323-325.

Govila, O.P. (1989). In Plant Breeding Theory and Practice. Ed.Chopra, V.L. Publ. Oxford and IBH Publishing Co. Pvt. Ltd., New Delhi. Pp. 213-223.

Hanson, M.R. and Conde, M.F. (1985). *Int. Rev. Cytol.*, 94:213-267.

Illyas, Aahmed, M. (1996). In "Hybrid Rice Technology" pp.55-65.

Jain, S.K. (1959). *Bibliogs. Genet.*, 18:101-166.

Jones, D.F. (1950). *Genetics*, 35:507-512.

Jones, H.A. and Davis, G.N. (1944). *USDA Tech. Bull.*, 874.

Jones, H.A. and Clarke, A.E. (1943). *Proc. Amer. Soc., Hort. Sci.*, 43:189-194.

Karper, R.E. and Stephene, J.C. (1936). *Heredity*, 27:183.

Kaul, M.L.H. (1988). In Monograph on Theoratical and Applied Genetics Vol.10. Ed. Frankel, R., Crossman, M., Liskens, H.F., Maliga, P.; R.Riley. Springer-Verlag Berlin. pp. 775-797.

Kinman, M.L. (1970). New developments in USDA and State Experiment Stations Sunflower breeding programme. Proc. 4th Intern. Sunflower Conf. Meniphis, T.N. 181-184. Intern. Sunflower ASSOG, Toowoomba, Australia.

Krug, C.A. (1932). M.S.Thesis. Cornell Uni., Ithaca, N.Y.

Lasa, J.M. and Bosemark, N.O. (1993). Male-sterility in plant breeding. Principles and Prospects, Edited by M.D. Hayward, N.O. Bosemark and I. Romagosa. Chapman and Hall. Pp.231-228.

Laser, K.D. and Larsten (1972). *Botanical Rev.*, 58:425.

Laughman, J.R. and Gabay-Laughman, S.J. (1983). *Annu. Rev. Genet.*, 17:27-43.

Leclercqm, P. (1969). *Ann. Amelior. Plant*, 19(2): 99-106.

Levings, C.S.III (1990). *Science*,: 250:942-947.

Levings, C.S.III and Pring, D.R. (1976). *Science*, 193:158-160.

Mariani, C.; Beuckeleer, M.; Truettner, J.; decmans, J. and Goldberg, R.B. (1990). *Nature*, 347:737-741.

Maunder, A.B. and Pickett, R.C. (1959). *Agron. J.*, 51:47.

Mital, S.P. and Thomas, T.A. (1961). *Indian J. Hort.*, 18:150-151.

Pour, A.K.; Gracen, V.E. and Everett, H.L. (1981). *Genetics*, 98:379-388.

Reddy, V.S.; Green, J.M. and Bisen, S.S. (1978). *Crop Sci.*, 18: 362-364.

Rhoades, M.M. (1931). *Science*, 73 : 340-341.

Rick, C.M. (1984). Plant Germplasm Resources. In: Handbook of Plant Cell Culture, Volume 3. Crop Species P.V. Ammirato, D.A., E., Vans, W.R. Sharp and Y. Ymada (eds) Mcnullah Publishing Co., New York.

Rick, C.M. and Butler, L. (1956). *Adv.Genet.*,8:267-382.

Richmond, J.R. and Kohel, R.J. (1961). *Crop Sci.*, 1:397-401.

Roever, W.E. (1948). *Science*, 107-506.

Sasakuma, J.; Maan, S.S. and Williams, N.D. (1978). *Crop Sci.*, 18:850-853.

Schweisguth, B. (1973). *Ann.Amer.Plant*, 13:221.

Sears, E.R. (1947). Yearbook Agr.USDA. pp.245-255.

Shen, Hin-Hwa (1980). Rice breeding in China. In: Rice Improvement in China and other Asian Countries, International Rice Research Institute, Los Banos, Laguna, Phillippines, pp.9-36.

Shi, M.S. (1981). J.Hubei *Agric.Sci.*1-3.

Shi, M.S. and D eng,J.Y. (1986). *Acta Genet.Simia*, 13(2): 107-112.

Shifriss, O. (1956). *Genetics*, 41: 265-280.

Siddiqu, E.A. (1996). "Hybrid Rice Technology" Ed. Illyas Ahmed, M.; B.C. Viraktamath; Ramesha, M.S. and Vijay Kumar, H.M. Diirectorate of Rice Research, Hyderabad, pp.1-26.

Sheep, J. and Hendriksen, A.J.T. (eds) (1979). Plant Breeding Perspectives. Pudoc, Wageningen, Netherlands.

Sheep, J., Murty, B.R.; Utz, H.P. and Vranceau, A.V. (1979). In: Perspectives, J.T. Hendriksen, J., Sheep and Holbek (eds) Centre for Agricultural Publishing and Documentation, Wageningen, pp.216-223.

Stephens (1937). *J.Amer.Soc.Agron.*,29:690-696.

Stephens, J.C. and Holland, R.F. (1954). *Agron.J.*,46:20.

Virmani, S.S.; Khush, G.S. and Pingali, P.L. (1996). In: "Hybrid Riced Technology" Ed. Illyas Ahmed, M.; B.C. Viraktamath, M.S. Romesha and C.H.M. Vijaykumar". Directorate of Rice Research, Hyderabad (A.P.). pp.27-50.

Vrqnceanu, A.V. (1974). Editura Academica Republici Socialiste, Bucharest, Romania (Sunflower). P.322.

Weaver, J.B. and Ashley, T. (1971). *Crop Sci.*, 11:596-598.

Yuan, L.P. (1996). In: "Hybrid Rice Technology". Pp.51-54.

CHAPTER - 18

SCREENING OF COTTON GENOTYPES FOR FIBRE QUALITY

Suman Bala Singh and Phundan Singh

1. INTRODUCTION

Cotton occupies an important position in the world fibre market. The crop is cultivated in more than 80 countries the world over in about 32.2 million hectares annually with an annual production of around 20 million tonnes of lint and 40 million tonnes of cotton seed. The average lint yield of the world is around 585 kg per ha. Seventy per cent of the global cotton production is concentrated in five countries namely China, USA, India, Uzbekistan and Pakistan. India ranks third in production with 1.5 million tonnes of lint and 3.98 million tonnes of cotton seed from 8.6 million hectares. India has not yet come out of the low productivity status, though in the last four decades it has risen from 88 kg. to 301 kg/ha. The low productivity is mainly attributed to rainfed cultivation, which covers about 70% of the cotton in India.

2. FIBRE PROPERTIES AND THEIR TECHNOLOGICAL SIGNIFICANCE

Fibre quality is one of the important factors that should be taken into account in cotton breeding programmes. Central Institute for Cotton Research on Cotton Technology was established at Bombay in 1924. The primary objective of CIRCOT is to evaluate cotton genotypes in terms of fibre quality parameters and help breeders in the development of cotton cultivars / hybrids which can compete in the global market for fibre quality.

The fabric quality is mostly governed by the yarn quality, which in turn depends upon the properties of the fibres. The quality of raw cotton is, therefore, an important factor to be taken note of and is judged from the physical characteristics of the fibres. Important fibre properties include fibre length, uniformity, fineness, strength and maturity. This chapter deals with various fibre properties, genetics of fibre quality, factors affecting fibre quality and sources of fibre quality traits.

2.1 Fibre Length

Fibre length is measured in three ways – (i) halo length, (ii) mean length, and (iii) 2.5% span length. These are briefly discussed as follows:

Halo Length: The length of the fibre with attached seed is referred to as halo length. It gives rough estimate of fibre length and is measured with the help of halo disc.

Mean Fibre Length: It is arithmetic mean of the length of all the fibres present in a sample. It is measured with the help of bolls sorter, comb sorter and AN sapling apparatus.

2.5% Span Length : It is the distance from the clamp on a fibre beard to the point upto which only 2.5% of the fibres extend. It is measured by digital fibrograph. Cotton fibre is grouped in five categories (Table 1).

Table 1. Categories of Fibre Length in India

Category	Mean Staple Length	2.5% Span Length
Short	<19	<20
Medium	20-21.5	20-24.5
Superior Medium	22-24	25.0-27.5
Long	24.5-26.0	28.0-31.5
Extra long	27.0	32.0

2.2 Uniformity Ratio

The ratio of 50% span length to 2.5% span length expressed as percentage is called uniformity ratio (UR%). Uniformity ratio gives an idea about length variability in a sample. Again, cotton fibre is classified into five groups based on uniformity (Table 2).

Table 2. Classification of Cotton on Fibre Based on Uniformity

Category	Range of Uniformity Ratio
Poor	42
Fair	42 to 43
Average	44 to 45
Good	46 to 47
Excellent	47

2.3 Fibre Fineness

This is another important quality of the fibre, which together with length is important for spinning value of cotton. Generally, short staple cottons are coarser than long stapled ones. It is closely related to fibre maturity because it depends upon both the diameter and the wall thickness of the fibre. An instrument, which is now used for this purpose, is the 'Micronaire' though there are several others. The test gives a direct reading of the weight in microgram per inch. Based on the micronaire value, cotton fibre is classified into five categories:

Table 3. Classification of Cotton Fibre Based on Micronaire Value

Category	Range in Micronaire
Very fine	3.0
Fine	3.0 – 3.9
Average	4.0 – 4.9
Coarse	5.0 – 5.9
Very Coarse	6.0

2.4 Fibre Strength

This is the most important property of cotton fibre and has direct correlation with yarn strength. The mean breaking load of single fibre or bundle of fibres under standard test length gives an index of strength of the fibres. Pressley Strength Tester gives the strength of the sample within a short time and expressed as pounds per milligram. This represents the average breaking load in pounds of fibres of one-milligram weight.

Bundle strength is generally measured at zero and 3.2 mm (1.8") gauge length. The strength at 3.2-mm.-gauge length is only about half the strength at zero gauge length. The reduction is due to the presence of "weak links" in fibre structure. The ratio of strengths at 3.2 mm. and zero gauge lengths is referred to as Strength Uniformity Ratio (SUR) has been found to range between 0.4 to 0.6 in different varieties.

$$\text{Tenacity (g/tex)} = \frac{\text{Breaking load in kg x 118}}{\text{Weight of bundle in mg}}$$

This formula is for zero gauge length i.e. without spacer between clamps. Here, 11.8 mm is the width of the clamp. When a space of 3.2- mm (1/8") is kept between the clamps to increase the gauge length, the length of fibres in the test becomes 15 mm, and hence tenacity at 3.2- mm gauge length is given by

$$\text{Tenacity (g/tex)} = \frac{\text{Breaking load in kg x 15}}{\text{Weight of bundle in mg}}$$

$$\text{PSI (LB/mg)} = \frac{\text{Breaking load in lb}}{\text{Weight of bundle in mg}}$$

PSI can be converted into tenacity (g/tex) by using the following relation:

Tenacity (g/tex) = 5.36 x PSI

When strength is measured at 3.2 mm Gauge

Tenacity (g/tex) = 6.80 x PR

$$PR = \frac{\text{Breaking load in lb}}{\text{Weight of bundle in mg}}$$

The bundle strength of cotton i.e. the strength of a group of fibres taken together gives a better correlation to yarn strength. Based on the bundle strength value, the cotton can be grouped into following categories.

Table 4. Categories of Fibre Strength in Cotton

Category	Range of bundle strength value		
	(LB/mb)	Tenacity (g/tex)	3.2 mm gauge (g/tex)
Very good	9.0 and above	47.5 and above	26.0 and above
Good	8.0 to 8.9	43.1 to 47.4	23.1 – 26.0
Average	7.0 to 7.9	37.5 to 43.0	20.1 – 23.0
Low	6.5 to 6.9	34.5 to 37.4	16.1 – 20.3
Very low	Below 6.5	Below 34.5	16.0 and below

2.5. Fibre Maturity

It is an important character contributing to fineness and strength and ultimately spinning quality. Maturity of a cotton fibre indicates the degree of secondary thickening of the cell wall in relation to its diameter. The deposition of the cellulose inside the fibre is not uniform in all fibres and varies widely in fibres on the same seed and even at different portions of a fibre. These fibres are classified as mature which are having thick cell wall and narrow lumen white half mature and immature fibres have no secondary thickening or extremely thin walls and a wide limen.

$$\text{Maturity coefficient (M.C.)} = \frac{M + 0.6H + 0.41}{100}$$

Where, M = Mature, H = Half mature, I = Immature fibres.

2.6 Percentage of mature fibres (PM)

This is the method described in ASTM standard and followed in USA. Procedure is the same. However, viewing the swollen fibre they are classified only into groups according to the following criteria:

Category	L / W
Mature	<2
Immature	>2

(L = Lumen width, Wall thickness)

From the total number of fibres and number of mature fibres, PM (Percentage of mature fibre) is calculated. Based on the maturity coefficient, the cotton is classified in five categories and based on percent mature fibres (Table 5).

Table 5. Categories of Fibre Maturity in Cotton

Categories	Range of M.C.	Percentage of mature fibres
Very high maturity	Above 90	Above 81
Good maturity	81 to 90	66 to 81
Average maturity	71 to 80	50 to 65
Immature	60 to 70	32 to 49
Very immature	60	Below 31

2.7. Genetic of Fibre Quality Traits

Fibre quality in cotton may be inherited by oligogenes with distinct effect or polygenes will small additive effects. The colour of fibre is governed by oligogenes. The inheritance of other fibre characters viz. Fibre length, strength, fineness and maturity is governed by polygenes. These characters are mainly governed by additive gene action with some degree of dominance. Seed oil and protein contents are also governed by polygenes.

Few mutants are known to influence fibre development. One such mutant is immature fibre, conditioned by the recessive allele 'im' (Kohel and McMichael, 1990). The 'im' mutant expresses on a normal seed, but has a fine, weak fibre typical of immature seed. Two dominant fibre mutants that affect fibre length are Ligon Lintless-1 and Ligon Lintless-2 (Kohel, 1972; Narbuth and Kohel, 1990).

Fuzz amount and distribution on seed coats varies widely in tetraploid species and probably is inherited quantitatively. Presence and absence of fuzz is a qualitative trait determined by two loci.

3. FACTORS AFFECTING FIBRE QUALITY

Fibre properties are largely hereditary and therefore are characteristics of the varieties. However, wide variations in these properties occur due to different environmental conditions of growth in the same variety. Some of the factors affecting fibre properties in cotton are discussed below:

3.1 Locality and Season

When seeds of any variety are grown in different localities, fibre properties of the lint produced vary considerably. Of these, fibre maturity is the most affected followed by fibre weight/unit length. In same cases, mean fibre length is also affected.

3.2 Fertilizers

The effect of fertilizers on the quality of cotton lint raised in a soil is complicated as it depends on several factors, such as the initial fertility, soil texture, moisture etc.

3.3 Picking

The quality of lint picked at different intervals varies considerably. Later pickings yield shorter and more immature lint than the earlier ones.

3.4 Insect Pests and Diseases

Cotton plants is attacked by many insects, like bollworms (*Heliothios*, pink and spotted), boll weevil, jassids, white-fly, etc. These insects have adverse effects on the quality and yield of the cotton produced. It is reported that a jassid attack leads to significant fall in the mean fibre length, fibre weight/unit length and maturity in susceptible varieties, while no such changes are noticed in resistant varieties.

4. SOURCES OF FIBRE QUALITY TRAITS

In cotton, there are four important sources of fibre quality characters, viz. cultivated varieties, germplasm collections, wild species and induced mutations. The first two sources are widely used for improvement of fibre properties in this crop.

Cultivated varieties are good sources of fibre quality e.g. MCU 5 of upland cotton is a very good source of fibre length, strength and fineness. This variety is widely used as a donor parent for improvement of fibre properties.

Germplasm collections are important sources of fibre quality. Large number of germplasm lines of all the four cultivated species of cotton viz. *G. hirsutum* (6018 lines), *G. barbadense* (1142), *G. arboreum* (1870) and *G. herbaceum* (528) are maintained at the Central Institute for Cotton Research, Nagpur. The germplasm of *G. barbadense* is maintained at CICR Regional Station, Coimbatore. Part of this germplasm has been evaluated for fibre properties and donor lines for fibre traits have been identified (Narayanan, 1985). A wide range of variability for fibre characters is available in the germplasm of cotton maintained at CICR, Nagpur (Table 6).

Table 6. Range of variability for important fibre characteristics in the cultivated species in India

Character	*G. hirsutum*	*G. barbadense*	*G. arboreum*	*G. herbaceum*
Mean halo length (mm)	16-36	22-32	12-30	16-27
Span length (mm)	15-37	22-35	13-27	19-27
Uniformity ratio (%)	34-52	40-52	44-55	42-52
Micronaire value	2.6-5.6	2.5-4.2	3.1-8.0	2.4-6.2
Maturity percent	55-79	52-88	70-93	53-78
Fibre strength (g/tex)	31-56	37-58	32-59	41-58

Wild species are also potential sources of fibre properties. There are some wild species, which possess good fibre quality as a latent feature e.g. American wild diploid species *G. thurberi* has fibre strength as a latent feature. There are some other wild species, which possess good fibre properties (Table-7).

Table 7. List of Wild Species Possessing Good Fibre Properties

Name of wild species	Quality characters found
G. thurberi	Fibre strength (latent feature)
G. armurianum	Fibre strength (latent feature)
G. anomalum	Fibre fineness, strength and maturity
G. raimondii	Fibre length and fineness
G. tomentosum	Fibre strength and fineness
G. sturtianum	Fibre strength
G. australe	High spinning
G. aridum	Fibre strength
G. stocksii	Fibre strength and elasticity
G. areysianum	Fibre strength and elasticity
G. longicalyx	Fibre fineness and strength

The latest parameters of fibre quality, as recommended by CIRCOT Mumbai, to meet the requirement of modern processing industries in India is presented in Table-8. The quality norms recommended by Indian Cotton Mills are different from the recommendations of CIRCOT, Mumbai (Table-9).

Table 8. Fiber Quality Norms for Cotton Breeding – CIRCOT Recommendation

Yarn	Counts2.5% span length (mm)	UR %	Mic.	Tenacity (g/tex)	Elongation (%)	Maturity (%)	FQI	CSP
6s-20s	22-25	50	4.0-5.0	20-22	6.0	80	200-270	2400
21s-30s	26-27	50	3.8-4.8	22-23	6.0	80	270-320	2500
31s-40s	28-30	50	3.8-4.5	24-25	6.0	80	320-380	2600
41s-50s	30-31	50	3.5-3.8	25-27	6.0	80	380-450	2800
51s-60s	31-33	48	3.3-3.5	27-28	7.0	75	450-510	2900
61s-80s	33-35	48	3.2-3.4	28-30	7.0	75	510-590	3000
81s-100s	36-38	48	3.2-3.4	30-32	7.0	75	590-680	3200
101s-120s	36-38	48	3.2-3.4	30-32	7.0	75	590-680	3400

Table 9. Fibre Quality Norms for Cotton Breeding – Mills' Recommendation

Yarn Counts	2.5% span length (mm)	U R %	Mic.	Tenacity (g/tex)	Elongation (%)	Maturity (%)	FQI	CSP
6s-20s	22-25	50	4.0-5.0	20-22	6.0	80	200-270	2400
21s-30s	26-27	50	3.8-4.8	22-24	6.0	80	270-320	2400
31s-40s	28-30	50	3.8-4.5	23-25	6.0	80	320-380	2600
41s-50s	29-31	50	3.5-3.8	25-27	6.0	80	380-450	2800
51s-60s	31-33	48	3.3-3.5	26-28	7.0	75	450-510	2900
61s-80s	33-35	48	3.2-3.4	28-30	7.0	75	510-590	3000
81s-100s	36-38	48	3.2-3.4	30-32	7.0	75	590-680	3200
101s-120s	36-38	48	3.2-3.4	30-32	7.0	75	590-680	3400

Fibre properties of some of the currently cultivated cotton hybrids and varieties are presented in Table 10 and Table 11.

Table 10. Important fibre properties of cotton hybrids

Name of the hybrid	Year of release	Yield (q/ha)	Ginning out-turn (%)	2.5% span length (mm)	Micro-naire value	Bundle tenacity g/t 3.2mm gauge length	Spin-ning counts	Area of cultivation
Gujarat								
H8	1989	35	36	29	4.5	25.0	50	Gujarat state
H10	1995	18R	35	27	4.0	22.5	40	Gujarat state
Karnataka								
DCH 32**	1981	35	36	38	3.0	30.0	80	South Zone & Maharashtra
DHB 105**	1996	30	34	32	3.2	25.0	80	South Zone
DHH 11	1996	30	35	28	3.9	24.0	50	South Zone
Tamil Nadu								
Savita	1987	30	34	33	3.8	26	60	Tamil Nadu and Andhra Pradesh
HB 224**	1989	30	33	35	2.9	30	80	Tamil Nadu
TCHB 213**	1989	30	32	35	3.6	25	80	Tamil Nadu
Surya	1997	25	38	33	3.6	26	60	South Zone
Sruthi**	1997	30	33	35	3.5	28	80	South Zone
Maharashtra								
PKV Hy-2	1981	12 R	36	25	4.2	24	40	Maharashtra, Vidarbha
NHH 44	1983	23	35	28	3.8	24	50	Marathwada
NHB 12**	1989	30	33	34	3.0	28	80	Marathwada
NHH 302	1991	20	35	28	3.8	24	40	Marathwada
Madhya Pradesh								
JKHY 1	1976	25	35	28	4.2	25	50	M.P. and A.P.
JKHY 2	1994	30 15R	34	28	4.2	25	50	A.P.
Andhra Pradesh								
LAHH 1	1987	28	35	28	3.6	25	60	A. Pradesh
LAHH 4	1997	30	35	32	3.8	25	40	All three zones
Punjab								
Fateh	1994	30	34	26	4.4	22	30	Punjab
LHH 144	1998	28	35	28	4.6	23	50	Punjab
Haryana								
Dhanlaxmi	1994	25	35	27	4.2	25	40	Haryana
Om Shankar	1996	28	34	27	4.5	20		

AAH 1*	1999	24	38	17	6.8	18	Below 10	Haryana
Rajasthan								
Maruvikas	1994	30	34	26	4.0	21	30	Rajasthan

* diploid hybrid, ** *hirsutum* x *barbadense* hybrids, rest are *hirsutum* x *hirsutum* hybrids.

Table 11. Important fibre properties of cotton varieties

Name of the Varieties	Year of relea-se	Yield q/ha	Ginn ing out-turn (%)	M.H.L	Spin-ning counts	2.5% span length	Micro-naire value	Bundle tenacity g/t 3.2 gauge length	Area of cultivation
Punjab									
LH 900	1985	27	34	22	30	25	4.6	23.0	Punjab
LH 1556	1996	21	34	27	40	28	4.4	22.0	Punjab
F 846	1993	26	35	23	30	25	4.4	22.0	Punjab
F 137	1997	30	35	23	30	26	4.3	20.0	Punjab
LD 327*	1987	20	41	16	8	18	7.1	16.0	Punjab
LD 491*	1996	14	39	20	10	19	7.3	15.0	Punjab
Haryana									
HS 6	1991	22	36	22	30	24	4.7	20.0	Haryana
H 974	1991	21	35	23	30	25	4.9	24.0	Haryana
H 1098	1995	19	35	25	30	24	5.0	19.5	Haryana
HD 107*	1996	26	38	18	10	18	7.5	18.0	Haryana
HD 123*	1997	23	38	18	10	19	7.4	18.4	Haryana
Rajasthan									
RST 9	1991	26	36	23	30	24	5.2	24.0	North
RS 875	1997	30	35	23	30	27	4.9	20.0	North
Ganganagar ageti	1978	14	33	23	28	24	4.0	22.0	Sriganganagar
RG 8*	1986	17	38	16	8	16	7.5	17.0	Sriganganagar
U.P.									
Vikas	1987	20	33	23	30	24	4.8	18.0	Western
Lohit*	1969	4R	38	18	10	18	7.0	20.0	Western
M.P.									
Khandwa 2	1971	8R	36	24	30	25	4.2	22	Nimar
Khandwa 3	1984	9R	34	23	36	23	4.8		

Vikram	1981	9R	33	24	35	23	4.2	24	Malwa
Maljari*	1954	6R	33	21	20	22	5.0	24	Malwa, Nimar
Jawahar Tapti*	1992	15	35	24	30	24	4.5	21.5	Nimar
Gujarat									
G.Cot 12	1981	6R	36	24	23	24	4.0	22.0	Wagad area
G.Cot 16	1985	16R	37	25	40	27	4.3	22.0	Middle
G.Cot 18	2000	30	36	25	30	24	4.8	21.0	Junagarh
G.Cot 15*	1988	16R	38	22	30	22	4.7	22.0	Methio tract
G.Cot 19*	1997	11R	34	23	30	24	5.0	20.0	Methio tract
G.Cot 13+	1981	8R	39	23	30	24	5.2	20.0	Wagad area
G.Cot 17+	1995	11R	37	23	40	24	4.8	20.0	Middle
G.Cot 21+	1998	11R	42	22	30	23	5.4	20.0	Wagad
Maharashtra									
Rajat	1994	12R	38	25	40	26	4.5	22	Vidarbha
AKH4*	1975	7R	38	24	30	25	5.3	20	Vidarbha, Marathwada

Name of the Varieties	Year of relea-se	Yield q/ha	Ginning out-turn (%)	M.H.L	Spin-ning counts	2.5% span length	Micro-naire value	Bundle tenacity g/t 3.2 gauge length	Area of cultivation
AKA 5*	1981	7R	38	24	30	22	4.2	22	Vidarbha, Marathwada
AKA 8401*	1989	10R	38	25	40	26	4.8	21	Vidarbha
PA 183*	1996	18	38	24	30	27	4.5	22	Vidarbha
Y1*	1962	5R	38	24	30	23	4.7	24	Vidarbha
AKA 7*	1998	10R	41	22	25	23	5.8	21	Vidarbha
Andhra Pradesh									
L 603	1997	23	35	28	40	26	4.1	21.0	NSP Tract
L 604	1997	26	36	26	40	26	3.8	22.9	Rice fallow areas
Kanchana	1984	24	34	26	40	28	4.2	25.0	AP
LK 861	1993	25	34	27	50-60s	30	3.8	24.0	AP
Mahanandi*	1978	5R	32	22	26	22	4.5	23.0	AP
NA 1325	1993	8R	36	24	30				

Karnataka									AP
Sharda	1981	12R	38	28	40	24	4.2	22.0	Rainfed areas
Abadhita	1988	25	37	27	50	28	4.4	24.0	Karnataka
Sahana	1986	20	40	26	40	27	4.6	23.0	South zone
DB 3-12+	1979	4R	33	22	30	22	4.5	20.0	N.E. east
Raichur 51+	1968	3R	34	21	26	21	4.4	22.0	Karnataka
Tamil Nadu									
MCU 5 VT	1982	20	34	29	50	32	3.3	25.0	Winter tract
LRA 5166	1982	10R	35	24	34	26	3.8	25.0	tn,ap,ms,ks
Surabhi	1997	35	35	29	55	30	4.0	26.0	South Zone
Sumangala	2000	25	37	25	40	26	4.5	22.0	South Zone
MCU 12	2000	30	37	27	40	28	4.4	25.0	Tamil Nadu
SVPR 12	1996	16R	36	25	30	25	3.9	20.0	Summer Tract
Suvin.	1974	15	30	26	120	36	3.2	38.0	Tamil Nadu
K 10*	1989	7R	38	24	30	24	4.4	34.0	Tamil Nadu
K 11*	1992	8R	35	24	30	24	4.4	23.0	Tamil Nadu

**G. arboreum*, + *G. herbaceum*, .*G. barbadense,* rest are *G. hirsutum*

Looking to the modernization and globalization taking place the world over, there is a need to make changes in the current available cultivars to meet the demand of global market. Still our pace of modernization is considered slower and lower than other countries like Malaysia, China, Taiwan and Pakistan in Asia-Pacific Region. Both quantitative and qualitative demands are going to increase in future and India has to gear itself to play a bigger role in global cotton and textile scenario as is known for years together.

REFERENCES

Ball (1915). The development and properties of raw cotton. A.C. Black Ltd, London.

Ball (1928). Studies of quality in cotton. Mac Millan and Co, London.

Iyer, K.R. Krishna and P. Bhama Iyer (1999). Cotton fibre quality evaluation. In : Handbook of Cotton in India. Publ. By : Indian Society for Cotton Improvement (ISCI), Mumbai, pp. 303-337.

Kohel, R.J. (1972). Linkage tests in upland cotton, *Gossypium hirsutum* L. II. Crop Sci, 12 : 68-69.

Kohel, R.J. and S.C. McMichael (1990). Immature fibre mutant of upland cotton. Crop Science, 30 : 419-421.

Narayanan, S.S. (ed.) (1989). A Catalogue of Cotton Genetic Resources in India. CICR Publ.

Narbuth, E.V. and R.J. Kohel (1990). Inheritance and linkage analysis of a new fibre mutant in cotton. J. Hered. 81 : 131-133.

Singh Phundan (1998), Cotton Breeding. Kalyani Publishers, New Delhi.

Singh Phundan (1999). Heterosis Breeding in Cotton. Kalyani Publishers, New Delhi.

CHAPTER - 19

FORMATION OF SEED AND ITS STRUCTURE

Dhirendra Khare and C.B. Singh

1. SEED FORMATION

Botanically seed is a ripened ovule containing an embryo in arrested state of development usually with a food reserve and a protective coat. Seed in most plants is formed by sexual reproduction (Syngamy). In plants, both male and female organs are present in flower.

1.1 Female sexual organ

The female sexual organ [also known as *gynoecium* or *pistil*] is a flask shaped structure made up of three major parts stigma, style and ovary. Stigma is the upper most part of the pistil with sticky and wet surface that releases a large amount of viscous exudates. Stigma receives pollen grains for fertilization. The exudate of stigma contains certain enzymes, which allow the germination of the compatible pollen grains. Style is a long slender narrow path, which connects the stigma with ovary. Length and width of the style varies from genera to genera. Ovary is the lower broad and hollow part of the pistil. One ovary may have one to several chambers (carpels), whereas, each carpel may contains one to several ovules depending on the genera. Number of style and stigma per flower indicates the number of chambers (carpels) present in the ovary.

Ovule is the main structure in which egg is formed and fertilization takes place. It is an oval or egg shaped body, attached to ovary with the help of a bridge known as *funiculus*. In the egg shaped body, a round mass of very soft juicy tissues known as *nucellus* is present. These tissues are enclosed by the double wall layer known as *integument*. It also formed the outer layer of the ovule. It leaves one opening on one side known as *micropyle* (*micro* = small;

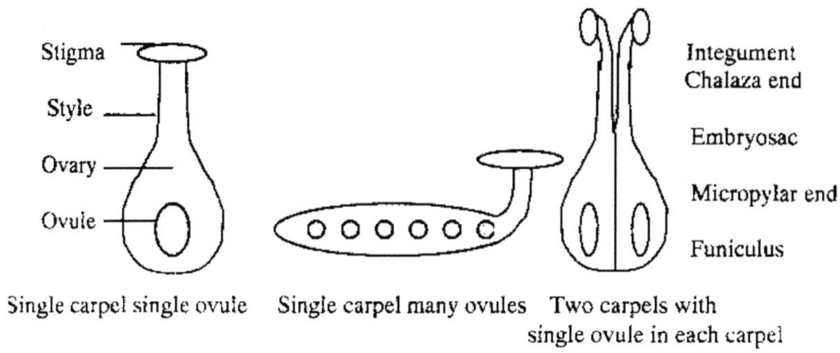

Fig. 1. Structure of female sexual organ

pyle = gate). Nucellus embedded one large cavity known as embryo sac with a single large nucleus. This nucleus through mega-sporogenesis forms the eight archesporium (*arche* = important) cells. These cells actively take parts in fertilization as female gametes. The end of the embryo sac towards the micropyle is known as *micropylar end* whereas end just opposite to micropyle is known as *chalaza* end (Fig 1).

1.2 Megasporogenesis

The process of reduction in chromosome number from diploid to haploid to form megaspore is known as megasprorogenesis. Female gametes are formed in embryo sac. One cell of embryo sac known as megaspore mother cell is enlarged in size. It gets food material from nucellus tissues (Fig 2). The megaspore mother cell under goes the reduction division [meiosis] and forms four haploid cells. Out of these four haploid cells, three consecutive cells are degenerated. The haploid cell, which remains alive, is known as megaspore.

1.3 Megagametogenesis

Megaspore divides thrice through eqational division [mitosis] and forms eight haploid cells known as *archisporium* cells. These cells are arranged in a particular fashion in the embryo sac. Embryo sac contains eight haploid cells arranged in 3:2:3 fashion formed through megagametogenesis (Fig 2). The three important cells towards the chalaza end are known as *antipodal* (anti = against; pous= foot). Two haploid cells at the centre are known as *polar nuclei*. Whereas, among the three haploid cells towards the micropylar end the middle

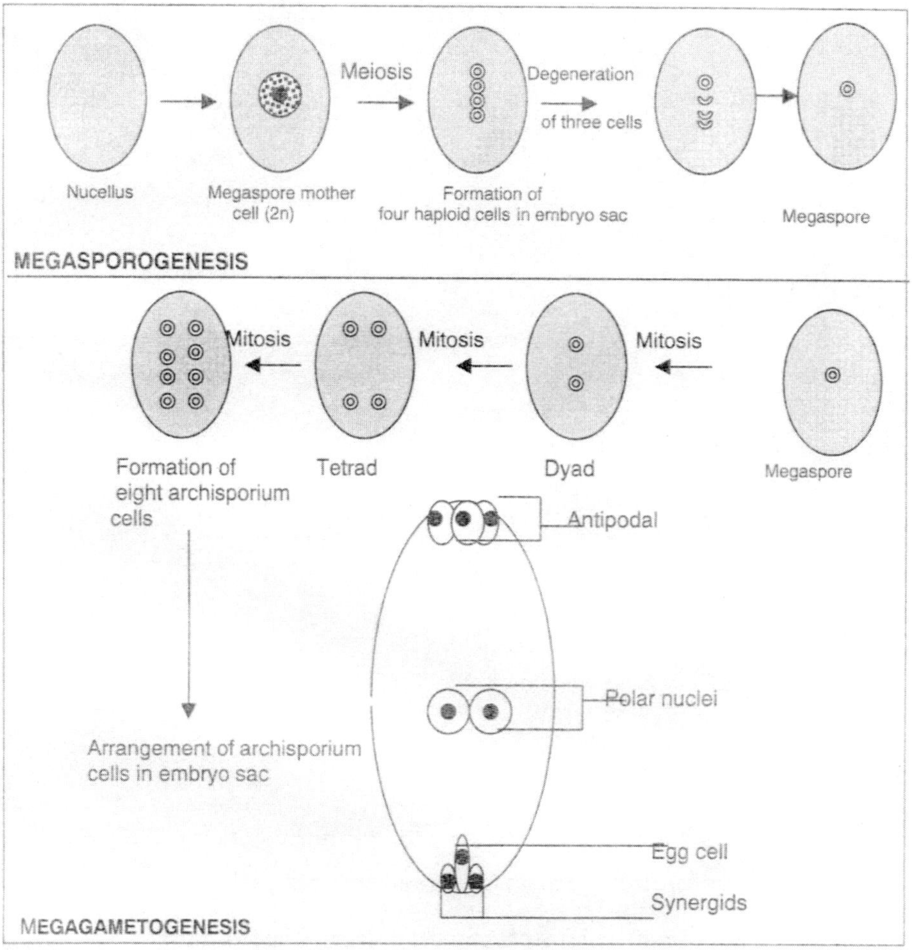

Fig. 2. Megasporgenesis and megagametogenesis

one of larger size is known as egg cell and the cells towards both the side of the egg cell are known as *synergids* (synergos= cooperator).

1.4 Male sexual organ

Male sexual organ of plant [also known as *androecium* or *stamen*] may present in one to thousand in numbers depending upon the genera and species. It is made up of three parts anther filament and connective tissues. Anther is an oblong inflated bilobed structure, with two cavities in each lob. These cavities produce large quantity of pollen grains through *micro-sporogenesis.* With the maturity of pollen the sac opens and discharges the pollen grains. Anther attached with elongated and narrow slender stalk known as filament through connective tissues (Fig 3).

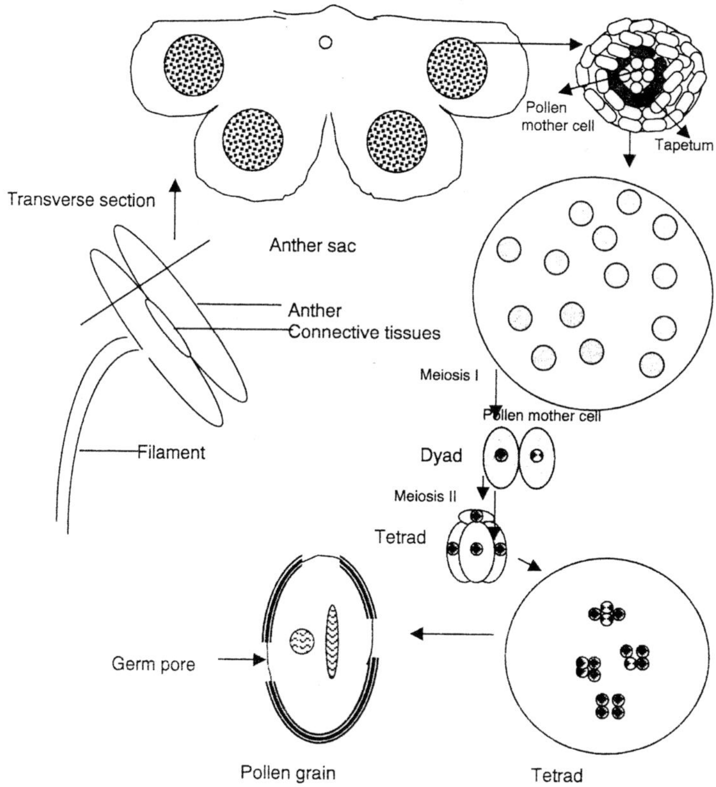

Fig. 3. Structure of male sexual organ and microsporogenesis

1.5 Production of pollen grains (Micro-sporogenesis)

The process of reduction in chromosome number from diploid to haploid to form pollen grain is known as micro-sprorogenesis. In most of the field crops stamen produces four microsporangia. The anther sac is first a mass of homogeneous cells covered by an epidermis. The hypodermal layer of these tissues becomes differentiated from the adjacent cells by their large size and larger nucleus. The number of these cells varies from crop to crop. These cells are surrounded by tapetum cells, which nurses the functioning mother cells. The mother cells and their nucleus usually increase very much in size, wall become thick, shape become round and they are separated from one another. These are known as *pollen mother cell* (2n). Pollen mother cell goes under the reduction division and forms a group of four haploid cells (enters in the gametophytic stage) known as *tetrad.* Each microspore is invested in a delicate wall differentiated into two layers. This structure develops as pollen grain.

The entire tetrad and each microspore within the tetrad is surrounded by a callose wall. Upon release from tetrad wall, the microspore increases rapidly in volume and changes its shape. In the microspore the haploid nucleus divides mitotically and forms two cells before dehiscence i.e., a large vegetative cell (n) with larger nucleolus and scanty chromatin net work and one small generative cell (n) with smaller nucleolus and dense chromatin net work. Both cells remain within the wall of the original microspore. This structure is known as pollen grain (Fig 3).

Pollen grain is a double walled unicellular structure, spherical to ovoid in shape with hard and rough outer layer known as *Exine* and thin and smooth inner layer known as *Intine*. The intine consists of pure cellulose, which in later stage forms pollen tube. The exine is cutinized and often become hard. Exine has spots known as *germ-pore*. The pollen tube emerges out from one of these pores. Pollen of monocot has only one germ pore whereas dicot may have two to many germ pores. After release from the anther pollen grains remain as a free haploid organism carried by wind, insect or other agents to the stigma of compatible flower. During this stage it protects the male gametes from environmental stresses and helps it in movement.

1.6 Pollination

Transfer of pollen from anther to the stigma is known as pollination whereas, release of pollen from anther is termed as anthesis (Fig 4).

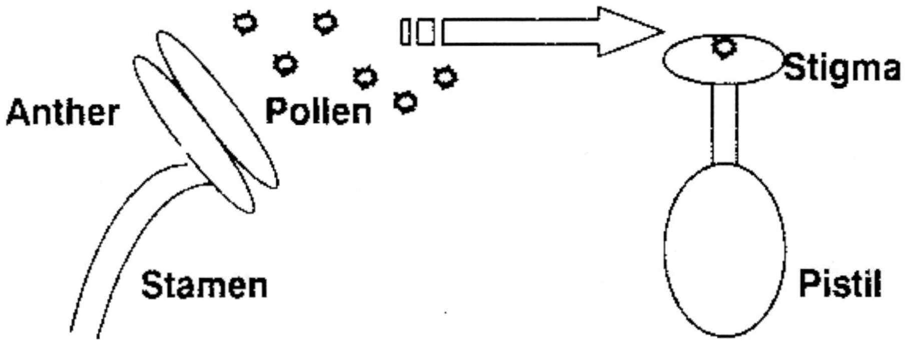

Fig. 4. Pollination

1.7 Micro gametogenesis

On the surface of the stigma, pollen germinates with the help of enzymes produced by the stigmatic surface. During germination, it produces a long slender delicate and thin walled tube known as *pollen tube* through one of the germ pores. The tube enters in the stigma and through style it reaches up to ovule. Generative nucleus and tube nucleus moves out of the pollen grain into the pollen tube. Initially pollen tube contains only two haploid nuclei i.e., vegetative and generative cell (Fig 5).

Tube nucleus: It is the vegetative nucleus with haploid number of chromosome without any cytoplasm. It occupies the space towards the tip of the pollen tube. The name tube nucleus was given because it was believed that it directs the growth of the pollen tube towards the ovule.

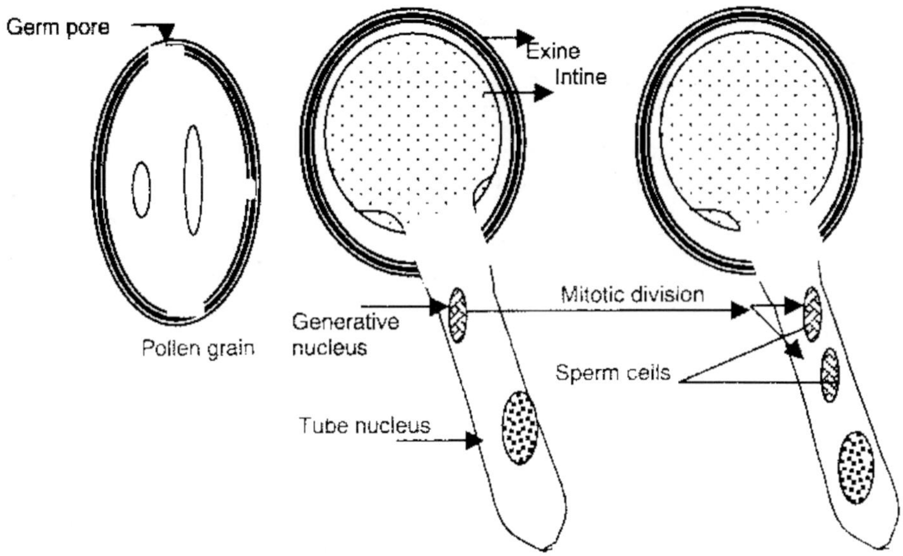

Fig. 5. Micro-gametogenesis

Generative cell: It is a haploid cell containing one set of chromosome from male parent with a small amount of cytoplasm. In the pollen tube it under goes in mitotic cell division and forms two haploid sperm cells (n) and(n).

The period from germination of pollen up to fertilization, ranges from few minutes to 48 hours and pollen tube growth rate as high as 35 mm per hour depending upon genera and species (Maheshwari, 1950).

1.8 Fertilization

The pollen tube enters in the embryo sac through micropyle with *male germ-unit* i.e., two sperm cells and one tube nucleus. The pollen tube penetrates one of the synergids, stops growing and discharges the sperm cells. The tube nucleus is degenerated at this place. Out of two sperm cells one combines with egg cell and forms a zygote (2n). Whereas, second sperm cell fuses with the two polar nuclei and gives rise to an endosperm (3n). Because of fusion of sperm cells at two places, it is known as *double fertilization* (Fig. 6). Out of two sperm cells, one is practically devoid of plastids whereas the other contains a large number of these organelles. The plastid rich but mitochondrial poor, sperm cell preferably fuses with the egg cell (Russell, 1984). It shows that there is specificity in gamete recognition and the process of double fertilization is not a random fusion event.

1.9 Development of endosperm

During fertilization, fusion of two polar nuclei (n+n) with one sperm cell (n) forms *primary endosperm nucleus* (3n). As a rule, the primary endosperm

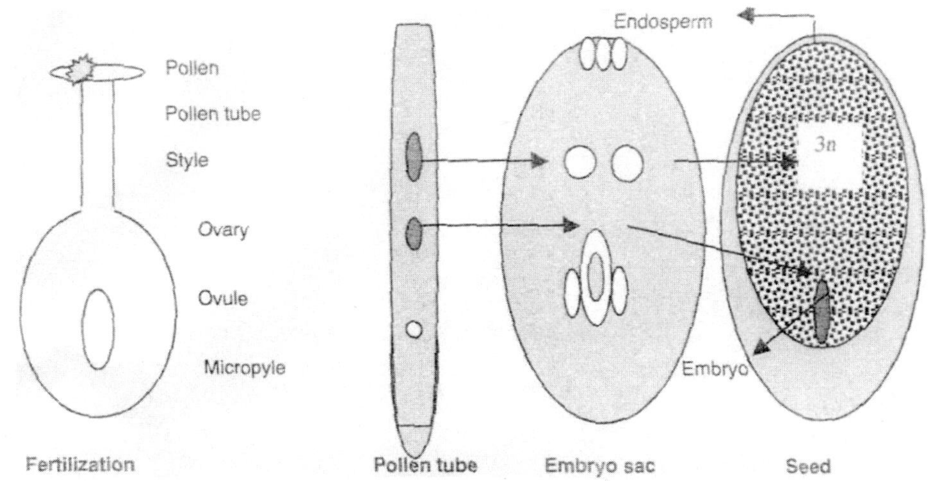

Fig. 6. Fertilization

nucleus began to multiply before multiplication of zygotic cell. It may divides in following ways:

Nuclear type: The nucleus of primary endospermic cell divides and forms many nuclei without wall formation. Initially these nuclei are concentrated at one place but later due to enlargement of the sac they migrate in every direction towards the wall of embryo sac. The nuclei are equally distributed and separated by lining of cytoplasm in later stage. This type of endosperm is common in monocot (Fig 7).

Cellular type: The primary endospermic cell divides and forms walls simultaneously. Thus, the chambers are formed in the sac. This division is continued until the embryo sac is filled with the 3n cells. This type of endosperm formation is common in dicot (Fig 7).

The cell of the endosperm (3n) multiplies mitotically and the tissue lacks the intercellular space. The cells are isodermic in shape, devised of pits with thin cell wall. The enlarged cells store the food. Normally, carbohydrates are stored in theses cells in the form of starch. In mature seed, the endosperm works as store of food with no life.

The outer most two layers of endosperm towards the periphery of embryo sac in the crops of family Gramineae are filled with aleurone grains composed of protein crystal. This layer of endosperm is known as *aleurone layer*. This part of endosperm is made up of viable cells.

1.10 Development of embryo

Embryo is formed by the zygote (2n). The zygote divides transversely and forms two cells by mitotic division i.e., basal and terminal. Basal cell is the larger cell towards the micropylar end, which is also known as suspensor whereas, Terminal cell is the smaller cell present above the basal cell towards the chalaza end, it is also known as embryo cell.

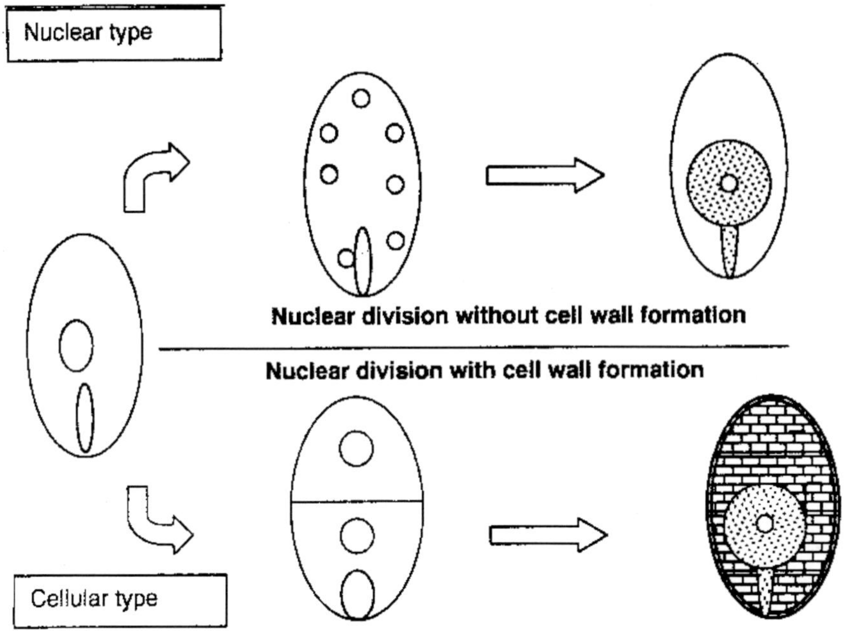

Fig. 7. Development of endosperm

The terminal cell divides vertically and forms two cells. Thus, a three celled structure known as *pre-embryo* is formed with one basal cell and two cells formed by the division of terminal cell. Lower cell formed by the division of terminal cell is again divided vertically and forms two cells just above the basal cell. At this stage the cell responsible for cotyledon, root and shoot formation are finalized. The cell present on the top forms cotyledon, the cell after the top cell forms shoot tip and the cell below this forms the root tip whereas, the last cell works as suspensor.

The top cell is always divide longitudinally and forms two cells. Then, It divides transversely and forms quadrant stage followed by the octant stage. The eight cells of the octant are arranged in two tiers of four cells each. The layers of cells running parallel to the surface of plant part (*periclinal division*) form ridges (*dermatogen*). Dermatogen first appears on the upper part and then proceeds towards the root end. The cell after terminal cell divides transversely and gives rise to stem tip whereas, the cell present above the suspensor also divides transversely and gives rise to root tip. The cell of suspensor always divides transversely. The central region of globose shape embryo is known as *Plerome,* gives rise to hypocotyl whereas the region between dermatogen (top two layers) and plerome is known as *periblem* gives rise to epicotyl. The stem tip is always present in between periblem and dermatogen whereas root tip is formed at the tip of the plerome (Fig 8). The early stage of embryo development does not show any fundamental difference between monocot and dicot.

1.10.1 Formation of monocot embryo

Initially the embryo is globose in shape, its cell starts unilateral growth and forms the single cotyledon *(scutelluem)* of cylindrical in shape. It occupies the terminal position in between endosperm and apical tip. At the base of the cotyledon, the cells responsible for stem tip formation are organized in a depression. In early stage, this depression is shallow which increase in depth with age. At the final stage, the extension appears on one side of the cotyledon with plumule present on tip in upward direction and radicle towards lower tip. This structure is surrounded by the sheath-like structure formed by the cotyledonary tissues. The basal cell formed initially by the division of zygote is not again multiplied in monocot (Fig 8).

1.10.2 Formation of dicot embryo

In dicot, the division of the terminal cell is same as of monocot. Nevertheless, the shape of the structure is heart instead of globose with two lobes. Throughout the period of development, it maintains bilateral symmetry. These two cotyledons are raised at apical end with plumule between the cotyledon and radical towards the micropylar end.

The basal cell of dicot seeds divides transversely and formed 6-10 celled filament with suspensor i.e., the large cell towards the micropylar end. This pushes the embryo cell deep into the endosperm. The cells of cotyledons absorb the food material of endosperm therefore the cotyledons are enlarged whereas the endosperm is disappeared. Cotyledons are the part of the embryo (2n) and they remain alive with the seed (Fig 8).

1.11 Structure of Seed

The embryo of dicot is made up of embryonic axis with two fleshy cotyledons. Endosperm may exist as a thin layer or it may disappear. Seed is round to oval embedded by a protective seed coat with scar of hilum and micropyle. Radicle portion of embryonic axis projects beyond the cotyledons. The other end of the embryonic axis is the plumule present in between both the cotyledons. In some species, wart like outgrowth called strophiole (pigeonpea) and caruncle (castor) are present below the hilum.

In monocot seed, endosperm is made up of non- living cells. A very small embryo made up of shield shaped single cotyledon known as scutellum and a short axis lies in a groove at one end of the endosperm. The epithelium layer separates the embryo and endosperm. Cotyledon forms a protective sheath known as coleoptile on plumule and coleorhiza on radicle.

Fruit wall is formed by the wall of the ovary (Fig 9) whereas seed coat is formed by the integument of ovule. Seed coat is a protective sheath made up of two layers i.e., *testa* (Latin word = bricks or tiles) and *tegmen* (inner thin membrane) present as an envelop on embryo and endosperm. It protects the seed from desiccation, mechanical injury, and damage due to insects, microorganisms and environmental fluctuations. It also helps in dispersal of

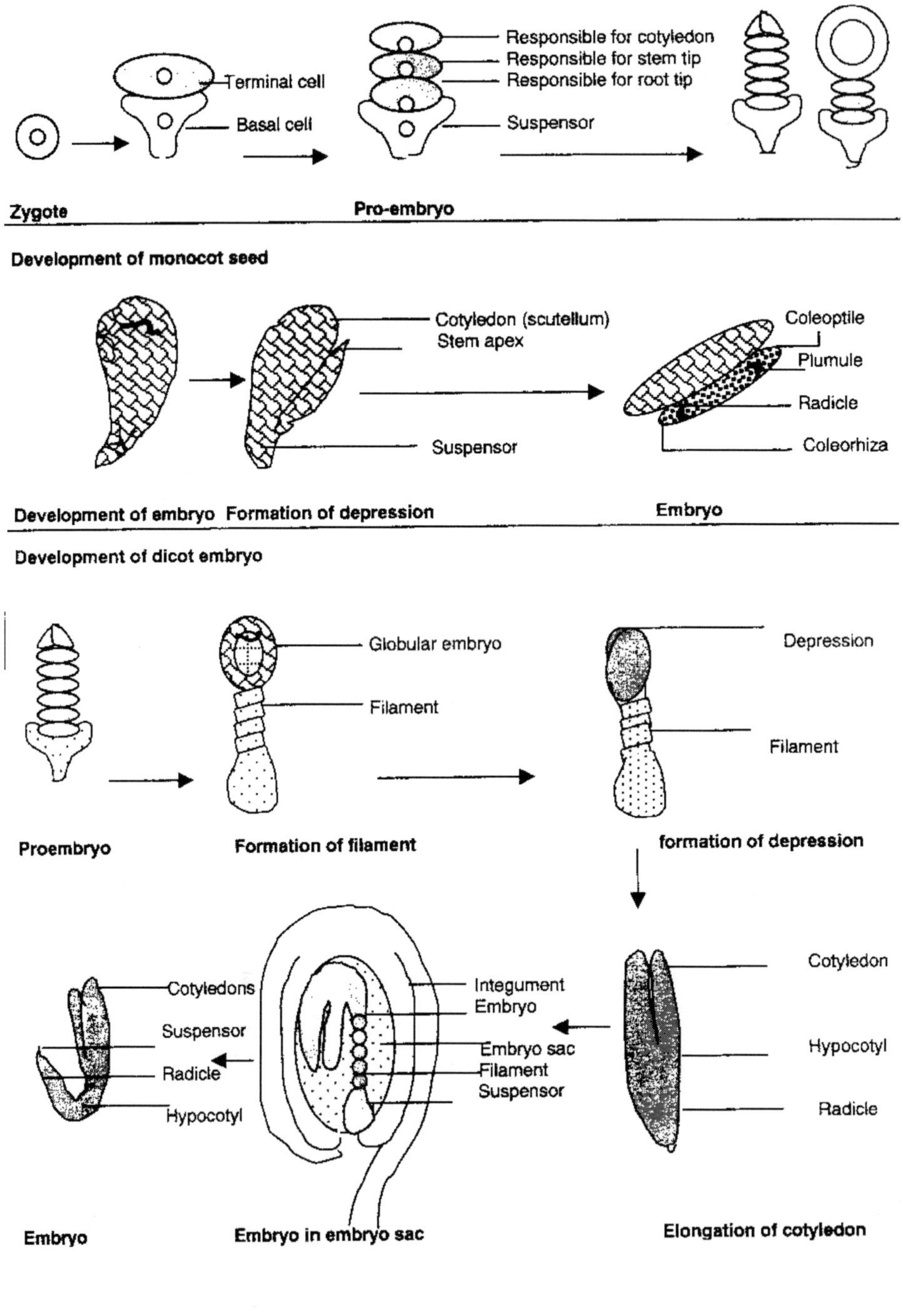

Fig. 8. Development of embryo

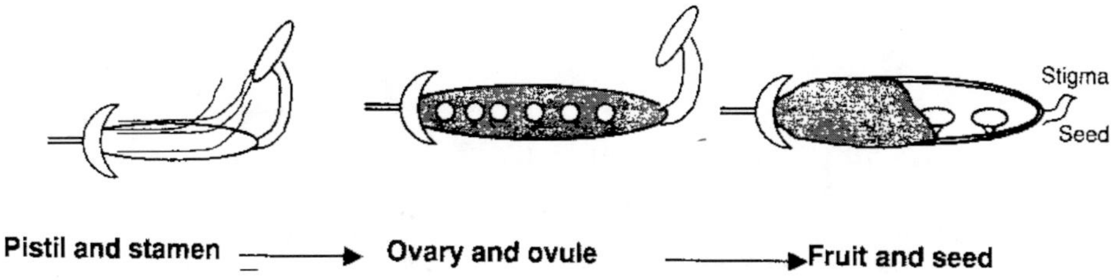

Fig. 9. Formation of fruit and seed

seed. A prominent scar on seed of dicot known as *hilum* (Latin word = trifle) is formed by the funiculus i.e. attachment of the ovule with ovary. The place left open by integument for entrance of the pollen tube (micropyle) (Fig 10) exists on seed coat as a microhole just above the hilum. In the seed of castor, the whitish out growth of integument known as *caruncle* is prominent.

Embryo is a rudimentary plant present in axis form with one tip known as plumule forms shoot portion and the other axis known as radicle forms the root (Fig 11). The portion of embryonic axis extended above the cotyledon is known as *epicotyl* whereas the portion extended below the cotyledon is known as *hypocotyl*.

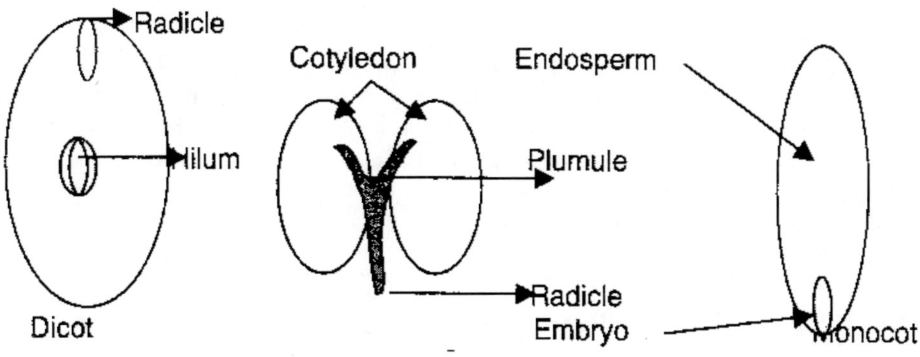

Fig. 10. Structure of monocot and dicot seed

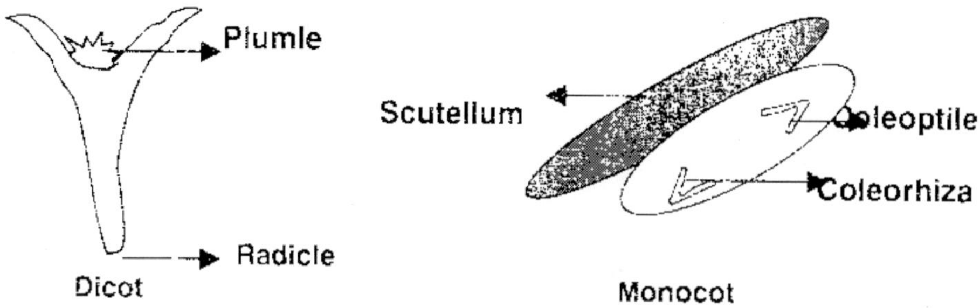

Fig. 11. Structure of embryo

Store of food

Endosperm (Greek word = with in and seed)**:** It is a thick and massive structure made up of elongated cells containing abundant starch. On the basis of presence or absence of endosperm, seeds are categorized into two groups : (i) Albuminous - endosperm present and (ii) exalbuminous - endosperm absent.

Cotyledon: It is the extension of the embryo and on the basis of number of cotyledons, the crop species are divided into two groups : (i) monocotyledonous - one cotyledon is present and (ii) dicotyledonous - two cotyledons are present. Whereas some gymnosperms have 2-18 cotyledons. In monocot, the scutellum situated in between the endosperm and the embryo is the single cotyledon.

Single seeded fruits those are considered as seed:

Achene: Single seeded fruit where pericarp can be removed from the mature seed. These fruits are used as such as seed for sowing. e.g., sunflower, safflower etc.

Caryopsis: Single seeded fruit where pericarp can not be removed from the mature seed, due to the fusion of fruit wall with seed coat. These fruits are used as such as seed for sowing. e.g. wheat, sorghum, pearl millet etc.

Schizocarp: A dry fruit, which is separated into two or more units at maturity, e.g. coriander, carrot etc.

Nut: Fruit with thick and hard shell like fruit wall, e.g. coconut.

1.12 Expression of qualitative genes in different parts of a seed:

Inheritance of the qualitative traits at seed level depends on trait on which expression has to be observed, genetics of the character and generation in which that character will be able to express.

Embryo : The embryo is made up of the chromosome received from both the parents. Therefore, the expression of the character on embryo of hybrid seed depends on the presence of dominant gene, irrespective of its source. If the dominant gene is from male parent it will also expresses in hybrid seed harvested from female parent. Therefore, attribute like embryo colour and seed shape (surface) should be observed in F_1 seed, i.e., harvested from female parent after hybridization.

Cotyledon: The cotyledon is the part of the embryo and is formed by the cells responsible for the formation of the embryo. Therefore, the inheritance of qualitative characters of cotyledon should be observed similarly as embryo. If green colour (GG) of cotyledon is dominant over yellow (gg) cotyledon colour. Then, the hybrid seed form by the cross between these two will be of green cotyledon colour in reciprocal cross also (Fig 12).

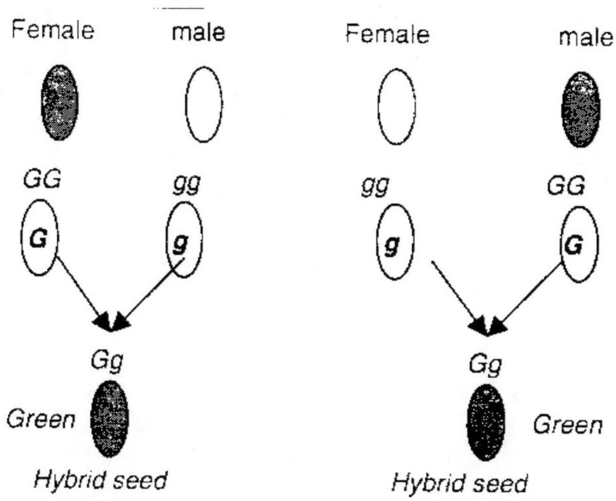

Fig. 12. Inheritance of qualitative traits of embryo and cotyledon

Endosperm : It contains the chromosome from both the parents but the dose from female parent (n+n) is double than the male parent (n). Direct effect of the pollen parent on the expression of the endosperm is known as *xenia effect*. The gene present in sperm cell contributes in the expression of endosperm colour in maize. (Table 1). If the yellow colour of endosperm (Y) is dominant over white (y) colour, then the colour of the endosperm of the maize seed will express differently based on number of dominant gene present.

Table 1. Xenia effect on maize seed

Polar nuclei	Sperm nucleus	Endosperm	Colour of endosperm
YY	Y	YYY	Yellow
YY	Y	YYy	Medium yellow
Yy	Y	Yyy	Light yellow
Yy	Y	Yyy	White

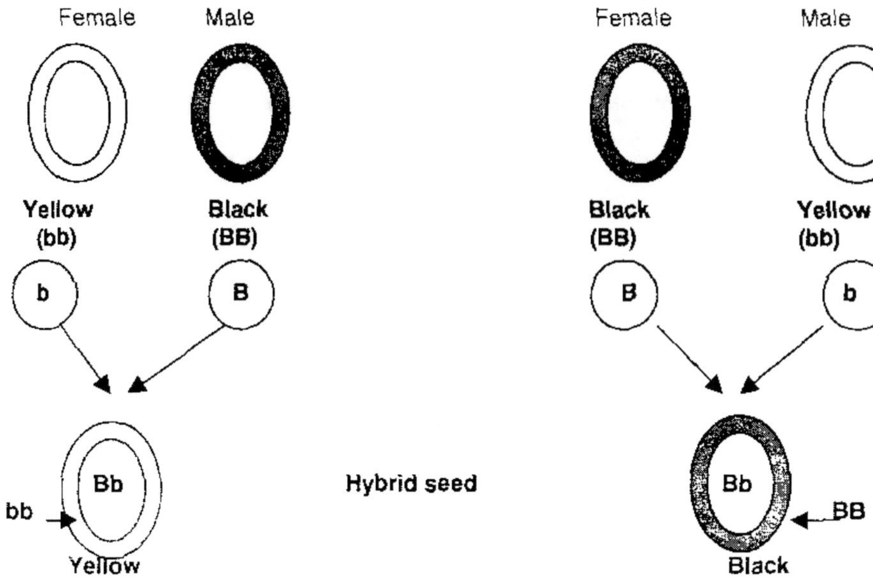

Fig. 13. Inheritance of seed coat colour

Table 2. Contribution of male and female parent in the formation of a hybrid seed structure

Part of seed	Female	Male	Chromosome
Embryo	Egg cell (n)	Sperm nucleus (n)	2n
Cotyledon	Egg cell (n)	Sperm nucleus (n)	2n
Endosperm	Polar nuclei (n+n)	Sperm nucleus (n)	3n
Testa	Outer integument (2n)	—	2n
Hilum	Scar of funiculus (2n)	—	2n
Micropyle	Micropyle (2n)	—	2n
Caruncle	Outer integument (2n)	—	2n
Strophiole	Wart like outgrowth (2n)	—	2n
Coleoptile	Egg cell (n)	Sperm nucleus (n)	2n
Coleorhiza	Egg cell (n)	Sperm nucleus (n)	2n
Scutellum	Egg cell (n)	Sperm nucleus (n)	2n
Pericarp	Fruit wall (2n)	—	2n
Coleorhiza	Egg cell (n)	Sperm cell (n)	2n
Coleoptile	Egg cell (n)	Sperm cell (n)	2n

Seed coat : To know the inheritance of this trait one has to observe seed of next generation i.e., for seed coat colour of F_1 seed, we have to observe colour of the seed coat present on F_2 seed. The seed coat colour of the "hybrid seed" is similar to the female parent irrespective of dominant gene present in the male parent because seed coat is formed by the integument of the female parent. If the black colour of the seed coat (B) is dominant over yellow (y) colour, then the colour of the seed coat of the hybrid seed will express similar as of female parent (Fig 13). The genetic constitution of other seed characters is presented in Table 2.

REFERENCES

Maheshwari, P. (1950). An Introduction to the Embryology of Angiosperms. New York, Mcgraw-Hill.

Russell, S.D. (1985). Preferential fertilization in *plumbago* : ultra-structural evidence for gamete level recognition in an angiosperm. Proc. Natl., Acad. Sci., USA 82:6129-34.

CHAPTER - 20

VARIETAL TESTING, RELEASE AND NOTIFICATION SYSTEM IN INDIA

N.D. Raut

1. INTRODUCTION

Identification of genotypes with higher and stable yield performance over the best ruling variety has been the common goal of all the crop improvement programmes. The ICAR-SAU system of research has pioneered in promoting varietal development and testing, by promoting interdisciplinary and intra-institutional collaborations. Most of the crop improvement work in India in recent years has been in the form of nationally coordinated programmes involving the SAUs and the Central Institutions of the ICAR. The major objectives of these programme is the development and release of large number of high yielding but genetically diverse varieties adopted to different agroclimatic zones of the country.

The release of a strain as a variety depends on conclusive demonstration of its superiority over the existing checks in yielding ability or in some other feature of economic importance like disease resistance, drought tolerance, quality attributes etc. Because of these reasons, the strains are extensively evaluated for their performance by combining any desirable agronomic, physiological, disease/ insect pest resistant factors and product quality features.

The plant Breeders located at various cooperating centres organize the need based varietal improvement programme for combining desirable attributes required by the growers in the region. For this purpose various plant breeding techniques/methodologies are used. These efforts are supplemented by the coordinated projects through providing source stock with desirable features identified through systematic multi-location screening of indigenous and exotic germplasm resources. The efforts of plant breeders are supplemented by scientists of other disciplines such as plant pathology, entomology etc. to identify resistant strains to diseases and pests. The coordinated crop improvement projects aims at:

Multi-locational variety testing and evaluation over a vide range of macro and micro- environments

Multidisciplinary research at regional and national level for assessment of the overall value of superior varieties

Promotion of exchange of materials, information and interaction among the scientists working on a crop or group of crop.

Strengthening of need based research activities at selected centres.

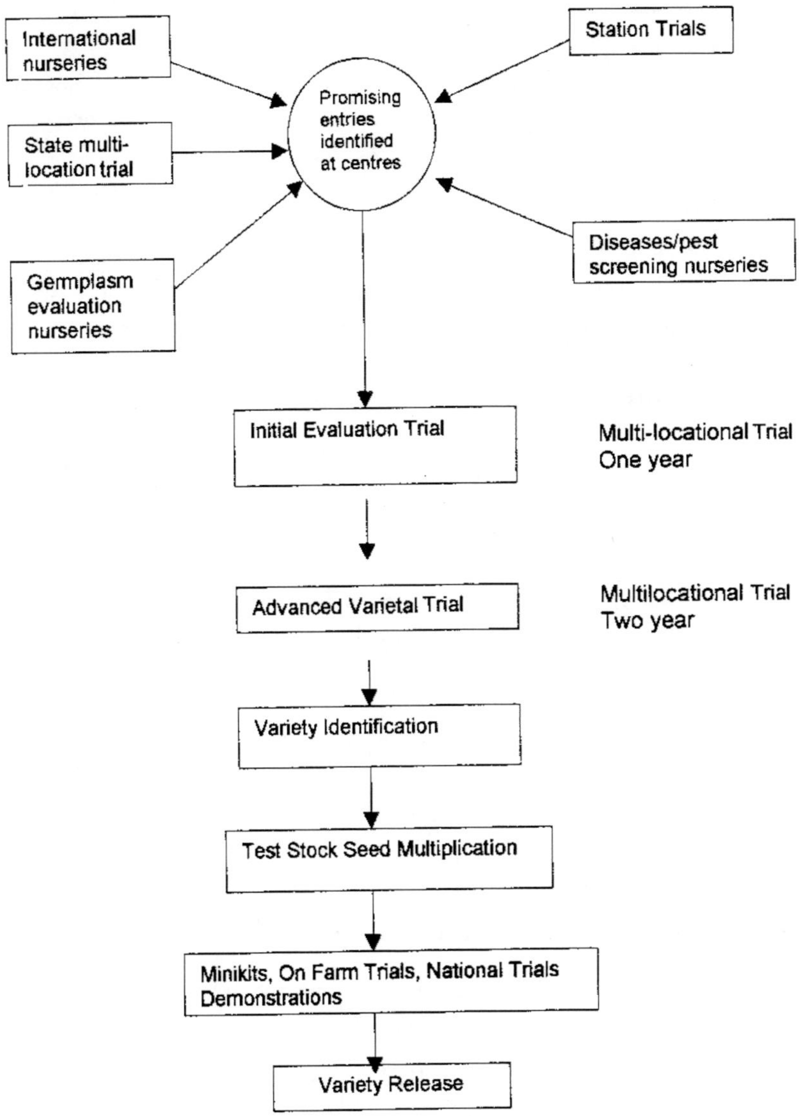

Flow chart — Variety testing

2. DEVELOPMENT OF IMPROVED VARIETY

The operational mechanism for evaluation of varieties by coordinated system, irrespective of the crop, can be broadly classified in to the following group for convenience

Evaluation

Identification

Release

Notification

The various activities and operations in the release of genotype as a variety consists of various trials and tests to determine its superiority over the best existing variety.

(a) Station trial

The new lines emerging from breeding programme are assessed for their performance in the station trials conducted at the concerned regional research centres. The objective of the station trial is to see that the new strains are superior in performance to the best available variety in the region. The breeder may assess the performance of their material at more than one location through common testing programs in collaboration with other centers. The assessment may be for yield and other desirable features for a year or more before the strain is given for testing in to the coordinated trials.

(b) Multi-location Trials

The objective of this trial is to evaluate the performance of newly developed strain at several locations distributed over region.

(c) Induction into coordinated testing

The material emerging from the pre-coordinated testing is sent for testing in the coordinated trials. The breeders are required to produce data generated from the station trial and multi-location trials. The information is discussed jointly by a group of breeders before these strains are accepted for coordinated testing.

(d) Coordinated yield evaluation trials

The various trials conducted under the coordinated projects normally run in three-tier system

Initial Evaluation Trial

Uniform Regional Trial

Agronomic Trial (Model agronomic experiments)

(e) Initial Evaluation Trial (IET)

A strain included in the IET or URT is commonly known as *entry*. The IET is conducted at 10-12 locations within zone. The system involves the most dynamic approach in which only the most promising entries succeed in being tested in subsequent years, whereas, poor performers are dropped at each stage. IET may constitute the first level of testing and may have 25-49 entries for the important environments. The IETs are normally laid out at the research centers and are grown in small plots with four replications. The best performing entries having other desirable attributes/features is advanced for evaluation in the regional trials and the one, which are not performing well are discarded. The objective of IET is to eliminate relatively inferior entries so that their number is reduced to a manageable size under URT evaluations.

(f) Uniform Regional Trials (URT)

The URT are organized at a larger number of locations, involving not only the research stations but also testing centres and regional stations of the universities and state department of agriculture. The URT is laid out in a plot of larger size and with more replications. The URT is conducted under the same conditions under which the concerned entry was evaluated in IET. After evaluation for one year in URT the data from all the centres are statistically analysed and compiled in appropriate form at the project headquarters. The outstanding entries are included in the agronomic trials, and their evaluation in URT is continued for one more year.

Entries, which have completed two years of testing in URT and with three years of overall coordinated testing, are eligible for identification/release

If the entry is identified for release, it is evaluated in adoptive trials, Model Agronomic Trials and Minikit Trials.

The criteria for promotion of an entry from IET to URT (CVT), for the retention for evaluation in the second/third year URT/CVT, and for the identification of entries by the workshop vary from one crop to the other crop.

Information generated by other disciplines

An entry showing superior performance in URT during the first year is included in the Agronomical Trials of the respective All India Coordinated Crop Improvement Projects for agronomical testing viz. varietal x cultural variable trials such as date of sowing/fertiliser levels and irrigation levels. Herbicidal sensitivity tests are also carried out.

Model Agronomic Experiments

In Model Agronomic Trials, only those entries are included that has been identified by the workshops of the respective coordinated projects. These trials are conducted at all the centres of the Coordinated Agronomic Research Project in the zone

(g) National Trials

The entries in the National Trials consist of one entry from each zone that ranked first in the zone in URT of the previous year. The purpose of the National Trials is to evaluate out standing entries of one zone in the other agroclimatic zones to see the performance in other zone.

(h) Adoptive Research Trials

Adoptive Research Trials are conducted on research stations or farms of the state Governments. The entries identified by the workshop of the concerned Coordinated Project are included in these trials. The data from these trials are considered for release

(i) Minikit Trials

After the entry is identified for release, these trials are conducted at the farmers field. The objective of the minikit trial is to popularize the variety among the farmers of the zone, secondly the seed of a good new variety reach to the farmers in advance prior to the availability of certified seed.

(j) Disease and Insect Tests

The entries under evaluation in IET and URT are assessed separately in plant pathological screening nurseries. Diseases and insect resistance is tested under both natural epidemic as well as artificial epiphytic (epidemic) conditions.

The value of disease tests can not be overemphasized. In an extreme case, susceptibility of released variety to diseases may spell disaster to the farmer, and some times, may even cause health hazards to the consumer as well.

(k) Quality Tests

In each crop, a different set of quality tests has been developed in order to determine the suitability of new strain of the crop for various uses.

3. IDENTIFICATION OF ENTRIES FOR RELEASE

All the data generated under the Coordinated testing are compiled in the form of proposal for varieties suitable for identification in a prescribed proforma by the breeder who has developed the concerned strain. The proposal for identification are considered in an open forum by the participants in the workshop or all the data are deliberated upon in a special committee constituted by senior level ICAR scientists, representatives of seed production and seed certification agencies and state departments of agriculture. Principal investigator of each research discipline in the project also participates.

All the information is discussed very critically by the varietal identification committee and decisions are arrived at whether a strain should be identified for release of otherwise. Normally the decisions are taken unanimously and announced in the plenary session of the workshop.

4. VARIETAL RELEASE

After identification, a variety is tested for at least one year in Adoptive Research Trials. For this purpose the breeder is required to supply ten quintals of seed (Sufficient to grow ten hectares) to the State Farm Corporation of India for multiplication as test stock seed, which is used for conducting the adoptive trials at the state farm and minikit demonstrations at the cultivators fields. Based on the data from adoptive trials, disease and quality tests, and the breeder who has developed the strain submits proposal for its release as new variety for consideration by the Variety Release Committee.

Proforma for submission of proposal for the release of the seeds of a variety to the Variety Release Committee

1.	Name of the crop and species		:
2.	(a)	Name of the variety under which tested	:
	(b)	Proposed name of the variety	:
3.	Sponsored by		:
4.	(a)	Institution/agency responsible	:
		for developing variety (With address)	:
	(b)	Name of person who helped in the development of variety	:
5.	(a)	Parentage with details of its pedigree	:
	(b)	Source material in case of introduction	:
	(c)	Breeding method	:
	(d)	Breeding objectives	:
6.	State the variety which are most closely resembles proposed variety in general characteristics		:
7.	(a)	Whether recommended by- seminar/ conference/workshop/State Seed Sub-Committee	:
	(b)	Specific area of its adaptation	:
8.	Recommended ecology		:
9.	Description of the variety/hybrid		:
	(a)	Plant height	:
	(b)	Distinguishing morphological characters	:
	(c)	Maturity /range in number of days. Seedling/ transplanting to flowering. seed to seed.	:

(d) Maturity group (early, medium & late wherever such classification exists) :

(e) Reaction to major diseases under field and controlled conditions (reaction to physiological strains/races/biotypes to be indicated wherever possible :

(f) Reaction to major pests (Under field and controlled conditions including store pests :

(g) Agronomic features (e.g. resistance to lodging shattering, Fertiliser responsiveness, suitability for early or late sown conditions, seed rate etc. :

(h) Quality of produce of grain forage including nutritive values where relevant :

(i) Reaction to stress :

10. Description of the parents of the hybrid

(a) Plant height :

(b) Distinguishing morphological characters :

(c) Days to 50 % Flowering :

(d) Maturity (range in number of days) Seed to seed

(e) Is there any problem of synchronization? if yes, method to overcome it :

(f) Reaction to major pests (under field and controlled conditions reaction to physiological strains/races/biotypes to be indicated wherever possible :

(g) Reaction to major pests (Under field and controlled conditions including store pests :

(h) Agronomic features (e.g. resistance to lodging shattering, Fertiliser responsiveness, seed rate etc. :

(i) Reaction to stresses :

11. (a) Yield data in region/inter regional/district trials year wise (levels of Fertiliser application density of plant population and superiority over local control Standard variety to be indicated :

(b) Yield data from national demonstration/large scale demonstration. Average yield under normal conditions (Yield in kg/ha) :

12	(a) Agency responsible for maintaining breeder seed	:
	(b) Quantity of breeder seed	:
13	Information on the acceptability of the variety by farmers/consumers/industry.	:
14	Specific recommendation, if any for seed production	:
15	Any other pertinent information	:

Signature of the Breeder

Signature of Head of Institution

Proforma for submission of proposal for the notification of a variety Under Section 5 of the Seed Act. 1966 (180 copies of the proposal are submitted to the Ministry of Agriculture and Irrigation, Govt. of India)

1.	State	:
2.	Crop	:
3.	Name of the variety under which released	:
4.	Year of release	:
5.	Parentage with details of its pedigree	:
	(a) Source material in case of introduction	:
	(b) Breeding method	:
	(c) Breeding objective	:
6	State the variety which most closely resemble the proposed variety in general characteristics	
7.	(a) Breeder/Institute responsible for maintaining breeder stock	:
	(b) Quantity of the breeder seed of the variety available (In kgs)	:
8.	Description of the variety/hybrid	:
9.	Description of the parents of the hybrid. Is there any problem of synchronization? If yes, methods to overcome it.	:
10	Describe at least two identifiable and distinguishable morphological characteristics of the variety. In case of hybrid, please describe also at least two identifiable and distinguishable characteristics of both the parents.	:

11	Maturity group (Early, medium and late, wherever such a classification exists)	:
12	Disease and pest resistance (give details of any resistance (Give details of any resistance to pests or diseases including races)	:
13	Recommended ecology	:
14	Yield (in kg/hectare)	:
	(a) Commercial product	:
	(b) Seed	:
15	Current approximate percentage of the area of crop (Kind) under this variety in the state	:
16	Recommendation of the All India Workshop about the variety	:

Signature of the Chairman/Convenor
State Seed Subcommittee
Name
Designation

5. WEAKNESSES

Since the contributing breeders themselves are involved in the testing procedure, there is likely hood of some biasness in the data generated, which may affect the objective of assessment of varietal performance in the trial.

When the new strains are included in the advance trial, distinctness, uniformity and stability (DUS) are not properly tested.

Since norms adopted for advancement are required to be followed very rigidly, chances of discarding very promising material can not be ruled out.

Suitability of variety under off-season condition needs to be tested for seed production during crop failure in the regular season.

Seed production technology should be worked out before a variety is released for general cultivation.

Testing should be done for different purposes such as fresh harvest, processing and off-season seed production.

CHAPTER - 21

MAINTENANCE BREEDING OF SELF POLLINATED CROPS

S.K. Rao

1. INTRODUCTION

Seeds of assured quality can be expected to respond to fertilizers and other inputs. Attention must be given to the maintenance of genetic purity and other qualities of the varieties in order to exploit the yield potential. The crop varieties are influenced by the environmental and ecological conditions. It is essential to go for rouging operations, as it is the only major field activity, which distinguishes the seed crops from that of grain. This has been a regular practice in seed production systems and also had theoretical basis in population genetics with special reference to change in population structures due to population size and sampling errors in the selected population in the seed production chain. These things have a greater relevance on the maintenance of genetic purity and genic balance of the varieties at the time of testing, identification, release, notification and registration.

Pollination behaviour in addition to the knowledge of chromosome number and information on fertilization habits play an important role for the breeders involving in the maintenance breeding programme. The varietal differences with respect to floral biology of a crop species have a major influence on practical aspects of maintenance breeding. Even in self-pollinated crops, the extent of out-crossing is varying with high magnitude depending upon the crop species e.g. pigeonpea and sesame. In general, unwanted out crossing is greater in crops growing under dry and hot weather conditions than the cooler and humid environment e.g., cotton, sorghum, sesame, pigeonpea etc. Hence, a particular attention is required in the maintenance breeding of these crops varieties for production of quality seeds.

The cross-pollination occurs due to different type of floral mechanisms that prevent self- pollination i.e. self-incompatibility of pollen (rye and beets), protogyny (Brassica), monoecious (maize) and dioecious (papaya). These pollination habits are taken into account in breeding new varieties and maintenance breeding of released varieties. In case of self-pollinated crops, the maintenance breeding is simple. However, special measures are required to check the cross-pollination in the form of isolation distances etc.

2. NEED FOR MAINTENANCE BREEDING

Providing good quality seeds of improved varieties to the farmers through various seed supply systems is a major task in the developing countries, which had a strong agriculture base. Self-pollinated crop varieties are generally homozygous and continuously subjected to influence of environmental conditions resulting in developmental variations. In most of the cases, the original variety changes in the course of few generations in such a way that the uniformity of variety characteristics especially height and maturity is replaced by certain amount of variation e.g. soybean varieties JS 72-44, JS 71-05 and JS 80-21. This ultimately results in deterioration of variety by loosing its performance.

At present the crop varieties are being released notified for seed production without testing for their Novelty (N), Distinctness (D), Uniformity (U) and, Stability (S). Due to novelty, the varieties released are having different gene combinations at a particular level of "genic balance" with respect to combination of characters. In seed production chain the variety loses its genetic make up and come up as some other variety with differential genetic composition due to maintenance breeding procedures to meet the requirement of certification standards of genetic purity at foundation and certified seed levels. This has been a major problem of the seed production programmes with respect to several varieties in the country.

It is essential to maintain the genetic purity of seed stocks in the large-scale multiplication and ensuring the conformity to the original types. Maintenance involves strict selection in the variety. Minor difference has to be identified and removed. This can be done through positive or negative mass selection or through line selection in self-pollination. The maintenance breeding of released and notified varieties for seed production in an organized and systematic way is a big task, because it has to maintain the relation ship of small amount of nucleus seed with that of originally selected by plant breeders. To meet this objective there are two types of systems under operation with respect to public sector plant breeding programmes. The first system is the maintenance breeding and is carried out in the institution that developed the variety. In this system, the major problem is that the breeder who developed the variety in a public sector organization continues with the same crop improvement after his retirement some body has to take up this maintenance breeding work in the same project. Now a days GOI has specified the notification of varieties for a period of time i.e. 14 years or less than that depending upon the importance/popularity of varieties with the farmers.

The another system of maintenance breeding of notified varieties is through functional breeder seed production units at SAU/ICAR Institutes. At these centres, qualified plant breeder takes up the responsibility of maintenance breeding work of all crop varieties under seed production chain. This can be better managed provided the breeders engaged in the quality control systems in the seed production depends on these basic principles of the variety maintenance, limited generations and official-quality standards. The functional breeder seed production units should be acquainted with the genetic back ground of development of varieties under seed production. They should also be acquainted with the initial genetic make up of variety through originating breeder. This help in maintenance breeding programme in a better way to maintain the initial genetic make up of varieties. In this system, if originating breeder of a variety is available at a particular station where maintenance breeding programme is going on, he can advise/guide the maintenance breeder.

The maintenance breeding programmes should be in the hands of qualified plant breeder and skilled technical personals for production of high quality seeds particularly in the production of nucleus and breeder seed. Systematic variety maintenance breeding procedures may be effective in retaining the allele frequencies in the original population selected by plant breeder. Genotype x environment interactions may also effect the maintenance of genetic purity of a variety. If the crop varieties are grown other than the environment where the varieties have been selected associated with natural selection and management practices may affect the genetic purity of a variety. This unplanned change in allele frequency is often referred to as a genetic drift. Growing environments also have major affects in population behaviour and will have important consequences for varietal maintenance. Preventing this deterioration of stocks during multiplication is a major concern in organized seed production.

The occurrence of offtype in a given genotype in the maintenance-breeding programme depends on —

Rate of addition of offtype in each generation and number of generations of multiplication

Proportion of progeny of offtype which are also offtype (i.e. stability of these offtype in subsequent multiplication)

The relative rate of multiplication of offtype

Effectiveness of removal of offtype by rouging at each generation.

Careful management of nucleus seed production programmes and restriction on the number of subsequent generations of multiplication is the major considerations of maintenance breeding. It is to counteract the changes if there is no control in the multiplication of variety. In order to carry out this job effectively, it is essential to know the typical causes of variety deterioration viz., developmental variations, mechanical mixtures, natural crossing, spontaneous mutations, loss of genetic components, minor genetic variations and premature release of varieties.

The significance of each of these factors varies depending on the nature of crop, the variety, physical environment and seed production procedures. It also includes isolation of crops from other sources. The deterioration of variety caused due to spontaneous mutations, natural selection, loss of genetic component and minor genetic variation do not become apparent until the next generation is grown after their induction. Such impurities can be controlled by initial purification and definition of varieties, cleaning of equipment, careful handling, crop-rotations and rouging operations in the seed plots. The breeder responsible for maintenance breeding and for production of nucleus and breeder seed is not responsible for these variations. However, the impurities of this sort are to be detected only in the maintenance breeding plots of the variety as within progeny row variations. The origin and detection of offtype are important in any of population and management of early investigations showed that the early multiplication of varieties may create problems in multiplication chain. The basic causes of deterioration of varieties so for known are discussed below —

1. Developmental variation: These variations may arrive due to either external or intrinsic factors. The variation in normal growth and development of seeds occur due to climatic, edaphic, biotic and abiotic factors. However, intrinsic factors responsible for variation are changes in anatomical structures, stomatal behaviour etc. These changes can be seen apparently in the subsequent generations of multiplication and should be removed after detecting them in seed plots. These effects can not be carried over to the next generation.

2. Mechanical mixtures: Impurities in the population can be ascribed to mechanical intermingling from a variety of causes mainly due to result of negligence and lack of care. There are four ways of detection the causes of contaminants i.e. grow out tests, post control plots, monitoring and inspection of seed plots. The use of specific land preparation practices may reduce the population of volunteers. The extent of efforts made for the removal of such volunteers from seed crop is another important consideration to maintain the genetic purity. The combine harvester having self cleaning devices have control over the possibility of getting mechanical mixtures. There are several other possibilities of obtaining mechanical mixtures in the following areas.

(a) *Seed drills:* Special attention must be given for careful cleaning of seed drills when changing from one variety to another. Air compressors, point brushes and feather dusters have proved to be extremely suitable for cleaning operation. There should not be any holes in the frame of seed drills, seed boxes and enrute of the seed. Path of the drills must be perfectly cleaned from time to time.

(b) *Sacks:* Several types of sacks are being used for the seed storage. Before filling the seeds, the sacks should be cleaned perfectly by turning inside out and brushing down carefully.

(c) *Bulk seed handling:* During the transport of seeds from the threshing plots, from combine harvester and threshing yards the transport equipment must be cleaned before transporting the bulk seeds at the acceptance points of cleaning/processing plants.

(d) *Variety to variety basis:* During threshing transport and delivery to the acceptance point of seed processing the batches should always the deal as per individual seed categories in order to facilitate necessary machine cleaning work.

(e) *Seed cleaning machine:* These must be thoroughly cleaned when changing from one seed category to another or from one variety to another. Simply idle running of the machines for a lengthy period is not adequate to clean the machines. Hence, it is advised to use compressed air or powerful suction equipment is strongly recommended for cleaning.

(f) *Storage:* Careful handling of stocked and cleaned produce in order to ensure that the sacks do not burst. The sweeping to gather the seeds form burst sacks can leads to considerable amount of intermingling. This will be a grate problem particularly when different varieties are stored right next to one another in state.

(g) *Harvesters/ Threshers/ Reapers:* Among these equipment, combine harvester represents the major source of risk because its complete cleaning is almost impossible. All accessible parts such as cutter-bars, threshers etc. should be cleaned thoroughly with the help of suction pump before switching over to other variety.

(h) *Soil:* It can be source of origin for intermingling of varieties as the shattered seeds of all species germinate as soon as possible. Therefore, the germinated plants should be destroyed during the field operations for sowing a subsequent crop. However, the techniques used in the areas, which have a dry summer will be different from those employed in the regions with a humid climate. For production of foundation seed, it is necessary to have fallow period to prevent extensive weed infestation and unwanted multiplication of shed seeds during the fallow period. In the regions of extremely dry summer, field selected for seed production should not be ploughed in the normal manner because it hinders the germination of the seed due to burring of threshed seeds. When the ploughing operations are carried out shortly before sowing, possibly after the fallow year, then the seeds are exposed to well-aerated soil layers near the surface. In favourable condition, they emerge as a source of contamination for subsequent crop.

(i) *Grazing:* Cattle, sheep and goats in the semi-arid regions give rise to number of problems due to grazing of crop stubble. Although, a large number of shed seed are eaten and utilized by these animals particularly if they are allowed for a long grazing periods. Nevertheless, the dung of these ruminants contains undigested seeds, which are in most of the cases capable of field emergence. Therefore, the lands reserved for seed production should not be grazed at all. The intermingling of seed due to grazing animal and soil could readily be distinguished from mechanical factors as in case of machine unwanted plants are always found in drilled rows.

(j) *Irrigation water/floods:* Special care should be taken while irrigating the seed plots especially in dry sowing conditions because water passed from one field may carry seed to the adjoining fields.

3. Natural Crossing: All self-pollinated varieties have certain percentage of chasmogamous flowers permitting cross-pollination. The tendency of the flower on individual plants is determined by a variety i.e. genetical as well as environmental factors. Drying, during flowering period considerable increases the tendency of plants for open blooming. In the semi-arid regions, out-crossing can occur upto 10% in case of wheat and barley varieties. Similarly, incase of pigeonpea and sesame the extent of out crossing goes up to 50 percent. The crops which has been categorized as a self-pollinated, due to intensification of breeding efforts for the development of high yielding varieties have been brought from diverse sources form the available germplasm collections. This has a tremendous impact on the change in floral biology of some of the cultivated crops resulted in increasing the extent of out-crossing. This indicates the need of physical isolation for multiplication of seed. This is significant for the production of nucleus seed, which requires extremely high levels of uniformity. The isolation distances required for certification should be based on scientific data rather than minimum required for certification.

In case of fodder oats, the seed production plots should be free from the wild oats *Avena sativa* and *A. fatua,* as these have ability to pollinate the cultivated oats because of same ploidy level. The effects of unwanted crossing are less conspicuous in case of cross-pollination but their yield reducing impact become serious after a few generations of seed multiplication. In case of maize, the differences in the seed colour is apparent on the seed forms due to out crossing because of Xenia effect. These seeds can be picked out in the actual year when the unwanted pollination took place.

Isolation of varieties from other varieties of the same species as well as isolation of various seed categories stages of variety from one another is the promising way of counter acting the risk of unwanted crossing. This isolation is always of a physical nature. Self-pollinated species require few meters of isolation whereas long distance is required for cross-pollinated crops. The isolation standards are as per the provision of the seed act in different countries. The prevailing wind direction during the flowering period must also be considered to determine the isolation distance.

Maintenance breeding in suitable cabins is possible only for initial generations because only small number of plants is to be multiplied or species have high multiplication rate (e.g. sesame/ mustard, production of hybrid seed incase of tomato's). This is the future thrust area for seed technological research under Indian conditions.

4. Spontaneous mutations: Spontaneous changes in a plant genes or genome caused by environmental influences such as effect of temperatures, radiation of cosmic rays and earth bound origins and intentionally released radiation from the radio active substances. It increases the mutation rate depending upon the plant species depending upon

(a) Rate of mutation is considerable increases in the seed stored for several years.

(b) Variety specific differences in the tendency to mutation.

(c) Due to grater plasticity of some specific genotypes exhibits higher mutation rate than the old varieties.

(d) The mutants appeared by micro-mutations can be over looked in the field populations because after subsequent generations of multiplication micro-mutations may interface with the genetic purity of variety. These mutations are generally recessive. Mutants derived out of micro-mutations are generally dominant in nature. Majority of mutations has a detrimental effect on the performance and quality for genes, a reduction in seed yield can be expected. The mutants derived out of the spontaneous nature such as early flowering mutant in linseed variety *Jawahar linseed 23* earlier by 10 days flowering is one such example. Careful observation on the populations of seed production chain can easily be detected so that its effects on genetic purity can be reduced.

5. Loss of genetic components: High yielding varieties of crop plants have varied genetic make up. It is very difficult to discriminate the genotypes based on visual observation. The maintenance breeding programme should be restrict to the defined diagnostic characters, and too much genetic/phenotypic uniformity in the populations of the genotype adopted to extreme situation is gradually disappeared. It was happened incase of one of the most popular variety of soybean JS 72-44 in the Madhya Pradesh. This has become popular due to inherent genetic variations resulted in wider adaptability. The maintenance breeding work of JS 72-44 resulted in the 90% rejection of genetically variable progenies. Rest of the similar progenies constituted as JS 72-44 for further seed multiplication. This newly created JS 72-44 has reduced yield potential and wider adaptability. Hence, this type of situations should be avoided by maintenance breeding programme by maintaining the initial genic balance of the variety at the time of release and notification. This is very important consideration to harvest the fruits of crop improvement research. In this way, several crop varieties loss its original suitability of adaptation and fails due to lack of stability. The choice of unsuitable locations for maintenance breeding of crop varieties can concentrate major risks. An early maturing high yielding soybean variety JS 71-05 was recommended for cultivation in Malwa/ Nimar region of Madhya Pradesh. Though, this variety was recommended for this region but the maintenance breeding was undertaken at Jabalpur location where this variety was not recommended for cultivation due to sensitivity to high temperature during field emergence and not having adaptability of cultivar to this region. This has created a problem even in the initial process of maintenance breeding until certified seed production chain. Later the maintenance breeding of this variety has been shifted to adaptable environments (Indore) where the variety has been recommended for cultivation. In this way, selection of unsuitable location for maintenance breeding can create major risks. The most important generations of maintenance breeding i.e. Nucleus and Breeder seed production stage should always be grown in the place where variety was breed. Therefore, the environment can maintain the performance and expression of the variety during the maintenance breeding. Due to adverse climatic influences diseases such as rusts/powdery mildew may occur in epidemic form e.g. occurrence of rust on soybean variety JS-335 during *Kharif* in Satpura hills of Madhya Pradesh. This phenomenon effects the self-fertilized crop varieties to a grater extent.

6. Selective influence of diseases and pests: Effect of epidemic form of diseases such as rusts, *Rhizocotinia* and pod blight in case of soybean and loose smut and rust in wheat has created a problem related to seed development and seed quality. There has been drastic reduction in seed size due to biotic stresses. Similarly, the seed yields were badly affected due to selective influence of pest viz., pod borer in chickpea breaks the seed multiplication chain and the process of maintenance breeding during the year 1995-96 in Madhya Pradesh. Influence of diseases and pests on the seed yield loses and quality are well recognized in the seed industry. This also time with respect to cotton boll worm in the different years of Andhra Pradesh. Incidence of foliar diseases in Mung and Urid bean has affected the maintenance breeding and seed production chain. It needs to be considered in the formulation of the maintenance breeding programmes.

7. Premature release of varieties : The present system of release and notification of crop varieties does not require DUS tests. In the self-pollinated varieties developed through hybridization programme, a lot of residual heterozygocity exist even at the time of release of varieties. The maintenance breeding work of the promising lines should be started at the final stage of advanced varietal trials. This will assure the availability of genetic pure seed for seed production chain at the time of release/notification of varieties. This permits easy maintenance breeding of improved varieties in subsequent generation.

8. Maintenance breeding procedure: Four basic types of populations that differed in amount of variability expressed in the growing of crop are clones, inbred-lines, hybrids and open pollinated populations. The maintenance breeding programme of varieties tends to be continues with individual character showing wide range of expression. Individual plant can be considered as off-type and removed but precise limits of acceptable variation may be difficult to judge under field conditions. The active participation of breeder is the main consideration to maintain the genetic purity in seed production. There is a greater need to manage the populations with some extent of wide-genetic base at the development of variety suitable for wider agroclimatic conditions.

A plant variety is continuously subjected to a large number of environmental influences after its release and notification resulting in the variation in expression as compared to the original characteristics. In most of the cases, it losses its standard performances acquired through breeding in both qualitative and quantitative terms when grown on for few generations.

Maintenance breeding attempts to counter act this tendency towards deteriorated performance by continuing individual selection. To over come this, breeders are asked to submit the seed samples along with the photograph of the plant of the variety at the time of release or notification. These seed samples can be grown for few cycles of maintenance breeding to check it's identity.

Self-pollinated species have a major share in the seed production programme in India in terms of quantity of seed produced at the start of maintenance breeding. A large number of individual ears/panicles/plant are harvested from a large population of a variety, threshed in a suitable manner and assessed with regards to the external seed characteristics. The number of the ear/plants harvested is determined by the seed multiplication ratio in concerned location.

The individual ear progeny is often designated as G_0 and is the most important generation in the maintenance breeding of self fertilized population. Depending upon the economic importance the quantity of produce of single ear/ plant few hundreds to thousands of progenies are grown. The row length may be variable for different crops.

The two rows method followed in cereals must be modified for legumes, oilseeds and fodder crops. The progenies are kept under continuous observation during vegetative period and all the progenies starting from emergence deviating from true to the type are removed entirely when the row contains deviating plants. For technical reason mearly marked rows should be removed by snapping the located long with the appraisal path. Rejected rows must be eliminated from the populations to avoid mixing up of crop during harvesting.

A large number of single ears plants should be harvested from the entire population of remaining progenies or nucleus seed plots of considerable size. These are then used for new G_0 cultivation representing the beginning of maintenance breeding cycle. While selecting plants/ears for maintenance breeding special attention must be given to the size of the base population from which the selection are being made. The selection should also be done based on systematic random sampling so that the subsequent plot of nucleus seed should represent the single ear/plant progenies grown in the maintenance-breeding block.

The maintenance breeding block should be in the centre of the breeder seed production unit so that high quality seed of the progenies can be projected from the cattle grazing etc. For the nucleus and breeder seed plots it is advisable to have a space planting of progenies row of single plant/ear produce. The spacing between the lines should be rouged in such way that the progeny sown should be district from each other for verification.

There are two distinct approaches of planting of progenies. The first approach in growing of progeny row in continues row in a compact block with adequate track distances between the tiers for the mobility. The isolation requirements should be considered when more than one variety of the same crop is grown in maintenance breeding block. The second approach is that the progenies of single plants/ears must be grown in single/double row based on produce available. In this method, after every 50 plots one plot of nucleus seed of the same variety should be grown with same plot size so that the characteristics of the variety under maintenance breeding can be checked from time to time. This is especially advantageous for the maintenance breeding of new varieties, if the programme is to be handled by a sponsoring breeders/producing breeder who is not associated with the development of variety under multiplication. However, at least once the sponsoring breeder/producing breeder involved in the maintenance breeding programme should bring the nucleus/breeder seed from the originating breeder so that the initial genetic make up of the variety can be subsequently maintained in seed production chain. It will be advantageous for the sponsoring breeders if they are acquainted with the details of development of variety by originating breeder. This will avoid the complications in seed production chain with respect to genetic purity.

The general scheme of maintenance breeding of self-fertilizers with straight growing habits are given in the single plants / ear are cut with stem length around 20-30 cm. It is desirable always to combine them into bundles of hundred plants/ears. This makes it easier to assess the total plant/ ear requirement of the next cultivation process without the need for great deal of counting. Each of bundles of plants/ears must be suitably labelled. The bundles of the same variety are hanged up together in a room/hall, protected against the rain, and allow them to dry until they are required for further use. The additional advantage of keep this may helps in preventing the damage or breaking off. It must be ensured that the plants ears are protected against the winds. Once the ears/plants for the next breeding cycle have been harvested, the reminder to G_0 should be threshed by properly cleaned plot combines/ threshers. If such machines are not available crop must be cut with sickles and threshed in a suitable manner. The seed threshed from G_0 are then sources G_0 during the next sowing period.

In order to increase the multiplication rates, the seed rate should be 10% lower than normally used in the locality. G_1 is attended and harvested in a normal manner. The populations must be kept under continues observations and roughing carried out starting from the field emergence. It has been observed that some of the plant emergence was poor in certain cases the special of seedling emergence was very low due to low vigour of the progeny. These things should be removed in between the 20 to 25 days after sowing. In this way, less vigorous progenies will be eliminated. During the vegetative phase, crop plants showing extremely tall plants should also be removed. All types of diseased and insect rouge types and off-types should be removed from the multiplication field. Track gaps are essential for management of seed crops and should be provided in all the seed production programmes. Similarly, G_2 can be a specialized crop like climbers and tendular plants grown out of G_1 with the normal recommended seed rates recommended in the area of multiplication. Because of several risks involved, it is advisable to grow this stage at number of places. This G2 stage of multiplication is considered as breeder seed. G3 follows the same way from G2. Proper roguing of the progeny of breeder seed i.e. G3 is essential. From the G3 generation of multiplication certified seed can be produced. This is the seed marketed to the farmers for commercial cultivation. Farmers generally grow the purchased certified seed for 2 or 3 generations to meet his own requirements. There will be a yield loss of 5% after first year growing of certified seed harvest and during second year it goes up to 12% and in third year the yield reduction goes up to 25% in self-pollinated varieties. A comparative assessment of performance of farmers saved seed with certified seed lots revealed that there has been increase in the yield potential up to 25% by simply changing their own farm saved seed with the certified seed and rest of the inputs and management practices are common. The farmers are achieving the substantial returns. This is a clear proof of importance of maintenance breeding.

Structure of maintenance breeding for self-pollinators

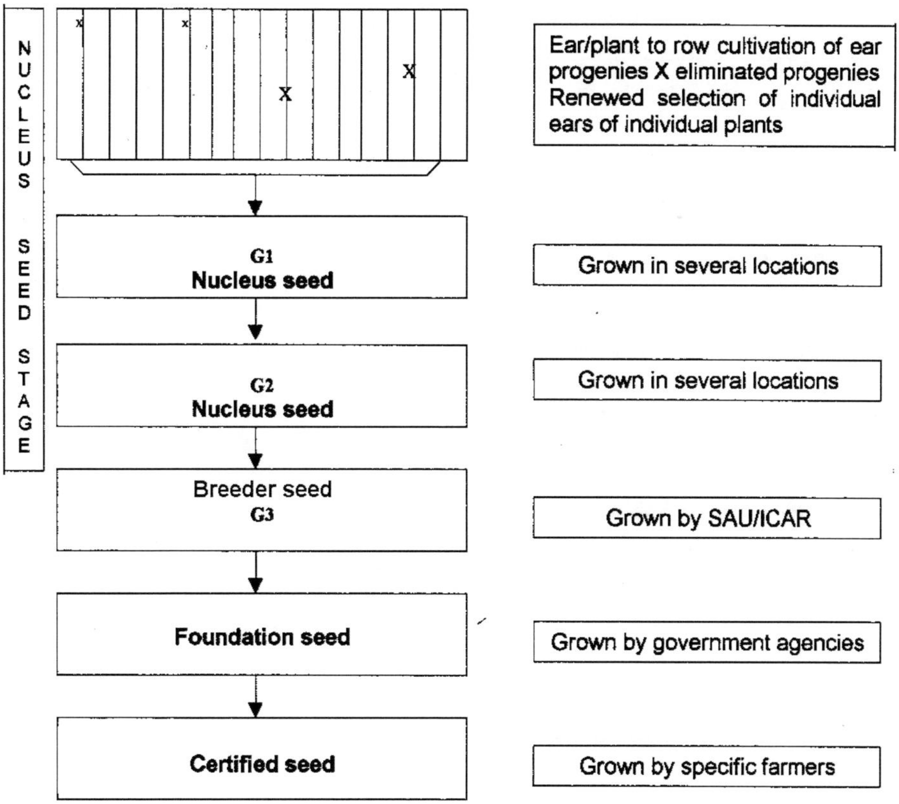

It could be concluded that maintenance breeding of self-fertilizers maintain the level of performance originally attained through breeding and selection. The above mentioned procedure of maintenance breeding applicable to all self pollinated crops except some specialized crops like climbers and tendular plants.

CHAPTER - 22

APOMIXIS AND ITS APPLICATION IN CROP IMPROVEMENT

Dhirendra Khare and C. B. Singh

Most plants reproduce through seed, which is normally formed by sexual reproduction i.e., union of egg cell (n) with sperm cell (n) derived from meiosis. It provides variability in next generation due to segregation and genetic recombination during micro-sporogenesis and mega-sporogenesis. It always follows alteration of generation (Fig 1). Few plant species are able to reproduce asexually through vegetative parts or through seeds formed without fertilization i.e., without alteration of generation. The asexual reproduction through seed or vegetative part is known as **apomixis**. The word is derived from two words *Apo* = "away from" and *mixis* = "act of mixing".

Formation of seed without fertilization is termed as **agamospermy**. In general, the term 'apomixis' is used to designate the seed, which is formed without combination of male and female gametes.

Apomixis is a genetically controlled process that bypasses —

Female meiosis

Fertilization

Therefore, the produced seed has embryo genetically identical to the female parent. This behaviour is reported in more than 300 species distributed in 35 families (Bashaw 1980; Marshall and brown 1981). Among the field crops, it is reported in species of families Gramineae, Compositeae, Crucifereae, Leguminoseae and Solanaceae (Johansen, 1950 and Davis, 1966).

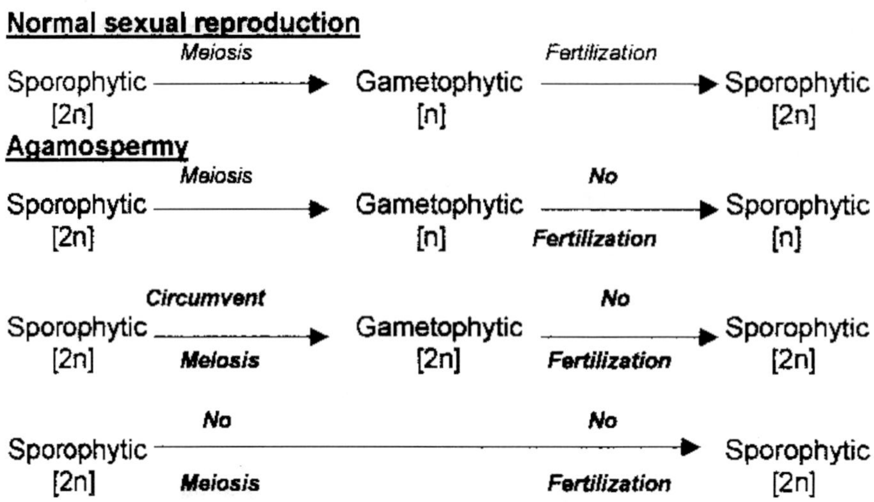

Fig. 1. Alteration of generation

1. TYPES OF APOMIXIS

1.1 Classification on the basis of the cell responsible for embryo formation

On the basis of the cells which are responsible for the formation of embryo, the apomixis is divided in two groups —

1. Sporophytic apomixis
2. Gametophytic apomixis

1.1.1 Sporophytic apomixis

Diploid embryo develops directly from the somatic cell without fertilization either out side of the embryo sac or in the embryo. Based on the place of its formation it is of two types —

1.1.1.1 Adventitious embryony

The seed is formed outside of the embryo sac from a sporophytic cell without pollination. It has no alternation of generation. It may arise from individual cell of the two different somatic tissues i.e., nucleus or integument but not from the megaspore mother cell or archisporium cell. In such cases, the seed is not surrounded by embryo sac. In adventitious embryony, archisporium cells are not formed.

Example: Citrus, mango, *Allium* sp., *Hosta* sp., *Ochana* sp. and *Eugenia jambos.*

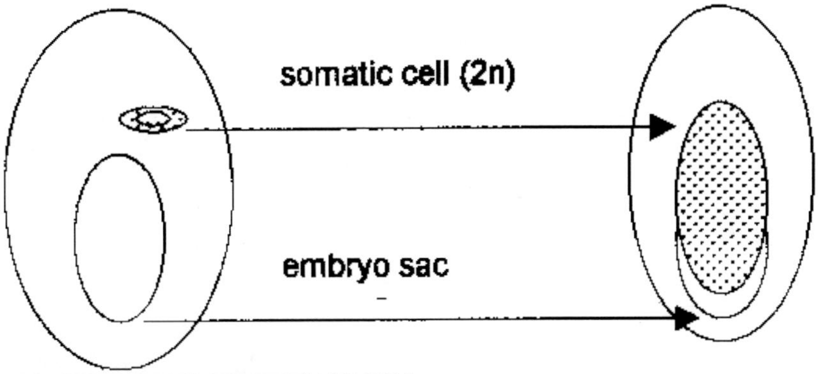

ADVENTITIOUS EMBRYONY

1.1.1.2 Apospory

Development of a diploid embryo from a somatic cell derived from nucellus or inner integument by a series of somatic division in an embryo sac. It may occur in following ways —

The MMC goes through the usual reduction division but at this stage a somatic cell situated in the Chalaza region begins to enlarge and become vacuolated. This cell gradually increases in volume. It encroaches the space of MMC and finally crushing it. Then a somatic cell multiplies with non- reduction division forms a diploid embryo without fertilization.

In some cases, the MMC degenerates and a number of nucellar cells become much enlarged. One of theses cells located near the micropyle gets vacuolated and then developed into uninucleate embryo-sac. In some apospory pollination is required for the development of endosperm.

Example : *Pennisetum, Cenchrus, Poa* etc.

1.1.2. Gametophytic apomixis

Seed is directly formed from gametophytic cell with or without pollination. The gametophytic cells may be formed by meiosis or mitosis. Based on cell division it is divided into following two groups —

1. Agamogyny
2. Diplospory

1.1.2.1 Agamogyny

Seed is formed from unreduced gametophytic cells due to irregularities in meiosis. The egg cell develops parthenogenetically, giving rise to embryo. It is more common in polyploids. It is of following types —

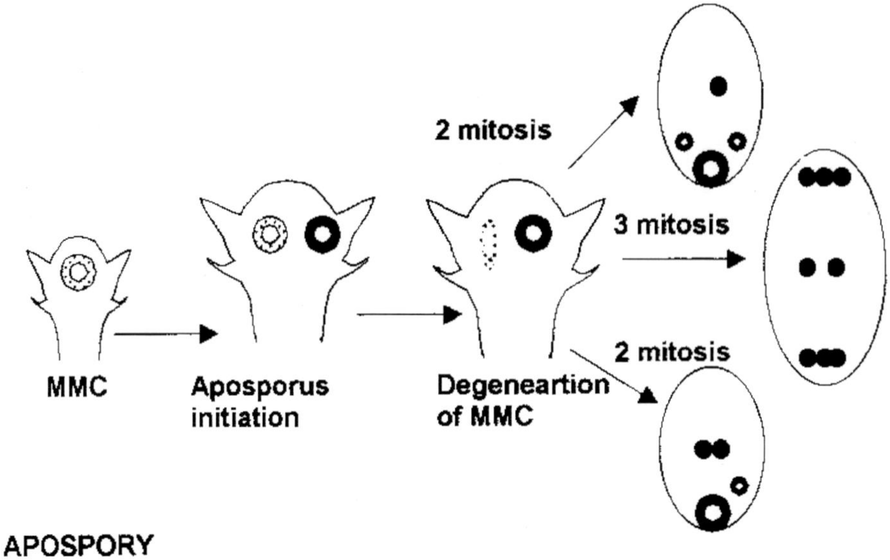

1.1.2.2 Diplospory

The diploid embryo arises from megaspore mother cell (MMC) after cell division with following abnormalities —

No meiosis

Mitotic cell division: Direct mitotic division of MMC resulting in the formation of diploid archisporium cells.

Reduction division modified in such a way that pairing and reduction of the chromosome does not occur.

The MMC initially follow meiotic prophase but due to *asynapsis* normal chromosome pairing does not occur. The *univalents* are scattered over the spindle at metaphase I. A restitution nucleus is formed with diploid chromosome number. This will form the archisporium cell with diploid chromosome number.

Example : *Paspalum* sp., *Arabis* and genera of Compositeae.

Formation of restitution nucleus due to abnormality in meiosis-I followed by normal Meiosis-II division but without cytokinesis. Therefore all the archisporium cells are formed with diploid in nature. The unreduced egg cell develops into an unreduced embryo sac without fertilization. The endosperm develops from the unreduced polar nuclei without fertilization.

Example : *Eragrostis, Rubus, Tripsacum, Parthenium, Allium tuberosum* and *Potentilla*

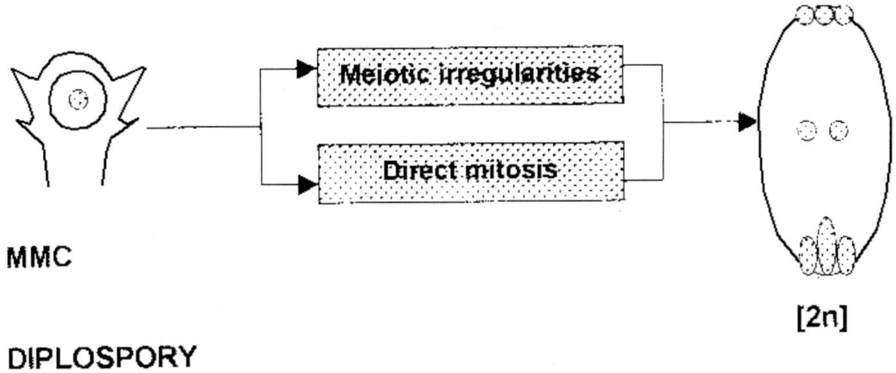

1.1.2.3 Apogamy

Embryo formed by the haploid nuclei synergids or antipodal (other than egg cell) without fertilization. Normally haploid embryo sac is formed. It is of two types haploid apogamy and diploid apogamy. Diploid apogamy is formed by fusion of two synergids or antiopodals.

Example : *Allium* species.

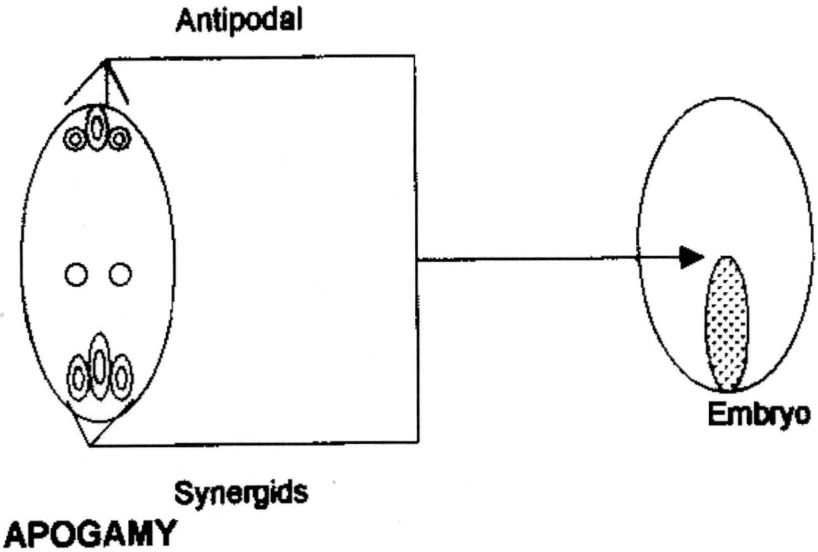

1.1.2.4 Pseudogamous

In some plant species, seed is formed apomictically but the process starts after pollination and pollen tube formation. Pollination provides a stimulus effect only.

Example: Mango and citrus.

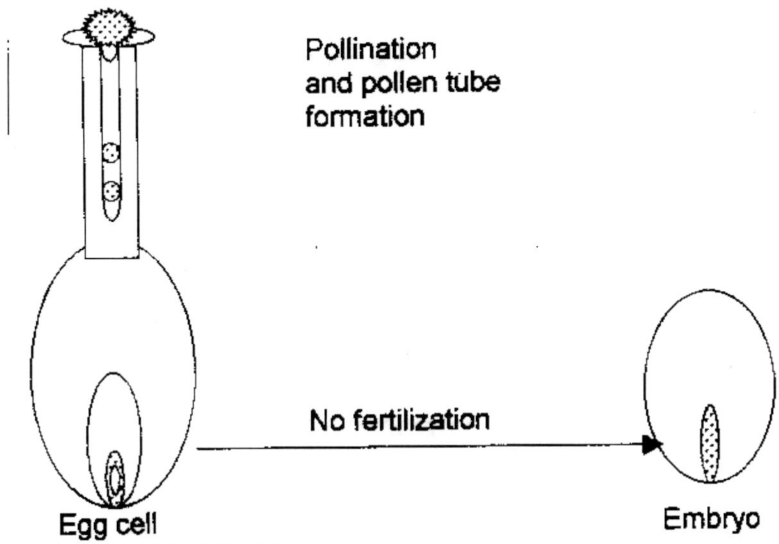

PSEUDOGAMOUS

1.1.3 Parthenogenesis

Haploid archisporium cells are formed by normal reduction division. Embryo is formed form haploid egg cell without fertilization whereas endosperm is formed by normal fertilization. It is of following types —

1.1.3.1 Gynogenesis

Formation of embryo from egg cell without fertilization. Pollen tube enters the embryo sac but instead of double fertilization, single fertilization takes place i.e., one sperm cell combines with two polar nuclei but egg cell form a haploid embryo without fertilization. Finally a hapliod embryo with normal 3n endosperm is formed.

Example: Maize and tobacco.

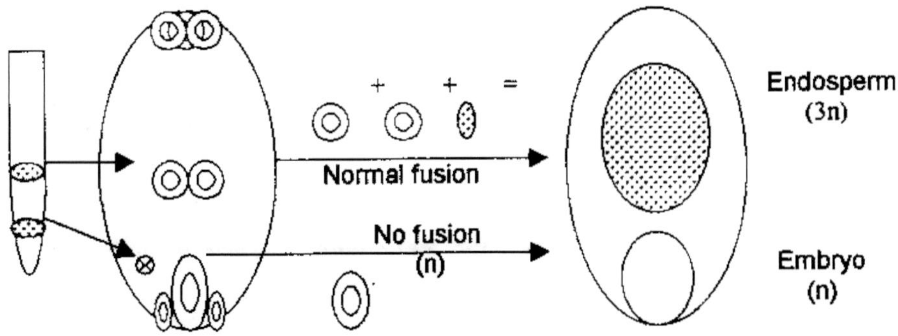

GYNOGENESIS

1.1.3.2 Androgenesis

Formation of haploid embryo sac by male gamete with normal triploid endosperm. Its average frequency of androgenesis in maize is 0.00125 percentage. It is also termed as ovule androgenesis.

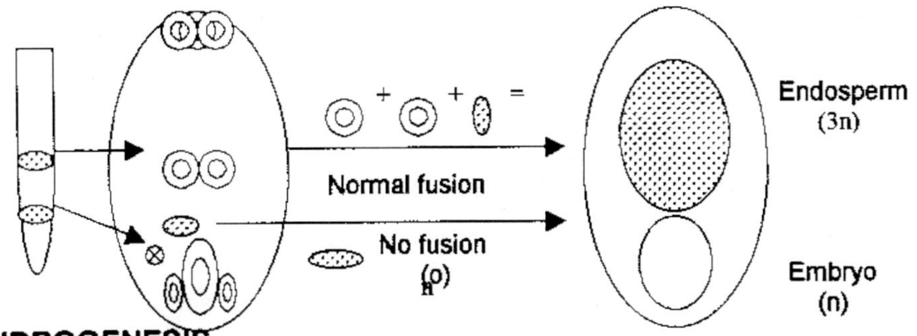

ANDROGENESIS

1.1.3.3 Semigamy

Pollen tube enters the embryo sac and releases both the sperm cells. One sperm cell combines with two polar nuclei and forms a normal 3n endosperm. Second sperm cell is not able to combine with egg cell. Both egg cell and sperm cell begins to divide synchronously. Thus, two haploid embryos are formed in one embryo sac with one endosperm. The developing embryos often give rise to plant chimera for the two parental genotypes. Although, occasionally one type either female or male gamete takes over to produce haploid plant.

Example: *Gossypium hirsutum, G. barbadense, G. tomentosum, Coix, Rudeckia, Zephyranthes* and *Arabidopsis.*

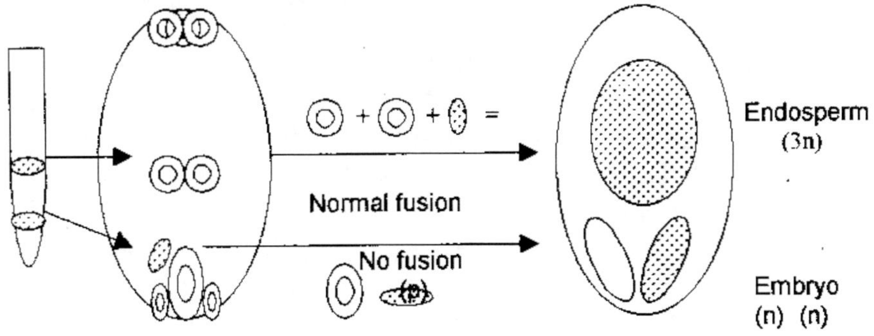

SEMIGAMY

1.2 Classification based on habit of genera for apomixis

On the basis of apomictic habit the genera are classified in the following two groups —

1.2.1 Facultative apomixis

The plant species those produce seed both apomictically and sexually are known as facultative apomict. Example; Kentucky grass, marvel grass, guinea grass sorghum and pearl millet.

1.2.2 Obligatory apomixis

Reproduction is strictly based on apomixis. Sexual reproduction is not possible in this group of plant genera.

1.3 Classification based on inheritance of apomixis

Based on repetition of the apomictic nature generation after generation apomixis is classified in following two groups —

1.3.1 Non recurrent apomixis

The nature of apomixis is not repeated from one generation to another. Normally haploid apomictic embryo, with only one set of chromosome is usually sterile and is not able to repeat this nature in next generation.

Example: Parthenogenesis and apogamey.

1.3.2 Recurrent apomixis

The phenomenon of apomixis is repeated from generation after generation. Embryo may arise either from a diploid archisporium cell or from other somatic cell. Because of no reduction division diploid embryo is formed. The genera with this type of nature repeat the phenomenon of apomixis from generation to generation.

Example: Adventitious embryony, apospory and diplospory.

2. IDENTIFICATION OF APOMIXIS

The seed is produced through sexual reproduction or asexual reproduction can be verified by following methods —

2.1 Indicators of apomixis

1. Unusual high seed set.
2. Multiple seedlings per seed.
3. Occurrence of multiple stigma and multiple ovules per floret.
4. Consistently high frequency of twin seedlings.

2.2 Progeny tests

1. Uniform progeny form a heterozygous plant.
2. Maternal phenotypes in crosses.
3. Absence of male dominant marker in crosses.
4. Emasculation should be carried out on a large scale and the stigma of the emasculated flower should be pollinated with pollen from a genotype bearing a dominant marker gene. If the seed produced from this repeatedly yield a consistent percentage of strictly maternal parent, apomixis can be suspected.
5. If the flower is very small and emasculation is not possible then pollen viability and pollen tube formation should be thoroughly examined. Uniform progeny from the seed harvested from bagged flowers may be due to self- fertilization because these two processes do not usually occur in the same group.

2.3 Cytological methods

Histology and cytology of the female gametophyte enables to identify.

Lack of embryo sac and endosperm confirms the adventious embryony during embryo development. Presence of proembryo those developed directly into embryos out side the embryo sac from somatic cell in the ovule also confirms the adventitious embryony.

Development of nucellus cell in the embryo sac and degeneration of megaspore mother cell confirms the apospory before pollination. Whereas, in later stage, lack of antipodal cells in embryo sac confirms the apospory.

Restitution of chromosome, absence of meiosis during megasporogenesis, during metaphase, univalents are scattered over the spindle and lack of linear tetrad of megaspores confirms diplospory at the cell division stage. However, in later stage it can be confirmed by Diploid cells of endosperm.

Normally eight archisporium cells in embryo sac confirms the sexual reproduction whereas, four nucleate embryo sac leads to apomixis.

More than one embryo sac per ovule leads to apomictic as the plants with sexual reproduction have generally only one embryo sac.

Archisporium cells of the plants with sexual reproduction are well distributed in the embryosac (3 : 2 : 3). Whereas, all the nuclei are concentrated at the same end in the apomictic embryo.

Uneven polyploidy or aneploidy, low pollen stainability and the presence of monomorphic self incompatible population indicates apomixis (Cowan *et al.* 1996).

The two basic features of apomixis in family Gramineae are plyploidy and heterogenity (Huang Qunce and Chen Qifeng, 1999).

2.4 Pistil clearing technique

Pistil collected at the time of antheis is fixed in Ethanol (70%) : Glacial acetic acid : Formaldehyde (37%) at the ratio of 18:1:1. This material is dehydrated by passing them through series of alcohol solution of different concentrations. After dehydration, it is washed by an aromatic easter, Methyle salicylate extracted from winter green. Methyle salicylate clears the ovule and embryo sac. The ovule is examined under the phase contrast micoscope (Young *et al.,* 1979).

2.5 Callose fluorescence technique

Pistil is fixed in Formalin-acetic acid- alcohol and then it is soften with the help of 8N NaOH. It is stained in the solution prepared with 1.36 mM aniline blue (pH 9.5) + 2.46M Sucrose +50mM K_2HPO_4. The stained pistils are observed in fluorescent microscope. Callose is deposited on the cell wall of megaspore mother cell during megasporogenesis, in the species produce seed with normal sexual reproduction. However, it is unable to deposit on the cell walls of diploid megaspore mother cell formed without sexual reproduction. This method is used to detect diplosporous embryo sac development (Peel and Carman, 1992).

2.6 Paraffin method of sectioning

It involves following steps —

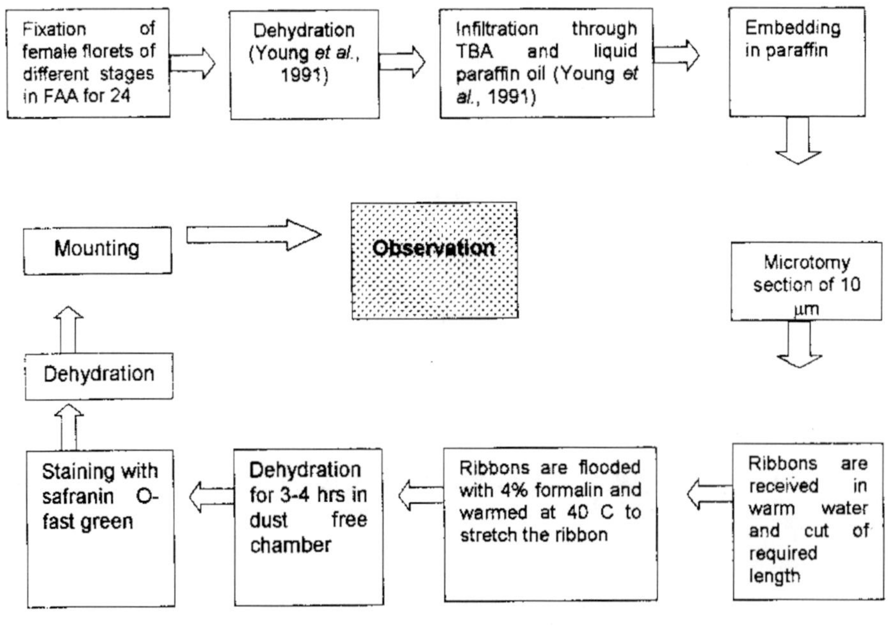

2.7 Biochemical method

Isozyme markers can be used to study gentic variation and detect the presence of apomixis. Zymogram of the hybrid formed through enzyme electrophoresis can be compared with both the parents. If the zymograph of hybrid is same to the maternal parent then it can be confirmed as apomictic.

2.8 Molecular techniques

Gene for apomixis can be tagged with molecular markers and confirmed at any stage. **Example:** UGT 197 and OPC-04 co-segregated with apomictic mode of reproductiuon in pearl millet (Ozais-Atkins et al., 1993).

3. ADVANTAGES OF ASPOMIXIS

1. Genotypes or hybrid will bred true regardless of heterozygosity.
2. It increases the opportunity for developing superior gene complex.
3. It provides a method for rapid incorporation of desirable traits.
4. Limitation of male sterile female for hybrid seed production will be over come.
5. The production of commercial hybrids in crops where male sterility system is not available.
6. It will broaden the germplasm base of hybrids because one would not be limited to female with the male sterility inducing system to produce commercial hybrids.
7. It will facilitate the commercial production of hybrid seed because of no need of isolation and rouging.
8. Every F_1 genotype with nature of apomixis has the potential of being a cultivar.
9. Progeny testing for genetic stability will not be required.
10. Comparable to other vegetative reproduction the spread of disease and viruses would be much less.
11. Farmer can use hybrid seed year after year.
12. Seed of any obligate hybrid could be increased through open pollination for unlimited number of generations without loss of vigour and change in genotype.
13. New apomictic genotypes could be produced through hybridization of sexual and apomictic plants because apomictic plants usually produce normal reduced male gamete.
14. The potential for rapid production of many new breeding genotypes might be strong force in preventing the genetic vulnerability due to wide spread use of only one or few superior genotypes.

15. It provides a way to incorporate and easily fix genetic markers in hybrid genotypes.
16. Haploids formed may be used for —
    (i) Production of 100 percent homozygous lines
    (ii) Production and identification of mutations
    (iii) To take the advantages of simpler genetic ratios and smaller population sizes at the haploid level.
    (iv) Direct use for ornamental crops.
    (v) Production of aneuploid stock for genetic research.
    (vi) Formation of somatic fusion.
17. It allows the establishment of the genetically stable seed-propagating clones of crops that can be propagated themselves across countless saprophytic generation (Pessiono et al. 1999).
18. It may prove to be unvalued biotech tool to improve crop yield (Pessino et al., 1999).
19. It can be apply to economize variety development.
20. It allows the establishment of the genetically stable seed-propagating clones of the crops.

4 USE OF APOMIXIS IN PLANT BREEDING

1. Sexual and apomictic crosses usually release a large amount of genetic variation because of the heterozygosity of the apomictic parents.
2. Heterozygosity in an obligate apomict is usually not released unless the apomict is crossed as a male parent with a sexual parent.
3. Obligate apomixis is the most desirable mechanism for plant breeding. Facultative apomixis is more difficult to manipulate and efforts such as inter crossing plant with highest levels of apomixis may be needed to increase the frequency of apomixis.
4. Different mechanisms and different levels of expression may exist in the same genus therefore, greatest number of genotypes with the expression of apomixis should be selected to enhance successful use of apomixis in crop improvement.
5. In rice 166 apomictic species with diplospory and apospory have been identified but adevntitious embryony have not yet been discovered (Huang Qunce and Chen Qifeng, 1999)

5. TRANSFER OF GENE CONTROLLING APOMIXIS

1. It involves manipulating ploidy levels of either the wild or cultivated species for production of hybrid.

2. Production of bridging hybrid is necessary to achieve higher crossability or to improve fertility.
3. It requires large population and efficient screening methods.
4. New molecular techniques such as vector transformation along with protoplast regeneration will be a helpful tool for transfer of apomictic gene.
5. *Inheritance of apomixis controlled by recessive gene(s)*: It can be introduced through male parent or through female parent heterozygous for method of reproduction. The resultant progeny will consist of both sexual and apomictic plants. A sexual plant heterozygous for method of reproduction can be selfed to produce more sexual and apomictic progeny. Superior apomicts at F_1 or selfed generation would breed true and will be ready for performance test.
6. *Inheritance of apomixis controlled by dominant gene(s):* It is easy to use in breeding programme because apomictic plant will always be a heterozygous for the method of reproduction. The cross-made between sexual and apomictic will always produce both sexual and apomictic F_1 plants. Apomictic plants from F_1 can be selected based on phenotype and tested in replicated trials in the event of obligate apomixis it can be released as a new cultivar.
7. *Development of apomictic hybrid* : Obligate apomictic genotype is used as female parent irrespective of dominant or recessive nature of apomixis for formation of hybrid. From the produce of F_1 in case of dominant inheritance or F_2 incase of recessive inheritance apomictic plants are selected (Fig. 1). These plants are tested for hybrid vigor and apomictic nature. Superior genotypes can be released for cultivation.

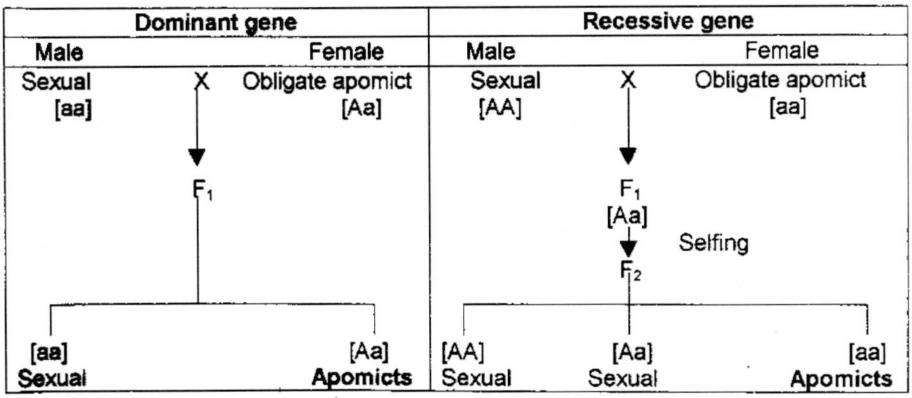

Fig. 2. Use of apomixis in development of hybrid

REFERENCES

Young, B. A., R.T. Sherwood and E.C. Bashaw (1979) Cleared-pistil and thick sectioning for detecting aposporous apomixis in grasses. Can. J. Bot., 57: 1668-1672.

Peel, M.P. and J.G., Carman (1992) Clearing procedure for detecting megasporocytes callose deposition during megasporogenesis. Am. Soc. Agron., Minnesota, USA, 100p.

Bashaw, E.C. (1980) Apomixis and its application in crop improvement In Hybridization of crop plants eds., W.R. Fehr and H.H. Hadley. American Society of Agronomy, Wisconsin, USA, page 45-63.

Marshall, D.R. and A.H.D., Brown (1981) The evolutuin of apomixis. Heredity, 47:1-15.

Johansen, D.A. (1950) Apomictic embryology. *In* Plant embryology. Chronical Botanica Co., Wattman, MA. Vol. 24:271-294.

Davis, G. I. (1966) Systematic embryology of the angiosperms. John Wiley and Sons New York.

Huang Qunce and Chen Qifeng (1999) Basic tactics of selecting and breeding apomctic germplasm. J. Fujian Agril. Univ., 28(2): 124-134.

Pessino, S., J.P.A. Ortiz, M.D. Hayward and C.L. Quarin (1999) The molecular genetics ofgametophytic apomixis. Hereditas (Landskrona) 130 (1):1-11.

Cowan, R., M.J. Ingrouille and M.D. Liedo (1998) the taxonomic treatment of agamosperms in the genus *Limonium* mill (Plumbaginaceae) Foloa. Gaeobotanica 33(3):353-366..